★ UNCOMMON VALOR SERIES ★

AIR WAR IN THE PACIFIC

THE JOURNAL OF GENERAL GEORGE KENNEY, COMMANDER OF THE FIFTH U.S. AIR FORCE

GEORGE C. KENNEY

FOREWORD BY STEVE W. CHADDE

Air War in the Pacific
The Journal of General George Kenney, Commander of the Fifth U.S. Air Force

George C. Kenney

with a Foreword by Steve W. Chadde

Foreword copyright © 2014 by Steve W. Chadde
Printed in the United States of America.

ISBN: 978-1951682743

Air War in the Pacific was first published in 1949 by Duell, Sloan and Pearce, New York, as *General Kenney Reports: A Personal History of the Pacific War,* by George C. Kenney. The original book is now in the public domain.

TYPEFACE: Calluna 10/12.5

FOREWORD

Air War in the Pacific is a classic account of a combat commander in action. General George Churchill Kenney arrived in the Southwest Pacific theater in August 1942 to find that his command, if not in a shambles, was in dire straits. The theater commander, General Douglas MacArthur, had no confidence in his air element. Kenney quickly changed this situation. He organized and energized the Fifth Air Force, bringing in operational commanders like Whitehead and Wurtsmith who knew how to run combat air forces. He fixed the logistical swamp, making supply and maintenance supportive of air operations, and encouraging mavericks such as Pappy Gunn to make new and innovative weapons and to explore new tactics in airpower application.

The result was a disaster for the Japanese. Kenney's airmen used air power—particularly heavily armed B-25 Mitchell bombers used as commerce destroyers—to savage Japanese supply lines, destroying numerous ships and effectively isolating Japanese garrisons. The classic example of Kenney in action was the Battle of the Bismarck Sea, which marked the attainment of complete Allied air dominance and supremacy over Japanese naval forces operating around New Guinea.

In short, Kenney was a brilliant, innovative airman, who drew on his own extensive flying experiences to inform his decision-making. *Air War in the Pacific* is a book that has withstood the test of time, and which remains as the definitive work on the Pacific air campaign in World War II.

ABOUT THE AUTHOR

At the time of original publication, General George C. Kenney was commander of The Air University of the United States Air Force, Maxwell Air Force Base, Montgomery, Alabama.

Born August 6, 1889, in Nova Scotia, where his family was vacationing, he was brought up in Brookline, Massachusetts. He attended grade and high schools in Brookline and spent three years at the Massachusetts Institute of Technology.

As a lieutenant pilot in World War I, General Kenney flew seventy-five missions, shot down two German planes, and was once shot down himself. He ended the war as a captain with the Distinguished Service Cross and the Silver Star.

As a peacetime officer, in the years from 1919 to 1939, he concentrated on aeronautical development and its application to warfare as he slowly worked his way up the Army promotion ladder to the grade of lieutenant colonel.

From 1939 to 1942 he served brief tours as Air Corps Observer with the Navy in the Caribbean; as Assistant Attache for Air at the American Embassy in Paris; as Commanding Officer, Air Corps Experimental Division and Engineering School, Wright Field, Ohio; and as Commanding Officer of the Fourth Air Force.

In August 1942, General Kenney assumed command of the Allied Air Forces in the Southwest Pacific. The story of the next three years of his career—from his transfer to combat duty to the surrender of the Japanese—is encompassed in *Air War in the Pacific*.

In December 1945 he was given an assignment with the Military Staff Committee of the United Nations, which took him from Washington to London to New York.

On April 1, 1946, he assumed command of the Strategic Air Command.

In October 1948, he was transferred to the command of The Air University at Maxwell Field.

General Kenney has been awarded the Distinguished Service Cross with cluster; Distinguished Service Medal with cluster; Silver Star; Distinguished Flying Cross; Purple Heart; Honorary Knight Commander, Military Division, Order of the British Empire; Croix de Guerre, with palm (Belgium); Grand Officer of the Order of Leopold with palm (Belgium); Philippine Star; Military Order of Merit First Class of Guatemala; Order of Orange-Nassau, degree of Grand Officer, with swords (Netherlands); William E. Mitchell Memorial Award (1948); Croix de Guerre, with palm (France), and the French Legion of Honor, Rank of Commander.

Although this array of awards and citations might seem to have set him apart in a rarefied atmosphere of military achievement, General Kenney has always been known, particularly among the officers and enlisted men of the Fifth Air Force, as "a soldier's general."

POST-WAR CAREER

In April 1946, Kenney became the first commander of the newly formed Strategic Air Command (SAC). In October 1948, Kenney became commander of the Air University, a position he held from until his retirement from the Air Force in September 1951.

During a career that spanned over 30 years, Kenney was awarded the Distinguished Service Cross with one oak leaf cluster, the Distinguished Service Medal with one oak leaf cluster, the Silver Star Medal, the Distinguished Flying Cross, the Bronze Star Medal, the Purple Heart and several foreign decorations. In addition to this book, Kenney wrote two books about World War II air campaigns in the Pacific: The Saga of Pappy Gunn (1959) and Dick Bong: Ace of Aces (1960). After his retirement, he lived in Bay Harbor Islands, Florida, where he died on August 9, 1977.

CLOSE-UP OF AN AIR GENERAL

(excerpts from TIME magazine cover story, January 18, 1943)

Last week was the most successful week of the war in the Southwest Pacific.... Last week belonged to the airmen. The center of the week's action focused first on a drab sedan which lurched over the pocked and pitted track that winds from Jackson airdrome to Port Moresby. The thick red dust of New Guinea blurred its windows, but not the three white stars on its license plate. Spying the stars, half-naked troops, Australian and American, grinned and threw casual salutes. One of their favorite brass hats was home again: General George C. Kenney, Commanding General of Allied Air Forces, Southwest Pacific Area, and Commander, Fifth U. S. Air Force.

The grins would have become cheers had the troops known what scrub-headed General Kenney was saying at that moment: "Good, Whitey! Let's smear 'em tomorrow."

"Whitey" was his deputy commander, Brigadier General Ennis C. Whitehead " 'Em" was the Japanese concentration at Rabaul....

In five months in the Southwest Pacific, the man chiefly responsible for the successes has yet to have a day off, or even to want one. General

Kenney's office is wherever he and his aide are at the moment. Places are always laid for General Kenney at two luncheon tables, one at Port Moresby, the other nearly 2,000 miles south in Australia. Most weeks he manages to have several meals at each of them. Last week he had three lunches at his mainland headquarters, two with MacArthur in New Guinea....

General Kenney was raised (to a height of 5 ft. 6 in.) in Brookline, Mass. He studied civil engineering at M. I. T., but left after three years to become an instrument man for Quebec Saguenay Railroad. Then he became a civil engineer and a contractor. In 1917 he enlisted in the U. S. Signal Corps as a private. He learned to fly under Bert Acosta, who was later to achieve fame as a transatlantic pilot. His first three landings were all dead stick, but he was notably successful once he got to France....

Between wars Kenney ... went through the routine which is designed to round out an air general: War College, Supply, Air Corps Engineering School, instructor in observation. In France in 1940 he riled other military observers by recommending that the U. S. throw its Air Force into the ash-can....

Between times he experimented. George Kenney was the first man to fix machine guns in the wing of a plane: back in 1922 he installed two .30-caliber Brownings in the wing of an old De Havilland. Kenney is the inventor of the parachute bomb, which enables bombing planes to fly lower, bomb more accurately. He invented this bomb in 1928....

"You've got to devise stuff like that," Kenney says. "I'd studied all the books ... and Buna was not in any of them."

The textbooks did not tell George Kenney what he would find in the Southwest Pacific.

THE BIG PICTURE IN THE PACIFIC
(excerpts from a paper by General Kenney describing the first phase of the war, up to July 1942)

On December 7, 1941, without warning and following the consistent pattern of her military tradition of dispensing with a formal declaration of war, Japan struck at Pearl Harbor.

Shortly after daybreak, possessing exact and accurate knowledge of its objective, and with each individual assigned a definite task and target, a superbly trained force of 360 fighter and bomber aircraft launched its bombs on the big Hawaiian naval base and its protecting Army and Navy air installations. In less than an hour our Pacific fleet was out of action, with most of its units heavily damaged or resting on the muddy floor of

Pearl Harbor. The Air Forces left to defend Hawaii were a mere handful. Shops, hangars, depots, supplies, and aircraft had gone up in smoke from the bombing and strafing of an air attack that was a model of perfection and precision.

Japan was then ready to begin the march to the rich empire to the south—the Philippine Islands, the Netherlands East Indies, Malaya, Melanesia, through Burma to India and perhaps even to Australia and New Zealand—without fear of interference by the United States Fleet, which up to then had been considered the strongest force in the Pacific. Nippon's dream of a Greater East Asia seemed ready for fulfillment.

The tide of Japan's conquest rolled on with a speed that seemed incredible to a paralyzed world. Almost coincident with the Pearl Harbor attack, a small American air force in the Philippines was knocked out of action and Japanese bombers appeared over Hong Kong and Singapore. Japanese fighter planes cleared the air of opposition and a flood of Japanese infantry, trained to the minute in amphibious warfare and schooled in the art of utilizing the jungle to aid them in overcoming the complacent and overconfident white man, poured into his Southeast Asia and East Indies holdings. Thailand became a satellite partner of Japan on December 8. Two days later, as Guam was being occupied by a Nipponese force that had sailed from Japan a week previously, the Prince of Wales and the Repulse, both first-line battleships of Britain's Far Eastern Fleet, went to the bottom off the Malay coast. Japanese torpedo bombers had proved to the Royal Navy something that it should have already known: surface craft cannot survive unless friendly aircraft control the air above them.

On the 25th, the white man's Christmas Day, Hong Kong, the Gibraltar of the East and a British colony for over a hundred years, fell to the Oriental conqueror. Wake Island had already gone on the 23rd. On January 2, 1942, Manila was occupied by the invaders, who had swarmed ashore on the main Philippine island of Luzon at Vigan and Aparri on December 10, 1941, at Legaspi on the 13th, and had landed in force at Lingayen Gulf on the 21st. That same month of January saw the Japanese take over Borneo, the Celebes, Ambon, New Ireland, New Britain, and move into the Solomons and the Gilberts. By February, Japanese bombs were falling on Darwin in northern Australia and on Port Moresby the capital of Papua and key port on the south coast of New Guinea. Singapore crashed ignominiously on February 15, with a British garrison surrendering to an invading force of half its size. Timor was seized on February 20. Portugal, the neutral owner of the eastern half of the island, was not consulted any more than were the belligerent

Dutch, who owned the western half. Sumatra, Java, and the rest of the Netherlands East Indies went early in March. Japanese occupation of Lae and Salamaua, strategically important ports and air bases on the north coast of New Guinea on March 8, followed a day after Rangoon was evacuated by the British. Resistance to the brown conqueror seemed to have stopped. Generals and their staffs evacuated themselves to India and Australia where they talked about last-ditch stands. Some of their troops got away with them. Most of them died in position or went to Japanese prison camps.

A gallant stand in the Philippines on Bataan held up the consolidation of the Nipponese empire for a while. On March 17, 1942, General Douglas MacArthur and a small nucleus of his staff arrived in Australia to take over command of the Allied forces there. It had taken repeated orders from the President of the United States to get General MacArthur to leave his mixed American and Philippine command. Twenty-three days later, out of food and ammunition, weakened by dysentery and malaria, his successor, Wainwright, evacuated Bataan and moved the remnants of his forces to the island fortress of Corregidor at the entrance to Manila Harbor, where he held out until May 6. America had lost everything in the Pacific west of Midway Island.

Flushed with victory, the Japanese pushed southward into Bougainville Island and the northern Solomons. A Nipponese naval task force, backed up with troop transports, sailed south from Rabaul, now the main enemy base in New Britain, into the Coral Sea. The goal was Port Moresby, the key not only to southern New Guinea but the jumping-off point for the invasion of the east coast of Australia itself. Between May 4 and 8, the Battle of the Coral Sea gave the Allies their first chance to take a real breath. Air reconnaissance had spotted the Japanese invasion force and had passed the word to a United States naval task force under Vice-Admiral Frank Fletcher which had been operating against the Japs occupying the important anchorage of Tulagi in the Solomons. Fletcher promptly moved into the Coral Sea to contact the enemy fleet. As soon as they were in range, both sides launched their carrier aircraft. Each side lost a carrier and some smaller vessels. The remaining carriers on each side were damaged sufficiently to put them out of action as far as this engagement was concerned. With one of his carriers sunk, one badly damaged, and most of his aircraft shot down, the Japanese Admiral Hinoue decided to abandon his mission and withdrew to Rabaul. For similar reasons our fleet withdrew south to New Caledonia. Tactically we had barely gained a draw, but strategically we had gained a real victory. Port Moresby had been saved from almost certain capture. A Japanese expedition had failed. It was the first one so far. Our most serious loss was the big aircraft carrier Lexington, but the price was small for the gain.

On June 3 more American territory changed hands when the banner of the Rising Sun was hoisted over Attu, Agattu, and Kiska in the Aleutian Islands off Alaska. Two days later the Allies took another breath. This was a deep one. Four aircraft carriers, a heavy cruiser, and 275 airplanes was the price the Japanese paid for their failure to occupy the tiny island of Midway, 1200 miles west of Pearl Harbor. A second Japanese expedition had failed. We lost the aircraft carrier Yorktown and a destroyer. Like the Battle of the Coral Sea, the Battle of Midway was an air show. No surface craft on either side fired a gun except at enemy airplanes.

July saw the tide reach its crest. Guadalcanal and Rekata Bay in the Solomons were occupied on July 6. On the 22nd a picked force of special Japanese landing troops seized the Buna-Gona area on the north coast of Papua and drove for the Kokoda Pass across the Owen Stanley mountain barrier protecting Port Moresby. Kokoda fell on the 29th and, driving a tired Australian militia brigade back, the invaders headed for Port Moresby in a furious headlong rush that carried them to within thirty miles of the port.

Here the tide turned.

UNCOMMON VALOR SERIES

Air War in the Pacific is part of a series entitled *Uncommon Valor,* taken from the quote by Admiral Chester W. Nimitz, U.S. Navy:

"Uncommon valor was a common virtue,"

referring to the hard-won victory by U.S. Marines on Iwo Jima. The intent of the series is to keep alive a number of largely forgotten books, written by or about men and women who survived extreme hardship and deprivation during immensely trying historical times.

Steve W. Chadde
SERIES EDITOR

As Commanding General of Allied Air Forces, Southwest Pacific Area, and Commander, Fifth U.S. Air Force, General George C. Kenney was a thinker who understood how to organize forces, mass them, achieve air superiority, and exploit the skies to achieve the broader purposes of a campaign.

PREFACE

Few men contributed more to the defeat of the Axis in World War II, or to the innovation in warfare known as air power, than General George C. Kenney. From the day in August 1942 when he assumed command of U.S. Army and Allied air forces in the Southwest Pacific Area to the final battles of the war against Japan in 1945, Kenney succeeded in providing the aerial support without which the land and naval campaigns could not have been won.

The obstacles-confronting him were formidable. His forces were thin and his commander, Douglas MacArthur, unrelenting in his determination to take the offense. Not only were the Japanese at first stronger, but distance, climate, logistics and the low priority of the theater in American grand strategy all conspired to make virtually every air operation a major undertaking. But George Kenney was a scrapper. He understood thoroughly the strengths and limitations of his men and equipment. He was an innovator, daring to modify aircraft and tactics to outwit and deceive the enemy. He was a leader, demanding but supportive of his commanders, encouraging to his aircrews, and consistent in his effort to provide his men some comforts against the rigor of combat and fatigue in a hostile clime. Most of all, he was a thinker who understood how to organize his forces, how to mass them, how to achieve air superiority, and then how to exploit the skies to achieve the broader purposes of a campaign.

In this fast-paced narrative first published in 1949, Kenney tells the story of the Southwest Pacific air war. First, he had to gain the confidence of his a boss. At their initial meeting, MacArthur recounted a long list of complaints against the air forces. After a half hour, Kenney interrupted. "I decided it was time for me to lay my cards on the table. . . . I told him that as long as he had enough confidence. . . the Air Force. . . would produce results. . . .I would be loyal, too. If at any time this could not be maintained. . . I would be packed up and ready for the orders sending me back home."

"George," MacArthur responded, "I think we are going to get along together all right."

Kenney then established the freedom to operate and support his forces as needed. He took the war to the Japanese. Almost immediately, he began experimenting with ordnance, changing fuzes to skip bomb Japanese ships, and wrapping bombs in wire so that air bursts would chew up Japanese planes parked in dirt revetments. For the next two years, he schemed day and night to wipe out the Japanese air forces, and he did it as much by guile and deception as by superior numbers. At the same time, his forces transported ground units and gave them fire support, pounded enemy shipping in order to interdict forces coming into battle, and provided crucial reconnaissance and intelligence to commanders at all echelons. It was a constant juggling act, not made any easier by geography or by the command arrangements in the Pacific, which in 1942-43 made the Solomons area a separate theater and which throughout divided the American effort between MacArthur's Southwest Pacific Area and Admiral Chester Nimitz's Central Pacific command. Nor was Kenney ever allotted enough in the way of resources. Three times he journeyed to Washington to plead for more aircraft, and three times he wheedled more out of a War Department reluctant to divert precious production from the main effort of the war: Europe. But Kenney made the most of what he got. He persevered and he succeeded. In the end, American forces—by means of control of the air (and thus of the sea)— were able to bypass enemy garrisons, and so weaken the enemy, that the Americans were poised to invade Japan itself without ever having had to fight the bulk of the Japanese Army.

The Office of Air Force History is reprinting General Kenney's memoir for use by the Department of Defense because of the book's value for today. While readers should ignore Kenney's characterization of the enemy (the book was written immediately following a war of unusual bitterness, by a man whose own personal involvement was intense), they should see in these pages a virtual textbook of theater command of air forces. How Kenney patterned his operations, how he related to ground and naval counterparts, his innovations, his daring to undertake the unconventional, his leadership style...all speak to issues of current importance to the military establishment. In addition, Kenney's experience provides a stimulating reminder that no war is ever as we expect it, and no commander is ever prepared quite perfectly for the conditions he will face.

The Introduction, based on wartime press releases, contemporary journalism, and General Kenney's own words, conveys the spirit of the time and places in wider context the observations in the narrative itself.

RICHARD H. KOHN

General George C. Kenney, LIFE Magazine, March 22, 1943.

CONTENTS

	PAGE
Foreword	iii
Preface	xi
1. Assignment to the Pacific, July 1942	3
2. With MacArthur in Australia, August 1942	19
3. New Guinea, August-September 1942	45
4. The Buna Campaign: I, September-October 1942	73
5. The Buna Campaign: II, November-December 1942	101
6. "Our Losses Were Light", January-February 1943	135
7. The Battle of the Bismarck Sea, March 1943	147
8. First Trip to Washington, March 1943	155
9. Slugging Match: Dobodura, April-May 1943	165
10. Air Supremacy: Marilinan and Wewak, June-August 1943	187
11. Nadzab and Lae September, 1943	209
12. Markham Valley September-October, 1943	223
13. Taking Out Rabaul, October-November 1943	233
14. Cape Gloucester, December, 1943	247
15. Second Trip to Washington, January 1944	254
16. Los Negros, February 1944	262
17. The Hollandia Operation, March-April 1944	277
18. Hollandia to Noemfoor, May-June 1944	295
19. Sansapor and Morotai, July-August 1944	309
20. Balikpapan, September-October 1944	325
21. The Battle for Leyte Gulf, October 1944	335
22. The Philippines: I. Leyte, November-December 1944	349
23. The Philippines: II. Mindoro, December 1944	371
24. The Philippines: III. Lingayen to Manila, January-February 1945	381
25. Third Trip to Washington, March 1945	397
26. Okinawa and the Kyushu Plan, May-July, 1945	413
27. The Japanese Surrender, August-September 1945	427

General Douglas MacArthur (center) with General George C. Kenney (right).

General George C. Kenney.

AIR WAR IN THE PACIFIC

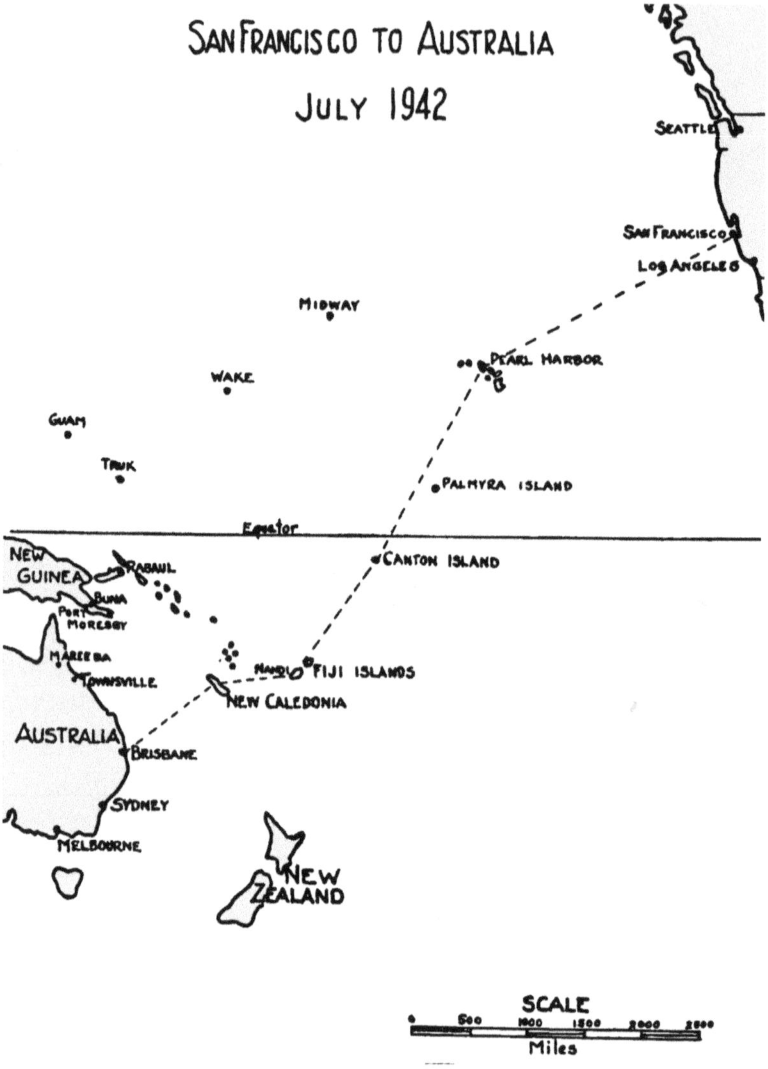
SAN FRANCISCO TO AUSTRALIA
JULY 1942
SEATTLE
SAN FRANCISCO
LOS ANGELES
MIDWAY
PEARL HARBOR
WAKE
GUAM
TRUK
PALMYRA ISLAND
Equator
CANTON ISLAND
NEW GUINEA
RABAUL
BUNA
PORT MORESBY
MAREEBA
TOWNSVILLE
NANDI
FIJI ISLANDS
NEW CALEDONIA
AUSTRALIA
BRISBANE
SYDNEY
MELBOURNE
NEW ZEALAND
SCALE
0
500
1000
1500
2000
2500
Miles

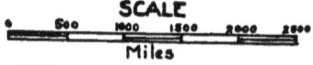

1. ASSIGNMENT TO THE PACIFIC
July, 1942

It was ten o'clock on the morning of July 7, 1942, at my headquarters in San Francisco where I was commanding the Fourth Air Force. I had just finished reading a long report concerning the exploits of one of my young pilots who had been looping the loop around the center span of the Golden Gate Bridge in a P-38 fighter plane and waving to the stenographic help in the office buildings as he flew along Market Street. The report noted that, while it had been extremely difficult to get information from the somewhat sympathetic and probably conniving witnesses, there was plenty of evidence proving that a large part of the waving had been to people on some of the lower floors of the buildings.

A woman on the outskirts of Oakland was quoted as saying that she didn't need any help from my fighter pilots in removing her washing from the clotheslines unless they would like to do it on the ground.

Considering the mass of evidence, it was surprising that more complaints had not been registered, but in any event I would have to do something about the matter. Washington was determined to stop low-altitude stunting and had put out some stringent instructions about how to handle the budding young aviators who broke the rules. The investigating officer had recommended a General Court-Martial.

I had sent word to the pilot's commander that I wanted to see the lad in my office, and I was expecting him at any minute. My secretary opened the door and said, "Your bad boy is outside. You remember—the one you wanted to see about flying around bridges and down Market Street."

I said to send him in. I heard her say, "The General will see you now, Lieutenant," and in walked one of the nicest-looking cherubs you ever saw in your life. I suspected that he was not over eighteen and maybe even younger. I doubted if he was old enough to shave. He was just a little blond-haired Norwegian boy about five feet six, with a round, pink baby face and the bluest, most innocent eyes—now opened wide and a bit scared. Someone must have just told him how serious this court-martial thing might be. He wanted to fly and he wanted to get into the war and do his stuff, but now he was finding out that they really were tough about this low-altitude 'buzzing' business and it was dawning on him that the commanders all had orders really to bear down on young aviators who flew down streets and rattled dishes in people's houses. Why, he might be taken off flying status or even thrown out of the Air Force! He wasn't going to try to alibi out of it, but he sure hoped this General Kenney wasn't going to be too rough. You could actually see all this stuff going on in his head just behind those baby-blue eyes. He didn't know it, but he had already won.

I let him stand at attention while I bawled him out for getting himself in trouble, and getting me in trouble, too, besides giving people the impression that the Air Force was just a lot of irresponsible airplane jockeys. He could see that he was in trouble just by looking at the size and thickness of the pile of papers on my desk that referred to his case. But think of all the trouble he had made for me. Now, in order to quiet down the people who didn't approve of his exuberance, I would have to talk to the Governor, the Mayor, the Chief of Police. Luckily I knew a lot of people in San Francisco who could be talked into a state of forgiveness, but I had a job of looking after the Fourth Air Force and I should spend my time doing that instead of running around explaining away the indiscretions of my wild-eyed pilots.

"By the way, wasn't the air pretty rough down in that street around the second-story level?" I was really a bit curious. As I remembered, it used to be, when I was first learning to fly.

"Yes, sir, it was kind of rough," replied the cherub, "but it was easy to control the plane. The aileron control is good in the P-38 and—" He paused. Probably figured he had said enough. For a second, the blue eyes had been interested more than scared. He was talking about his profession and it was more than interest. It was his life, his ambition. I would bet anything that he was an expert in a P-38 and that he wanted to be still better. We needed kids like this lad.

"Lieutenant," I said, "there is no need for me to tell you again that this is a serious matter. If you didn't want to fly down Market Street, I wouldn't have you in my Air Force, but you are not to do it any more and I mean what I say. From now on, if I hear any more reports of this kind about you, I'll put you before a General Court and if they should recommend dismissal from the service, which they probably would, I'll approve it."

I began slowly to tear up the report and drop the pieces of paper in the waste basket. The blue eyes watched, a little puzzled at first, and then the scared look began to die out.

"Monday morning you check in at this address out in Oakland and if that woman has any washing to be hung out on the line, you do it for her. Then you hang around being useful—mowing a lawn or something—and when the clothes are dry, take them off the line and bring them into the house. And don't drop any of them on the ground or you will have to wash them over again. I want that woman to think we are good for something besides annoying people. Now get out of here quick before I get mad and change my mind. That's all."

"Yes, sir." He didn't dare to change his expression, but the blue eyes had gone all soft and relieved. He saluted and backed out of the office. The next time I saw Lieutenant Richard I. Bong was in Australia.

I was still chuckling to myself over the look on that kid's face as he watched me tearing up the charges against him and thinking how wonderful it would be to be twenty-two and a lieutenant flying a P-38 instead of fifty-two and a general looking after a whole Air Force, when the light flashed on the direct telephone line from my desk to General 'Hap' Arnold, the head of the Army Air Forces in Washington.

Hap believed in working directly with his commanders. He made quick decisions and he demanded immediate action. Once in a while, when his staff had given him insufficient information, his decisions would be wrong. If you had the real facts at your fingertips and could present the case briefly and correctly, you could argue with Hap, but your argument had better be good if you wished to emerge from the 'brawl' unscathed. On the other hand, while the interview might be, and generally was, exceedingly stormy, if you put across your point he would reverse his decision immediately and correct the situation.

I remember once hearing Hap say that if he had a hundred problems put up to him in a day he would make one hundred decisions that day. Fifty of them would be good decisions. The other fifty might range from good to fair to poor. Before the day was over he would find out about most of these and correct them, but some days he might make twenty-five that were not so good and he'd just have to take the blame for them later on.

While he and I have had lots of arguments, and the finish on a lot of desk tops has suffered in the process, I like to work for a man who will make decisions. If he is right three times out of four, his batting average is far better than most. Hap and I understood each other, we respected each other's judgment and were strong personal friends of over twenty years' standing. He called me almost daily about a multitude of matters, some big, some little, and sometimes, I suspected, just to blow off a little excess steam. Hap lived with the throttle well open most of the time. I wondered what was on his mind now.

"George, I'm keeping my promise to you. Pack up and be in my office at eight o'clock Monday morning. That's the twelfth. I'll give you all the dope then. I can't tell you any more on the phone. Tell me, who do you recommend to take over your job?"

I knew this meant that I was leaving the command of the Fourth Air Force for another job. I had three brigadier generals under me, but Barney Giles was my choice if one of them was to succeed me. I told Hap that I wanted to turn the job over to Giles but that he would have to get out some special orders from Washington as Barney was outranked by both the other brigadier generals.

Hap said okay and that he would get the orders out right away. We said goodbye and I told him I would be in his office on Monday morning.

Back in 1940, General Arnold had ordered me to the Air Corps Materiel Division at Wright Field, Dayton, Ohio, to look after the aircraft-production program, which was being stepped up to increase our air strength, and to coordinate the stepped-up output with the huge orders then being placed in this country by the British and French. At that time I had remarked to him that, the way things were going, it looked to me as though we would be getting in the war ourselves before very long. If we did get in it, I wanted him to let me take a combat outfit. I promised him that I would get the aircraft production speeded up but asked him not to make me stand around counting airplanes coming out of factories if we started shooting bullets for keeps.

Hap refused to give me any definite promise right then but hinted quite strongly that if the production machinery were set up and operating satisfactorily he would not keep me out of the combat action.

By February 1942, the program was getting into high gear. The aircraft output was already exceeding our fondest hopes and the curves on the charts were all spiraling upward. Hap called me at Detroit, where I was completing the final negotiations to put the Ford Company into the bomber manufacturing business. It was a typical Arnold telephone call.

"George, pack up and move to San Francisco and take over the Fourth Air Force. I want a lot of fighter and bomber units trained in a hurry, the offshore reconnaissance maintained, and I want to see that accident rate come down. Come in and see me tomorrow morning at ten o'clock and I'll give you the whole story. By the way, I hope you will notice that I'm starting to keep my promise to you. Oh, yes, another thing. You will be promoted to major general by the time you get to California. Goodbye—see you tomorrow."

A couple of weeks later, with two stars on my shoulders, I arrived in San Francisco and took command of the Fourth Air Force. Now, only four months later, another telephone call was putting me on the move again.

I sent for Brigadier General Barney Giles, who headed my Fourth Bomber Command, and my top staff officers and told them that I would be leaving the Fourth Air Force in a few days, that General Giles was the new boss, and that I expected them to work even harder for him than they had for me as I had no authority to take any of them along where I was going. I told Barney to select the best colonel in his command to succeed him and to put in a recommendation right away for his promotion which I would take to Washington with me; at the same time I would see what I could do to fix up Giles with a promotion to the rank of major general.

The next job was to call my other two brigadier generals and explain to them that, while I had nothing against them and I considered their work eminently satisfactory, I believed Giles was the best of the three and better qualified to take over when I left. I told them that I expected them to serve Giles as loyally and as faithfully as they had me but that, if they did not want to carry on in their present assignments, I would do what I could to get General Arnold to place them where they wanted to go. They both assured me that I need not worry about the matter at all, that they had jobs to do and not only liked Giles but would do everything they could to make the Fourth Air Force a credit to its new commander. I had known they would but I felt much better after I had talked to them. After all, no one likes to be "passed over" by someone junior in length of service.

Sunday afternoon, July 11th, I landed at Bolling Field with Major William Benn, my aide, and went over to Air Force Headquarters to see if I could find out anything. Arnold was not in, but I saw Major General "Joe" McNarney, an old friend of mine who was now Deputy to the Army Chief of Staff, General George C. Marshall. Joe told me I was going to Australia. MacArthur was not satisfied with Brett or with the way the Air Force was working. My name had been submitted and MacArthur had said I was acceptable. Joe wished me luck and remarked that, from the reports coming out of that theater, I was going to need it.

Lieutenant General George H. Brett had been General Wavell's deputy in the now disbanded American-British-Dutch-Australian Command. When the Japs ran the Allies out of Java in March 1942, Wavell had gone to India and Brett had taken over the Allied command in Australia. On MacArthur's arrival in that country a couple of weeks later, Brett, as the senior airman in the theater, became General MacArthur's Allied Air Force Commander. He had not had much to work with and his luck had been mostly bad. MacArthur's Chief of Staff, Major General Richard Sutherland, and Brett had seldom seen eye to eye on anything. Brett's own staff was nothing to brag about and that had not helped either. I had worked under Brett when he commanded the Air Materiel Division at Dayton a couple of years before and liked him a lot. I hoped they would give him a good job when he got back. General McNarney said he hadn't heard what Brett was lined up for.

I talked with McNarney at some length about the general situation. Europe was going to get the real play, with a big show going into North Africa that November to chase Rommel and his Afrika Korps out of the area before they grabbed Egypt and the Suez Canal on one of their tank drives.

The Pacific war would have to wait until Germany was disposed of before any major effort would be made against Japan. There simply were not enough men, materials, or shipping to run big-scale operations in Europe and the Far East at the same time, so the decision had been made to concentrate on defeating Hitler and to take care of the other job later on.

McNarney held out no encouragement that troops, aircraft, or supplies in any appreciable numbers or quantities would be sent to the Pacific for a long time to come. No wonder he wished me luck.

The next day General Arnold took me into General Marshall's office, where I was officially told about my new assignment. My instructions were simply to report to General MacArthur. Their analysis of the problems that I would be up against not only confirmed what I had heard from McCamey but sounded even worse. The thing that worried me most, however, was the casual way that everyone seemed to look at the Pacific part of the war. The possibility that the Japs would soon land in Australia itself was freely admitted and I sensed that, even if that country were taken over by the Nipponese, the real effort would still be made against Germany.

I gathered that they thought there was already enough strength in the Pacific, and particularly in Australia, to maintain a sort of "strategic defensive," which was all that was expected for the time being. Arnold said he had sent a lot of airplanes over there and a lot of supplies, but the reports indicated that most of the airplanes were out of commission and that there didn't seem to be much flying going on. He said that Brett kept yelling for more equipment all the time, although he should have enough already to keep going. I was told that there were about 600 aircraft out there and that should be enough to fight a pretty good war with. Anyhow, while they would do what they could to help me out, they just had to build up the European show first.

General Marshall said there were a lot of personality clashes that undoubtedly were causing a lot of trouble. I said I knew of some of them already and that I wanted authority to clean out the dead wood as I didn't believe that much could be done to get moving with the collection of top officers that Brett had been given to work with. They told me that I would have to work that problem out with General MacArthur. I said I wished that they would wash the linen before I got out there and save me the trouble, but I didn't get to first base with the suggestion.

Luckily I would have two good brigadier generals to work with me as Ennis Whitehead and Ken Walker had already been ordered to Australia. I had known both of them for over twenty years. They had brains, leadership, loyalty, and liked to work. If Brett had had them about three months earlier, his luck might have been a lot better.

I stayed around Washington for three more days, absorbing all the data I could find in regard to the Southwest Pacific Area. I knew Arnold didn't think much of the P-38 as a fighter plane, so it wasn't hard to get him to assign me fifty of them with fifty pilots from the Fourth Air Force. I intended to make sure that a Lieutenant Richard I. Bong was one of the pilots.

While looking around for anything that was not nailed down, I found that there were 3000 parachute fragmentation bombs in war reserve. No one else wanted them, so they were ordered shipped to Australia on the next boat.

Back in 1928, in order to drop bombs in a low-altitude attack without having the fragments hit the airplane, I had put parachutes on the bombs; the parachutes opened as the bombs were released from the airplane. The parachute not only stopped the forward travel of the bomb, but slowly lowered it down to the ground while the airplane got out of range of the fragments by the time the bomb hit the ground and detonated. With a supersensitive fuze, which kicked the thing off instantaneously on contact with anything—even the leaf of a bush, the bomb was a wicked little weapon. It weighed about twenty-five pounds and broke up into around 1600 fragments the size of a man's little finger. At a hundred yards from the point of impact these fragments would go through a two-inch plank. I had had a hard time getting the Air Corps or the Ordnance to play with the thing, in spite of a dozen demonstrations I had put on. It was actually 1936 before an order of about 5000 was made up for service test. Everyone that used them was enthusiastic, but somehow or other the 3000 remaining got hidden away in war reserve and people gradually forgot about them. I think the Ordnance Department was actually glad to get rid of them. But I was speculating about trying them out on some Jap airdrome and wondering if those fragments would tear airplanes apart—as well as Japs, too, if they didn't get out of the way.

Hap Arnold told me that Major General Millard (Miff) Harmon, who had been his Chief of Staff, had been ordered to the South Pacific Theater, where, under Vice-Admiral Robert H. Ghormley, he was to command all Army troops, air and ground. He was taking with him, to command his Army Air Force units, Brigadier General "Nate" Twining and several colonels. We would all go out together in a Liberator bomber that had been converted into a sort of passenger carrier. Sort of—because they had dispensed with such luxuries as soundproofing, cushioned seats, heating system, or windows to look out of. It might not be very comfortable, but I preferred it to spending a month on a boat. According to the present schedule, we would leave on July 21 from San Francisco for New Caledonia, via Hawaii-Canton Island-Fiji. Miff and his gang would get off in New Caledonia, and Bill Benn and I were to have the plane to ourselves during the remainder of the trip to Australia.

Among the multitude of people that I conferred with about the arrangements for sending airplanes, engines, propellers, and spare parts to Australia to keep my show going, was "Bill" Knudsen, the former president of General Motors, who had been called into government service to speed up production of war materials, particularly through mobilization of the automotive industry of the country. I had had a lot of dealings with him during the past two years, and besides admiring his methods of getting things done, I had gotten very fond of him. We liked each other.

His ability in his field was unquestioned. His simple honesty, his sincerity, his unselfish patriotism, and his unfailing sense of humor endeared him to everyone. The country owed a lot to William S. Knudsen, who had been made a lieutenant general a few months previously, with the additional title of Director of War Production for the War Department.

Bill had just come back to his office from a long conference which had consisted of a lot of talk but no decision. In that charmingly thick Danish accent of his, he told me about it and then suddenly said, "George, do you know what a conference is?"

I said, "Go ahead. I'm listening." He grinned, hesitated a little, and gave me his definition. I still consider it a gem. It still fits most of them.

"A conference is a gathering of guys that *singly* can do nothing and *together* decide that nothing can be done."

My favorite of all Bill Knudsen's sayings, however, came one day in Washington when we were listening to an efficiency expert give us a lecture on speeding up the process of letting contracts to industry. In the course of the talk, he kept using the phrase, "Status quo." After about ten repetitions of the two words, Knudsen nudged me and whispered, "Georgy, do you hear that 'status quo' stuff?" I nodded. Bill waited a few seconds and continued, "That's Latin for what a hell of a fix we're in."

The last thing I did before leaving Washington was to get Major William Benn fixed up with orders to go along with me as my aide.

I inherited Benn as an aide when I took over the Fourth Air Force. It wasn't long before I found out that he was much more than the generally accepted version of a general's aide. I could open my own doors and put on my own overcoat without any help, so I "fired" Benn as aide and ordered him to take over a heavy-bombardment squadron that needed a leader to put it on its feet. In a week that squadron was the best in the Fourth Air Force. Benn had leadership, energy, new ideas, and enthusiasm to burn. In addition, he was a lot of help as a personal staff assistant. That was the kind of aide I really needed, so when Arnold told me I was leaving San Francisco, I recalled Benn from command of his squadron and now I insisted on taking him with me to Australia. A general is supposed to have an aide, so Arnold said I could take him along. Benn grinned when I told him the news and asked how long I thought he would hold the job this time. I told him not to buy any more aide's insignia until I told him to. He would be a lot of help to me in lots of ways but I had a hunch that I was going to need some young commanders out there and Bill Benn had the makings of a real star performer.

On July 16, I said goodbye to Hap Arnold, and Benn and I flew back to San Francisco.

During the next five days, I wound up my affairs in San Francisco and packed for the move. I was not sorry to leave but my job on the West Coast had been a pleasant one. General DeWitt was a prince to deal with. He believed in this loyalty business working both ways. He definitely had a mind of his own and a temper that was a joy to watch in operation, but he was a square shooter and his decisions were sound. He told you what he wanted done and then let you alone to produce results. We got along together fine, in spite of my complicated assignment. I was responsible to DeWitt for the air defense of the West Coast and the offshore reconnaissance. At the same time, I was training fighter and bomber groups under General Arnold. We combined the two missions as far as possible and worked out a fairly respectable solution. The bomber crews learned navigation, instrument flying, and reconnaissance by performing offshore patrols to distances seaward up to 500 miles. If a submarine or a suspicious wake or a trail of oil were sighted, the airplane passed the word ashore, giving the position, and then circled the suspicious spot until a destroyer or some other type of antisubmarine vessel arrived to take up the hunt with its sound-detecting apparatus and depth-bombing charges. Once in a while the bomber would get a chance to drop its bombs, but most of the time the coup de grace was left to the surface vessel. This scheme of handling the submarine menace, later known as the "hunter-killer" method, became standard procedure throughout the war in both the Atlantic and the Pacific.

With all flying, both military and civilian, in the states of California, Oregon, and Washington under my control, we had plenty of opportunity to train the fighters in interception of other aircraft in the air. Fighter control centers were established in Seattle, Washington, Portland, Oregon, San Francisco, Los Angeles, and San Diego; into them was fed a continuous stream of information from thousands of volunteer civilian ground observers, who telephoned aircraft sightings in to the nearest control center. At these centers, where all air traffic was plotted by more civilian volunteers working in shifts twenty-four hours a day, Air Force, Navy, and civil airline controllers identified the aircraft from flight schedules. Markers representing the airplanes were kept moving over a huge flat map of the area by the plotters, in accordance with the information coming in from the ground-observer stations. All flight schedules had to be furnished us and approved before an airplane could fly in our area of responsibility. If an airplane was sighted which could not be identified by any of our controllers, fighters from one of my airdromes were dispatched to intercept the intruder and escort him to the nearest landing field for identification. The scheme worked with surprisingly little interference with either military or civil air traffic. Once in a while someone would neglect to notify us of a flight and would find himself forced down. I remember one time "Tom" Girdler, the steel tycoon, who had a private plane and who had just taken over the management of Consolidated Aircraft at San Diego, decided that he could cross the Arizona-California line and visit his own factory without asking any permission from anyone. Our fighters picked him up, called his pilot on the radio, and ordered the plane landed at a nearby airdrome. A little hesitancy on the part of Girdler's pilot prompted the observation that the fighters' guns were loaded and the argument was settled immediately. Mr. Girdler didn't quite approve and so stated rather forcibly on landing, but our fighter pilot acted as a well-bred policeman should, explained the system and the fact that he was simply carrying out his orders. He then called the Los Angeles control center, received permission for Mr. Girdler to proceed, and escorted him all the way to San Diego. I saw Tom a couple of days later. Much to my gratification, he was loud in his praises of the youngster and the system. He said he had not understood the situation or he would never have tried to crash through and, with his plant so close to a possible raid from some clandestine base which Jap agents might have established in Lower California, he was glad that our surveillance and control were that good.

On another occasion Admiral McCain, in charge of the Naval Base at San Diego, found himself intercepted on the way to Los Angeles when his operations officer at the Navy field at San Diego neglected to notify either the Los Angeles or the San Diego control center of the flight. McCain was also a bit more than annoyed at first, but, when he found out whose fault it was, he promptly replaced the offending operations officer.

Our radar warning service was a bit sketchy. We had only six sets to cover the whole West Coast from Seattle to San Diego and some of our operators were still learning how to operate their equipment. Our fighters took off on many wild-goose chases to intercept "big formations of unknown aircraft" spotted out over the Pacific and presumably coming from a Jap aircraft carrier. We never found these "enemy" attackers, but it was good practice for the fighters, to see how fast they could get going, and for the bombers, who would load their bombs and go on the alert to take off as soon as the "enemy" carrier could be located. One morning it looked like the real thing. The radar scope clearly showed a lot of "spots" at fairly low altitude about fifty miles due west of the Golden Gate. In less than five minutes our whole available fighter force in the San Francisco Bay area, totaling about 35 P-38s, was roaring out over the Pacific for the big "kill." All fighter and bomber squadrons on the West Coast were alerted and told to stand by for orders. A half hour later, I called everything off and told the gang to go back to breakfast. An incoming vessel had unloaded its garbage at the point where the radar operator had observed the "enemy" formation, and a few thousand sea gulls, wheeling and diving at this welcome meal, had been picked up on the radar scope and caused the illusion. The kids were disgusted, but I was just as well satisfied. It was not time yet to send my green fighter pilots—who were just learning how to fly the P-38 and who knew practically nothing of aerial gunnery—up against the seasoned Nip veterans of the Pearl Harbor, Manila, and Hong Kong operations.

The day before I was to take off for Australia, I went over to General DeWitt's headquarters to say goodbye. He was very complimentary about my show under him and said how sorry he was to have me leave but that he knew how I felt about going to a real combat theater. Said he wished he were going with me. He showed me a message that he had sent to General MacArthur the day I had returned from Washington and told him where I was going. The message read: "Major General George C. Kenney Air Corps has received orders relieving him from duty Fourth Air Force this command and directing him report you for assignment stop Regret to lose him but congratulate you stop He is a practical experienced flyer with initiative comma highly qualified professionally comma good head comma good judgment and common sense stop High leadership qualities clear conception of organization and ability to apply it stop Cooperative loyal dependable with fine personality stop Best general officer in Air Force I know qualified for high command stop Has demonstrated this here stop Best wishes."

DeWitt then grinned and said, "Now read this," and handed me MacArthur's slightly "I'm from Missouri" answer: "Personal for General DeWitt stop Appreciate deeply your fine wire and am delighted at your high professional opinion of Kenney stop He will have every opportunity here for the complete application of highest qualities of generalship stop Wish you and the Fourth Army could join me here we need you badly."

We both chuckled over it, shook hands, and wished each other luck for the duration.

At 9:30, the evening of July 21, with Major General Miff Harmon and his staff, Bill Benn and I took off from Hamilton Field, about thirty miles north of San Francisco, and headed west for Hickam Field, Oahu, Hawaiian Islands.

Shortly after daybreak the next morning we circled Pearl Harbor and landed at Hickam Field. While Pearl Harbor still showed the scars of the Jap attack of last December 7, with the Oklahoma still lying capsized on the bottom and the Arizona a tangled mass of twisted wreckage sticking up out of the water, the whole island of Oahu was bustling with activity. Both Army and Navy had certainly become "shelter" conscious. Everybody seemed to be digging in. Fuel storage tanks were being installed underground. A complete underground headquarters was in operation by the Army Air Force and a huge air depot and engine overhaul plant was being constructed by tunneling into the side of a mountain.

Hickam Field and Wheeler Field, the two Army airdromes at the time of the big Jap attack, still showed the effects of the disaster. Hangars, barracks and warehouses, burned and wrecked during the bombing, had not yet been rebuilt. There was no doubt about it, Hawaii had been in the war and, from all the digging going on, they had not forgotten it either.

When I asked about the news of the war out where I was going, I learned that, the day we left San Francisco, the Japs had made a new landing on the north coast of New Guinea at a place called Buna and had started driving the Australians back along the trail over the Owen Stanley Mountains toward Port Moresby. MacArthur's communique spoke of the landing being opposed by our aircraft, which had sunk some vessels but had not stopped the invasion. I figured that it probably meant more trouble for me. Anyhow I'd find out in a few days.

On the 24th, we flew to Canton Island and spent the night at the Pan American Hotel which had been taken over by the Navy, who had done a real job blasting a channel through the coral reef to get an anchorage for vessels and building an excellent flying field on the atoll itself. A squadron of Army Air Force P-39 fighters was stationed there for protection from Jap air attack, although the Nips had no airdromes within operating range. If an aircraft carrier should get close enough to threaten the place, this one P-39 squadron couldn't do anything about it except be the gesture of defense that it really was. As a matter of fact, if the Japs had really wanted Canton Island, they wouldn't have needed much of an expedition to take it. It was practically defenseless against any decently organized attack.

The next day we arrived at Nandi in the Fiji Islands about eleven o'clock in the morning. We taxied up in front of a hangar, stopped the engines, got out, and looked around to see if anyone lived in the place. We had been in radio communication, but there certainly was no welcoming party for General Miff Harmon, the new commander of the South Pacific Theater, which included Fiji. After about fifteen minutes, a young lieutenant hurriedly drove up in a jeep and asked for General Harmon. Miff identified himself and was handed a message from the colonel commanding the Nandi Base, saying that he would see him at mess at 12:30, as at the moment he was taking a sun bath. Miff handed me the note without saying a word. I read it and handed it back to him, also without a word. Miff put it in his pocket and said, "Let's go find where we are going to stay and where they serve that mess the colonel mentioned."

I suspected that there would be a new commander there shortly. (Note: There was, although he did hold the job for another week.)

Harmon's new command included an American Infantry Division, the 38th, which was relieving the New Zealand troops who were pulling out. We went over to their headquarters and got briefed on how they would defend the place if the Japs should try to take over Fiji. It sounded too much like the old textbook stuff to be very impressive. I didn't get the impression that they realized that World War Two differed radically from World War One. However, while we may get pushed around rather roughly at first, we generally learn fast. The bad thing about it is that, in war, it costs lives to learn that way.

To take care of the air situation, Harmon had a couple of P-39 fighter squadrons and two squadrons of B-26 medium bombers stationed at Nandi to defend the Fiji Islands, along with a few New Zealand reconnaissance planes which were helping out on antisubmarine patrol. It wasn't much more than a token force, at best.

So far, I had seen nothing to indicate that our air line of communication across the Pacific was very secure. Canton was wide-open and any one of the Fiji Islands could have been taken easily, including Viti Levu, the one that Nandi is on, and the only one defended at all. If either of these links in the chain were taken out, the air route would be gone and the ship distance to Australia increased by at least another thousand miles to keep out of range of Jap bombers that would then be based at our former Canton or Fiji airdromes.

Benn and I had been discussing low-altitude bombing all the way from San Francisco. It looked as though there might be something in dropping a bomb, with a five-second-delay fuze, from level flight at an altitude of about fifty feet and a few hundred feet away from a vessel, with the idea of having the bomb skip along the water until it bumped into the side of the ship. In the few seconds remaining, the bomb should sink just about far enough so that when it went off it would blow the bottom out of the ship. In the meantime, the airplane would have hurdled the enemy vessel and would get far enough away so that it would not be vulnerable to the explosion and everyone would be happy except the Japs on board the sinking ship.

The more we talked about the scheme, the more enthusiastic we got, so finally we borrowed a B-26 from the boys at Nandi, loaded on some dummy bombs, and tried the idea out against some coral knobs just offshore. It was quite evident that it was going to take quite a bit of experimental flying to determine the proper height for release of the bomb and how far from the ship it should be released. From this first experiment it looked as though 100-feet altitude and a distance of about 400 yards would be somewhere near right. We bounced some bombs right over the targets, others sank without bouncing, but finally they began skipping along just like flat stones. Benn and I both agreed that we would have to get some more firepower up in the nose of the bomber to cover us coming in on the attack if the Jap vessels had very much gun protection on their decks, but it looked as though we had something. The lads at Fiji didn't seem to think much of the idea but I decided that as soon as we got time after we got to Australia I would put Benn to work on it. He was really enthusiastic about it, particularly after we began to score some good "skips" against the coral knobs.

On July 28 the whole party hopped over to Plaine des Gaiac, a new field that American engineers had built on the west coast of New Caledonia, a couple of hundred miles northwest of Noumea where Miff's headquarters had been located. Another plane flew up from Noumea to ferry Miff and his party to their new home, so Benn and I pushed off right after lunch for Brisbane, Australia, on the last leg of our transpacific joy ride. My trouble-shooting job was about to start.

2. WITH MacARTHUR IN AUSTRALIA
August, 1942

WE HAD excellent weather all the way from New Caledonia, which gave us a chance to appreciate the beautiful blue water of Australia's east coast and the startling green colors around the pink and white coral reefs offshore. The land itself made you feel at home. Here were well-kept farms and villages, beaches that looked as though they were used as summer resorts, and rivers with wharves and boats on them. After all the water, atolls, and jungle-covered islands I had seen for the past week I was glad to be arriving among people and civilization again.

We crossed the coastline, flew inland across the city of Brisbane, Australia's third largest city, sprawling on both sides of a river and looking to be about the size of Cincinnati, Ohio. With its multi-storied business district, apartment houses, and the outskirt fringe dotted with bungalow-type single houses, from the air it could have been any one of a dozen Midwest American cities.

Just before sundown, we landed at Amberley Field, about twenty miles west of Brisbane. As we got out of the airplane, a Royal Australian Air Force billeting officer at once took me in tow and in a few minutes I was on my way to town assigned to Flat 13 in Lennon's Hotel. Benn was also assigned there. It was explained that only the top brass and their aides lived there, as it was not only the best hotel in town but the only one that was air-conditioned.

The air-conditioned part didn't interest me very much as August is a winter month in Australia and, although it was only twenty-eight degrees below the equator or about the same distance south that Tampa, Florida, is north, the wind was chilly enough so that I was glad I had on an overcoat.

Flat 13, however, sounded all right. Although I don't claim to be superstitious, a lot of good things have happened to me with that number involved. Back in 1917 I passed my flying tests, which first entitled me to wear pilot's wings, on the thirteenth. I left New York to go overseas in World War 1 on that date; my orders sending me to the front were dated February 13, 1918. I shot down my first German plane on the thirteenth, and so on. That number on my door at Lennon's Hotel looked like a good start.

After dinner I spent a couple of hours talking to Major General Richard K. Sutherland, General MacArthur's Chief of Staff. I had known him since 1933 when we were classmates at the Army War College in Washington. While a brilliant, hardworking officer, Sutherland had always rubbed people the wrong way. He was egotistic, like most people, but an unfortunate bit of arrogance combined with his egotism had made him almost universally disliked. However, he was smart, capable of a lot of work, and from my contacts with him I had found he knew so many of the answers that I could understand why General MacArthur had picked him for his chief of staff.

I got along with him at the Army War College when we were on committees together and I decided that I'd get along with him here, too, although I might have to remind him once in a while that I was the one that had the answers on questions dealing with the Air Force. Sometimes, it seemed to me, Sutherland was inclined to overemphasize his smattering of knowledge of aviation.

It didn't take long for me to learn what a terrible state the Air Forces out here were in. Sutherland started out right at the beginning on December 8, 1941. He said that he advised General Lewis H. Brereton, the Air Commander in the Philippines at that time, to move his airplanes to Mindanao and that, as Brereton didn't do it, the loss of our planes on the ground to the Jap bombing and strafing attack was Brereton's fault. He claimed that, following the attack, our air officers and men were so confused and bomb-happy that the military police were busy for days rounding them up around Manila. He castigated all the senior air officers except Brigadier General H. H. George, Brereton's Fighter Commander in the Philippines, who was killed in May 1942 at Darwin, Australia, when he was hit by an airplane which swerved off the runway.

I had known "Hal" George ever since World War I. He was a good boy and, according to all the reports, he had been a thorn in the side of the Japs in the Philippines right up to the time he had gotten down to his last fighter plane. His loss by that unfortunate accident at Darwin was another piece of bad luck for George Brett.

The more I heard about those opening events of the war in the Philippines the more it seemed to me that the confusion that day had not been confined to the Air Force. These stories about everyone else being calm and collected were told and retold with so much emphasis that I began to suspect they were alibis.

Sutherland finally worked up to present-day conditions. Brett certainly was in wrong. Nothing that he did was right.

Sutherland said they had almost had to drive Brett out of Melbourne when MacArthur's headquarters moved from there a couple of weeks previously to the present location in Brisbane. According to Sutherland, none of Brett's staff or senior commanders was any good, the pilots didn't know much about flying, the bombers couldn't hit anything and knew nothing about proper maintenance of their equipment or how to handle their supplies. He also thought there was some question about the kids having much stomach for fighting. He thought the Australians were about as undisciplined, untrained, over-advertised, and generally useless as the Air Force. In fact, I heard just about everyone hauled over the coals except Douglas MacArthur and Richard K. Sutherland.

Come to think of it, he did say one thing good about the men of the Air Force. It seemed that up on Bataan, after all their airplanes were gone and they had been given some infantry training, they made good ground troops. . . .

This was going to be fun. Already it began to look like a real Number One trouble-shooting job.

The next morning I saw General Brett at Allied Air Force headquarters on the fifth floor of the AMP Building, a nine-story life-insurance building which had been taken over by Allied Headquarters and in which General MacArthur and his staff, together with the Air and the Navy commands, maintained their offices. We talked for a while and then I went upstairs to the eighth floor to report to General MacArthur. General Brett did not go with me. Sutherland said the "Old Man" was in, so I walked into his office and introduced myself. The general shook hands, said he was glad to see me, and told me to sit down.

For the next half hour, as he talked while pacing back and forth across the room, I really heard about the shortcomings of the Air Force. As he warmed up to his subject, the shortcomings became more and more serious, until finally there was nothing left but an inefficient rabble of boulevard shock troops whose contribution to the war effort was practically nil. He thought that, properly handled, they could do something but that so far they had accomplished so little that there was nothing to justify all the boasting the Air Force had been indulging in for years. He had no use for anyone in the whole organization from Brett down to and including the rank of colonel. There had been a lot of promotions made which were not only undeserved but which had been made by Washington on Brett's recommendations. If they had come to him, he would have disapproved the lot. In fact, he believed that his own staff could take over and run the Air Force out here better than it had been run so far.

Finally he expressed the opinion that the air personnel had gone beyond just being antagonistic to his headquarters, to the point of disloyalty. He would not stand for disloyalty. He demanded loyalty from me and everyone in the Air Force or he would get rid of them.

At this point he paused and I decided it was time for me to lay my cards on the table. I didn't know General MacArthur any too well as I had just met him a few times when he was Chief of Staff of the Army six years before, but if he was as big a man as I thought he was, he would listen to me and give me a chance to sell my stuff. If not, this was a good time to find out about it even if I took the next plane back to the United States. I got up and started talking.

I told him that as long as he had had enough confidence in me to ask for me to be sent out here to run his air show for him, I intended to do that very thing. I knew how to run an air force as well or better than anyone else and, while there were undoubtedly a lot of things wrong with his show, I intended to correct them and do a real job. I realized that so far the Air Force had not accomplished much but said that, from now on, they would produce results. As far as the business of loyalty was concerned, I added that, while I had been in hot water in the Army on numerous occasions, there had never been any question of my loyalty to the one I was working for. I would be loyal to him and I would demand of everyone under me that they be loyal, too. If at any time this could not be maintained, I would come and tell him so and at that time I would be packed up and ready for the orders sending me back home.

The general listened without a change of expression on his face. The eyes, however, had lost the angry look that they had had while he was talking. They had become shrewd, calculating, analyzing, appraising. He walked toward me and put his arm around my shoulder. "George," he said, "I think we are going to get along together all right."

I knew then that we would. I had been talking with a big man and I have never known a truly big man that I couldn't get along with.

We sat down and chatted for over an hour. I told him that General Marshall, Admiral King, and Harry Hopkins had gone to England to discuss with the British some operations to take the pressure off the Russians and that the next move would be to go into North Africa. He thought it a mistake and a poor gamble. If the Germans should move into Spain, Gibraltar would fall, closing the Mediterranean, and German aircraft operating from Spanish bases would wreck the whole show. Even if the Germans did not move into Spain, the results to be obtained by fighting our way east along the Mediterranean coast would not be worth the effort. It would be far better to land in France just as soon as the necessary force were built up and the proper degree of air superiority attained. Landing in Africa would not shorten the war anywhere near as quickly as landing on the European continent. Actually, by landing in Africa, our forces would be dissipated and the eventual landing in Europe thereby delayed.

As far as taking the pressure off Russia was concerned, he did not share the prevalent view that Germany was about to knock them out of the war. He said that, while the German was a better soldier than the Russian, Hitler was overextending himself without adequate rail and road communications and it had already been demonstrated that in winter the German offense could not move and even trying to hold their positions was bleeding them white. There weren't enough Germans to keep this process going forever, especially against the almost inexhaustible manpower reserves of Russia. Hitler was trying to conquer too much geography and in the long run the Germans would exhaust themselves.

The General told me that on the 7th of August, the South Pacific forces were to land at Tulagi and Guadalcanal in the Solomon Islands. The First Marine Division was to make the landing, supported by the available fleet units. He had been called upon to furnish naval and air support of the show and had already turned over a cruiser and a couple of destroyers from the naval forces under him which were commanded by Admiral Leary. He wanted to know what my recommendations were for the use of the Air Force. I told him that I didn't have any as I didn't know what there was to work with, but I intended to fly to New Guinea that night, look the show over there, and visit the airdromes in the Townsville area on the way back to Brisbane about the 2nd of August. I said we would do everything possible to help him carry out his commitments, but that I would prefer to tell him what that "possible" was when I got back. General MacArthur agreed and said, "As soon as you get back, come in and see me and we'll have another talk." We shook hands and I left.

MacArthur looked a little tired, drawn, and nervous. Physically he was in excellent shape for a man of sixty-three. He had a little less hair than when I last saw him six years ago, but it was all black. He still had the same trim figure and took the same long graceful strides when he walked. His eyes were keen and you sensed that that wise old brain of his was working all the time.

He wanted to get going but he hadn't anything to go with.

He felt that Washington had let him down and he was afraid that they would continue to do so. He had two American infantry divisions, the 32nd and the 41st, but they still needed training. The Australian militia troops up in New Guinea were having a tough time and they were not considered first-class combat troops, anyhow. The 7th Australian Division was back from the Middle East and they were about the only veteran trained fighting unit that the General could figure on for immediate action. His naval force, a mixed Australian and American show, was small and he had no amphibious equipment and little hope of getting any for a long time. All that could be spared was going to the South Pacific effort in the Solomons. The amount of transport shipping was barely sufficient to supply the existing garrisons in New Guinea. His Allied Air Force of Australian and American squadrons was not only small but what there was had not impressed him very favorably to date. No wonder he looked a little depressed.

I went back downstairs to see Brett. He took me around to meet all the Allied Air staff and tried to explain the organization to me. This directorate system of his, with about a dozen people issuing orders in the commander's name, was too complicated for me. I decided to see how it worked first, but I was afraid I was not smart enough to figure it out. Furthermore, it looked to me as though there were too many people in the headquarters. I thought I'd better see what could be done to cut down on the overhead and plow those people back into the combat squadrons.

In order to make it a truly Allied organization, the Americans and the

Australians were thoroughly mixed everywhere, not only in the staffs but even in the airplane crews. An Australian pilot on a bomber, for example, might have an American co-pilot, an Australian bombardier, an American navigator, and a mixed collection of machine-gunners. It sounded screwy but maybe it was working. That was another thing I'd have to find out about.

Air Vice-Marshal Bostock of the RAAF (Royal Australian Air Force) was Brett's Chief of Staff of the Allied Air Forces. He looked gruff and tough and was very anti-GHQ, like all the air crowd I'd talked to so far, but he impressed me as being honest and I believed that, if he would work with me at all, he would be loyal to me. A big, red-faced, jolly Group. Captain Walters of the RAAF was Director of Operations, and Air Commodore Hewitt, also RAAF, was Director of Intelligence. Both had several American officers as assistants.

Only an advanced echelon of the American part of Brett's air organization was in Brisbane. The rear echelon was still back on the south coast of Australia, 800 miles away, at Melbourne, under Major General Rush B. Lincoln. It was just as if the headquarters were trying to function at Chicago with its rear echelon in New Orleans, except that we would have had much better telephone service. All personnel orders were issued through Melbourne, the personnel and supply records were all down there, and, as far as I could learn, the boys were really bedded down to stay. That would be another thing for me to look into.

Brett told me of his troubles with Sutherland. They just didn't get along. Brett said he had so much trouble getting past Sutherland to see MacArthur that he hadn't seen the General for weeks and he just talked to Sutherland on the telephone when he had to.

Brett said it would take him several days to pack up and that he expected to leave about the 3rd of August. I told him I was going up to Port Moresby that night to look things over and he said to take his airplane, an old B-17 which had been named "The Swoose." The Swoose was supposed to be half swan and half goose; the plane had been patched and rebuilt from pieces of other wrecked B-17s.

I asked about Brigadier Generals Walker and Whitehead. Brett said he had sent both of them to Darwin to look things over there and become familiar with the situation in Northwest Australia and that he had intended to have them do similar tours in Townsville and New Guinea before bringing them back into his own headquarters. He said he thought Whitehead had returned to Townsville, so I asked him to have "Whitey" meet me there and accompany me to New Guinea. Brett said he would send a wire to Major General Ralph Royce, the Northeast Area commander at Townsville, to that effect and have Royce go to New Guinea with me to show me around.

I took off for Townsville at eleven o'clock that night. Four hours later we landed, picked up Royce and Whitehead and an Australian civilian named Robinson, who, according to Royce, owned a piece of everything in the country and, as a leading industrialist, had a big drag with the government.

On the way to New Guinea over the 700-mile stretch of the Coral Sea between Townsville, Australia, and Port Moresby, New Guinea, Robinson spoke of knowing President Roosevelt and Winston Churchill and I gathered that he acted as a sort of unofficial ambassador for the Premier, Mr. Curtin. He had been quite helpful to Brett and was quite pro-American. He gave you the impression that he was a big shot and I believed that he was. He asked me to look him up in Melbourne and to call on him for help any time I needed it. I decided I'd remember him. I would probably need all the help I could get from any source before I got through with this job.

Soon after daybreak I got my first sight of New Guinea. The dark, forbidding mountain mass of the backbone of the country, the Owen Stanley range, rising sharply from a few miles back of the coastline to nearly twelve thousand feet and covered with rain-forest jungle was truly awe-inspiring. You didn't have to read about it to know that here was raw, primitive, wild, unexplored territory.

We picked up the now familiar line of coral reefs offshore, crossed the two-mile-wide calm lagoon, and flew over the beautiful little harbor of Port Moresby with the town and wharves on one side and perhaps a hundred native huts on stilts bordering the water's edge on the other. A few flat areas along river courses could be seen to the north and east of the harbor, where the bulldozers were at work raising huge clouds of dust but the general impression was that even the foothills of the Owen Stanleys that rimmed the harbor and separated the landing areas were rough and rugged. I figured that living in New Guinea would be like that, too.

About seven in the morning we landed at Seven Mile strip, just north of the town of Port Moresby. The airplane immediately took off for Horn Island on the northeast tip of Australia, in order to be out of danger from the daily morning Jap air attacks coming from Lae, Salamaua, and Buna on the north coast, where the Nips had air bases. Seven Mile strip was used only as an advanced refueling point for the Australian-based bombers. It was defended by some Australian antiaircraft artillery and by the American 35th Fighter Group, equipped with P-39s.

Brigadier General "Mike" Scanlon, who was in command of the Air Forces in New Guinea, met us and accompanied us on an inspection of the camps and the construction of several other fields in the area. I had known Mike ever since 1918 and liked him immensely, but he was miscast in this job. He had been an air attache in Rome and London for the best part of the last ten years, with a tour as intelligence officer in Washington. I don't know why they sent him up to New Guinea; he was not an operator and everyone from the kids on up knew it.

The set-up was really chaotic. Missions were assigned by the Director of Bombardment at Brisbane; Royce at Townsville passed the word to the 19th Bombardment Group at Mareeba, which was a couple of hundred miles north of Townsville; and the 19th Group sent the number of airplanes it had in commission to Port Moresby, where they were refueled, given their final "briefing" on weather conditions along the route to the target and whatever data had been picked up by air reconnaissance. The fighter group at Port Moresby sat around waiting for the Japs to come over and tried to get off the ground in time to intercept them, which they seldom did, as the warning service rarely gave the fighters over five minutes' notice that the Nip planes were on the way. Scanlon had the authority to use any aircraft in New Guinea in an emergency, if targets of opportunity such as Jap shipping came within range, and this authority had to be exercised occasionally when the bombardment planes arrived without knowing exactly what they were supposed to do.

The briefing was done entirely by Australian personnel. The system did not contemplate utilizing the squadron or group organization but dispatched whatever number of airplanes was able to get off the ground. No one seemed to be designated as formation leader and even the matter of assembly of individual planes was ignored. The aircraft might or might not get together on the way to the target. On the average, seven to nine bombers usually came up from Australia. Six would probably get off on the raid. The take-off generally took about one hour until they had all cleared the airdrome and usually within another hour two would be back on account of motor trouble. Others came back later on account of engine trouble or weather conditions, so that from one to three usually arrived at the target. If enemy airplanes were seen along the route, however, all bombs and auxiliary fuel were immediately jettisoned and the mission abandoned. The personnel was obsessed with the idea that a single bullet would detonate the bombs and blow up the whole works. The bombs, of course, were not that sensitive but no one had explained it to the kids, so they didn't know any better.

The weather service was not even sketchy. Weather forecasts were prepared by the Australian weather section at Port Moresby from charts of preceding years which purported to show what would happen this year. During July, the youngsters told me, three quarters of all bomber missions had been abandoned on account of weather. We could have done better than that by tossing a coin to see whether to go or not.

During the morning I attended a briefing down at Seven Mile Airdrome. The target was listed as Rabaul, the main Jap base in New Britain. Before the war Rabaul was a town of about 15,000. It had an excellent harbor and was defended by a heavy concentration of antiaircraft guns and two airdromes at Lakunai and Vunakanau. No particular point was assigned as target and I found out afterward that nobody expected airplanes to get that far anyhow, but, if they did, the town itself was a good target and there were always vessels in the harbor and the two airdromes were also considered worth while. The Australian weather man, in giving the weather conditions along the route and the forecast, mentioned "rine" clouds. I heard one of the youngsters in the back of the room turn to his co-pilot and say, "What are rine clouds?" The co-pilot answered, "I think he said 'rime.' Probably the kind of clouds that have rime ice." The pilot remarked that he did not think that they had ice this close to the equator but then remembered that he had once picked up some ice at 20,000 feet over the Mojave Desert in California, and guessed that that was what the Australian weather man probably meant, after all. Several of the crews who overheard this conversation then became a little bit worried as they had no de-icing boots on their wings or fluid for the slinger rings to keep ice off of the propeller. I relieved their apprehensions by explaining to them that in the Australian accent "rain" became "rine."

I then told Whitehead to see that in future an American staff officer was present at all briefings by the Australians and that as soon as I took over officially he was to form an American staff for operations in New Guinea and that this staff would be in charge of or at least supervise all briefings. Furthermore, that from then on we would have a definite primary, secondary, and tertiary target assigned for every mission.

About noon the Japs raided Seven Mile with twelve bombers escorted by fifteen fighters which came in at twenty thousand feet and laid a string of bombs diagonally across one end of the runway and into the dispersal area where our fighters and a few A-24 dive bombers were parked. A couple of damaged P-39s didn't need any more work done on them, three A-24s were burned up, and a few drums of gasoline were set on fire, shooting up huge flames that probably encouraged the Jap pilots to go home and turn in their usual report of heavy damage to the place. A few other aircraft were holed but not seriously. Luckily the last of the five bombers that had come in from Mareeba at daybreak had left for a raid on Rabaul about ten minutes before the Japs arrived. One bomb hit on the end of the runway but that damage was repaired in about twenty minutes. I had a grandstand seat from a slit trench on the side of a small hill just off the edge of and overlooking Seven Mile, although I did not enjoy the spectacle as much as I might have under better conditions. The trench was full of muddy water and there were at least six too many people in with me.

There was no combat interception. The fighter-group commander seemed a bit proud of the fact that he had gotten ten P-39s off the ground before the Japs arrived overhead. With forty P-39s on the airdrome, I didn't think that was a very good percentage but I found that most of the rest were out of commission because they couldn't get engines, propellers, and spare parts up from Australia to replace stuff that had worn out or been shot up in action. Due to the lateness of the warning, which only gave four minutes' notice, the fighters, of course, could not climb to twenty thousand feet to engage the Jap bombers, which did their stuff and left for home without interference. Even the antiaircraft fire was ineffective. It looked, from where I was, short of the twenty-thousand-foot level and scattered all over the sky.

One thing was certain. No matter what I accomplished, it would be an improvement. It couldn't be much worse.

Throughout the whole area the camps were poorly laid out and the food situation was extremely bad. There was no mosquito-control discipline and the malaria and dysentery rates were so high that two months' duty in New Guinea was about all that the units could stand before they had to be relieved and sent back to Australia. Everyone spoke of losing fifteen to thirty pounds of weight during a two months' tour in Port Moresby. The *pièce de résistance* for all meals, which was canned Australian M-and-V (meat-and-vegetable) ration, was cooked, or rather heated, and served in the open as there was no mosquito screening anywhere. Swarms of flies competed with you for the food and, unless you kept one hand busy waving them off as you ate, you were liable to lose the contest. Even while walking around, the pests worked on you and it didn't take long to figure out what the kids meant when they referred to the constant movement of your hands in front of your face as "the New Guinea salute." Now, just throw in a continuous choking dust that seemed to hang in the hot, humid atmosphere and you began to realize that, in spite of the truly gorgeous mountain scenery and the heavenly blue of the ocean, it would be hard to sell the glamour of this part of the South Seas to the GIs. If you were thinking of dusky but fair sarong- or grass-skirt-clad girls, flitting among the palm trees, you were in for another shock. There were a few Papuan women in evidence around the native villages. They were dusky, in fact quite so, and they wore grass skirts, but from there on the dream faded. They forgot to give the poor girls any beauty to start with and the tattooing of tribal marks on their faces didn't help any. The prevalent mode of hairdressing called for the naturally fuzzy, kinky black hair to stand out from the head. This was accomplished by thoroughly applying pig grease, which got results apparent to the eye and, in the down-wind direction for a distance of at least a hundred yards, to the nose as well.

Both men and women worked around the camps building huts of bamboo framework thatched with palm leaves. The huts looked picturesque, kept out the rain, afforded shelter from the sun, and were nice things to stand in front of while you had your picture taken to send home to the folks, but after a short time they served as homes for a million varieties of spiders, scorpions, centipedes, lizards, and even birds. Luckily they were highly inflammable and accordingly, through accident or design, you didn't stay in the same house too long.

As New Guinea is close to the equator, darkness succeeded daylight with scarcely any twilight. You had just paused to admire the truly wonderful color display of the sunset, when blackness and the mosquitoes descended upon you. The mosquito problem was not being taken care of, with the result that the malaria rate was appalling. Most of the men had mosquito netting, but a large percentage of it was torn in places or full of holes and afforded little protection.

They really had mosquitoes up there. Big ones that dwarfed the famous New Jersey variety. I heard the usual yarn with a new locale when one of the kids tried to keep a straight face as he told me that one of them had landed over on the edge of the strip the other evening just after dark and the emergency crew had refueled it with twenty gallons of gasoline before they found out it wasn't a P-39.

The next morning I returned to Townsville with Royce and Robinson, the Australian "billionaire." I told Royce that he was going home as Arnold had a job for him running the First Air Force, which was a training show with headquarters at Mitchel Field in New York. He said he was glad to go and would like to have about ten days to straighten out his affairs and get packed up. I told Benn to make arrangements for space on a plane going back about August 10th. Royce said Walker was due in Townsville from a tour in Darwin, in a few days, and asked whether or not I wanted him in Brisbane. I said no, that as soon as Royce left I wanted Walker to take charge in Townsville as head of a Bomber Command which I would organize as soon as I got back to my headquarters in Brisbane. Whitehead would stay in Port Moresby, succeeding Scanlon. I asked Royce when Walker got in to show him the ropes and help him all he could before he left.

I decided to leave the Townsville show for Walker to straighten out. It turned out to be another scrambled outfit of Australians and Americans, with so many lines of responsibility, control, and coordination on the organizational chart that it resembled a can of worms as you looked at it. I made a note to tell Walker to take charge, tear up that chart, and have no one issue orders around there except himself. After he got things operating simply, quickly, and efficiently he could draw a new chart if he wanted to. The top Australian and Royce's Chief of Staff at Townsville was Air Commodore Lukis, a big likable individual who reminded me of Mike Scanlon. They even had the same kind of mustache. I didn't get much information out of him. In fact, the only one in Townsville who had the answers to very many of my questions was an Australian RAAF officer named Group Captain Garing. They called him "Bull," but he was active, intelligent, knew the theater, and had ideas about how to fight the Japs. He had flown and been all over the New Guinea country before the war and also knew New Britain, New Ireland, and the Admiralty Islands. I decided to keep my eye on him for future reference. I asked Garing about landing-field possibilities along the north coast of New Guinea between the eastern end of Alilne Bay and Buna where the Japs were. He told me that halfway between Milne Bay and Buna, at a place called Wanigela Mission, there was a good natural landing field, approximately a mile long and half a mile wide, normally covered with the native grass called kunai which grew six to eight feet high. A few weeks previously an Australian Hudson reconnaissance plane had a forced landing there. The crew got the natives to cut a path through the grass and, after a second Hudson had flown up from Port Moresby and made repairs on the first airplane, both of them took off and returned home. This looked like something for me to investigate as a possibility for the future, the next time I went to New Guinea.

That afternoon I flew to Mareeba, where the 19th Bombardment Group equipped with B-17s was stationed, and talked with Colonel Carmichael, the group commander, and the youngsters. Carmichael had not had command very long but he seemed to have taken hold of the show and was trying. He had a real job on his hands and it was apparent that I would have to help him a lot. Walker would have to help, too. The 19th Group had been kicked out of the Philippines and out of Java and kicked around ever since. They had had innumerable group commanders who had come and gone without leaving anything behind them. The crews were thinking only of going home. Their morale was at a low ebb and they didn't care who knew it. The supply situation was appalling. Out of the thirty-two B-17s at Mareeba, eighteen were out of commission for lack of engines and tail wheels. About half of them were old-model B-17s, without the underneath ball turret to protect the airplane from attack from below, and in addition they were pretty well worn out. Anywhere else but this theater they would probably have been withdrawn from combat, but they were all we had so I'd have to use them if we wanted to keep the war going. Airplanes were continually being robbed to get parts to keep others running. At that moment the group could not have put over four airplanes in the air if called on for immediate action. Requisitions for supplies and spare parts were submitted by the group through Royce's headquarters at Townsville to an advanced air depot at Charters Towers, about sixty miles south of Townsville. Charters Towers, for some reason, sent the requisitions to Melbourne. Melbourne forwarded them to Tocumwal, the main Air Force depot about one hundred miles north of Melbourne. An average time of one month elapsed from the time the requisition started until it was returned, generally with the notation "Not available" or "Improperly filled out."

I said I didn't believe it, but the kids made me eat my words when they showed me a whole filing case of returned requisition forms. I took along a handful for future reference and as evidence that some of the people in the organization were playing on the wrong team.

I told Carmichael to cancel all flying and put his airplanes in commission for a maximum effort about a week later. I said I'd be back in a few days to talk things over with him again. I got a list of all the bits and pieces he needed to put his airplanes in shape, called Royce and told him not to schedule anything except the usual reconnaissance, and pushed off for Brisbane. It looked as though I'd have to start at Melbourne to get anything moving. I decided to postpone bothering the Townsville-Charters Towers supply organization until I came back in a couple of days to brief Carmichael and see what I could do to fire those kids up a bit. Carmichael told me that no one had received any decorations for months and then just one or two had been passed out. He was told to prepare recommendations for every deserving case in the group and that I would see that they got them promptly. I knew that little bits of pretty ribbon had helped in World War I; maybe they would help in this one, too. Anyhow, it wouldn't hurt to try. The next morning at the office I got General Lincoln on the phone and read him a list of things to ship to Carmichael at once. I told him not to worry about the paper work but to load the stuff in every airplane he had and fly it to Mareeba without consulting Townsville, Charters Towers, or anybody. Also that I would be in Mareeba in a day or two myself and would check up on deliveries and, if I found that he had delayed for any reason, I would demote a lot of people and send them home on the slowest freight boat I could find. I told him to call me back as soon as he had seen his supply people, report what action he had taken and what the schedule of departures from Melbourne and Tocumwal looked like. A colonel named Frye, who was running a small air depot in Brisbane, came in and I told him to report what requisitions from my list he had in stock and also to give me a forecast right away on how many B-17s then undergoing repair and overhaul at Amberley Field and Eagle Farms, two airdromes a few miles out of Brisbane, would be ready that day, the next day, and each day for the rest of the week.

Brett came in and asked me if I had any objection to his taking The Swoose, his private B-17, for the trip home as far as Hawaii, as it was hard to get air transportation that far by air. I said, "No." He also said that he would like to take Brigadier General Perrin back with him to Washington on temporary duty to help him compile his reports. I said okay, he could have him. I knew Perrin and didn't know what I could use him for. After Brett left, I'd wire Arnold to give Perrin a job in the United States.

I checked in with General MacArthur and talked with him for about two hours. I told him frankly what I thought was wrong with the Air Force set-up in both Australia and New Guinea and discussed the corrective action that I intended to take immediately. I asked him to give me authority to send home anyone that I thought was deadwood. He said, "Go ahead. You have my enthusiastic approval." I then discussed the air situation and told him that I wanted to carry out one primary mission, which was to take out the Jap air strength until we owned the air over New Guinea. That there was no use talking about playing across the street until we got the Nips off of our front lawn. In the meantime, our reconnaissance aircraft would be constantly looking for Jap shipping that should be hit at every feasible opportunity, but we were not going to get anywhere until we had won the air battle. I told him I had called off flying all bombers, B-17s, B-25s, and B-26s, until we could get enough of them in shape to put on a real show; that about August 6, just prior to the coming South Pacific operation to capture Guadalcanal and Tulagi, I would send the maximum number of B-17s against the main Japanese airdrome at Vunakanau, just southeast of Rabaul. The Jap aircraft there would raise the devil with the Navy landing operations if they were not taken out. At the same time, the B-25s and B-26s with fighter escort should be given a mission to clean up the Jap airdromes at Buna, Salamaua, and Lae. The effort against the New Guinea airdromes should be continuous until the Jap airpower in New Guinea was destroyed and the runways so badly damaged that they even stopped filling up the holes. In the meantime, I would put everything else in support of the Australian drive along the Kokoda trail. General MacArthur approved this program and said to go ahead, that I had carte blanche to do anything that I wanted to. He said he didn't care how my gang was handled, how they looked, how they dressed, how they behaved, or what they did do, as long as they would fight, shoot down Japs, and put bombs on the targets. When I told him that I hoped to put between sixteen and eighteen B-17s on Vunakanau, he remarked that it would be the heaviest single attack in the Pacific war up to that time. It did seem like a lot of bombers, although the Japs were making it look rather small with their formations of twenty-five to forty bombers with about the same number of escorting fighters whenever they raided us at Darwin.

Back in my office, I asked Brett how many airplanes the Allied Air Forces had in the theater. He said he didn't know but called in the Director of Operations, Group Captain Walters, and Walters said that he would try to get the information for me right away.

Brigadier General Carl Connell, Lincoln's assistant at Melbourne, called me to give a report on how things were moving. They had a few spare engines for the B-17s but only two of their cargo planes would carry them, so only four engines could be flown to Mareeba the next day unless I could send down the one cargo airplane in Brisbane which could carry engines. I told him I would start it south right away. Propellers, tail wheels, and other spares, enough to fix up 12 B-17s, would also leave Tocumwal the next day. I told Connell to keep on shipping the rest of the list I had given them as the war would not stop after this operation. He wanted to know if he couldn't come up to see me about supply matters but I said not until the 19th Group had all the parts necessary to fix up every airplane in Mareeba.

I drove out to Amberley Field and talked with the mixed repair outfit, composed of Australian Air Force mechanics, American Air Force mechanics, and Australian civilians. I told them I wanted the place to operate twenty-four hours a day to finish up the five B-17s in work there. After a few pep talks and some arguments, I got a promise that four of them would be ready by the afternoon of the 5th.

In the evening I invited Benn and several of the lieutenant colonels and majors from Air Headquarters in Brisbane to eat, drink, and talk at Flat 13 at Lennon's. There were a couple of them, named Freddy Smith and Gene Beebe, who looked pretty good. They hadn't been in Australia very long and were anxious to work but didn't seem to be able to find out just what part they were to play in the war. So far they had spent most of their time shuffling papers from one directorate's office to another. There was a Lieutenant Colonel Bill Hipps who seemed to have a lot on the ball, too, but he was a Java veteran and a little tired. He did not say anything about going home but I figured I would use him for a few months and then send him back to the United States for a rest. I decided to clean out all the Philippine and Java veterans as far as possible. Most of them were so tired that they had almost lost interest. If we were to salvage them, they would need a long rest back home. I learned a lot from talking with those kids. Under the existing organization, so many people were putting out instructions, with or without the knowledge of the Commanding General, that no one could tell what the score was. The organization for getting supplies moving around the various gauges of the Australian railroad system and moving them up to the two fronts at Darwin and New Guinea was evidently so complicated that nothing moved. The whole service of supply was centered at Melbourne, which was 2500 miles away from the war in New Guinea. Even orders of the American part of the Allied Air Forces had to be issued from the rear echelon at Melbourne presided over by Major General Rush B. Lincoln. Most of the business was done by mail, although the telephone service was fairly good. The communication between Brisbane and Darwin was mostly by radio, as there was no telephone line between those two points. Communications to New Guinea were telephoned or mailed to Townsville and from there radioed or flown or sent on a boat across the Coral Sea to Port Moresby. The tendency seemed to be to keep everything in the south on account of the probability that the Japanese would soon seize Darwin and land on the east coast of Australia somewhere between Brisbane and Horn Island. There was a lot of talk about the Brisbane line of defense, which vas about twenty-five miles south of the city and upon which the Australians were supposed to rely for the last-ditch defense of Sydney and the country to the south. Everyone seemed to speak of the coming evacuation of New Guinea as a fairly good possibility. The organization was too complicated for me. I decided I would have to simplify it at least to the point where I could understand it myself.

Those youngsters following topside lead seemed to have an intense dislike of GHQ and all American ground organization. Sutherland was definitely unpopular with them. They said that he was dictating practically every action of the Air Force by his instructions to Brett. I told them to lay off the hate campaign. That MacArthur was okay, that he intended to let the Air Force run itself, carrying out general overall missions that he, of course, would approve. I also told them to quit wasting their time worrying about Sutherland, that I would take care of that situation myself, and that I would handle our dealings with the rest of the GHQ organization as well. These lads all wanted to work and I believed they could work. All they needed was somebody to lead them and we wouldn't have to get out of New Guinea or fight along the Brisbane line. However, someone had to go to work and it looked as though the finger were pointing at me.

Officially, Brett remained the Allied Air Force commander for the Southwest Pacific Area until he left for the United States. General MacArthur told me that for the present he was not going to release to the press the fact that I was even in the theater, as he didn't want the Japs to know that he was shaking up his command. However, Brett had already stopped issuing any orders and was helping all he could to get me acquainted with things before he left.

At noon each day, the commanders and top staff officers of the Air Force, the Navy, and the Australian Land Forces, together with a few representatives of GHQ, gathered in the Brisbane Air Force War Room, which was operated by the Air Force Directorate of Intelligence. There we were given the latest information from the combat zones, an analysis of enemy land, sea, and air strengths, locations of units and movements, with an estimate of probable hostile intentions. The briefing was done by reference to a huge map, about twenty feet square, of the whole Pacific theater from China to Hawaii on which colored markers and miniature airplanes and ships were correctly spotted to represent our own and enemy forces. The map was laid out on the floor in front of and a few feet below a raised platform on which we sat. The whole thing was very well done and served to keep us up to date on what was going on in our own and in adjoining theaters. Another valuable feature was that each day the heads and top staff officers of the three services saw each other and had a chance for brief discussion of the common problem: What were we all going to do about defeating the Japs?

General Sir Thomas Blamey, the Australian Land Force commander, took his seat regularly at the daily briefing. He was a short, ruddy-faced, heavy-set man with a good-natured friendly smile and a world of self-confidence. No matter what happened, I didn't believe Blamey would ever get panicky. He appealed to me as a rather solid citizen and a good rock to cling to in time of trouble. He was one of the few people I'd talked to outside of General MacArthur who really believed that we would hold New Guinea.

There was a Major General George Vasey who came in with Blamey that I liked the looks of. He was more like the Australian that we generally visualize, tall, thin, keen-eyed, almost hawk-faced. He was going to New Guinea soon to command the 7th Australian Division in combat and I thought we would hear a lot of him before this war was over. They told me his record with the 7th in the Middle East was excellent.

Vice-Admiral Herbert F. Leary, the Allied Naval commander, was another regular attendant. I'd never met him before but he was extremely cordial to me. There was some conflict between Leary and GHQ, but I had enough troubles of my own, so I didn't try to find out anything more. I gathered, however, that a successor to Leary would be on his way out here shortly.

The Australians seemed a little more than worried about being pushed aside by the Americans as soon as our forces and equipment would arrive in sufficient quantities to start offensive action. They are a young, proud, and strongly nationalistic people. So far, they had done most of the fighting in this theater and they wanted to be sure that their place in the sun was going to be recognized. I couldn't blame them and I promptly assured all the RAAF crowd that I was not going to be pro anything but Air Force and that I would fight to get airplanes, engines, and spare parts for both the Americans and the Aussies. I wanted aviators, regardless of nationality, who could fly, shoot down Japs, and sink Jap ships.

On the afternoon of the 3rd, Brett told me he was leaving the next day and then went upstairs to say goodbye to General MacArthur. He told me this would make the eighth time he had seen MacArthur since he had arrived in Australia last March.

Later he said the General had been extremely nice to him, had wished him all kinds of luck, and had presented him with the Silver Star Medal.

In the evening I went out to Brett's house where he was giving a farewell cocktail party. The house, called "Braelands," was one of the best in Brisbane. I understood that Robinson had secured it for Brett.

At the party were all the top rank of both the American and Australian forces, land, sea, and air, and a few representatives of General Headquarters. I spent most of the time talking to Sir Thomas Blamey. I discussed the question of Wanigela Mission as a landing-field possibility and asked Blamey if he would let me fly some of his troops in there to occupy it while we built an airdrome. He did not offer any objection but said that he would give us Buna in a few weeks. He stated that he intended to reinforce his force on the Kokoda trail, send a really aggressive commander up there, and order the Australians to advance, capture Buna, and drive the Japs out of Papua. The Australians had just withdrawn to the Kokoda Pass in the Owen Stanley Mountains and expected to hold there until reinforcements arrived. Blamey seemed a little overconfident of the ability of his troops to make the trip over that extremely difficult, alternately mountainous and swampy trail through the rain forests from Port Moresby to Kokoda and still have enough drive left in them to chase the Japs out of the territory between there and Buna itself. I didn't say so to General Blamey but I hoped that his confidence could be transmitted to his troops. From what I had been told in Port Moresby, those men up on the trail were worn out and it wasn't going to take much effort on the part of the Japs to crack the defenses of the Kokoda Pass.

On August 4th at daybreak, General George Brett left for Hawaii in The Swoose, Major Frank Kurtz piloting, taking Brigadier General Perrin with him, and I officially took over the job of Allied Air Force Commander of the Southwest Pacific Area.

Air reconnaissance at daybreak that morning showed a Jap concentration of approximately 150 aircraft, mostly bombers, on Vunakanau airdrome and over fifty fighter planes at Lakunai. The strength at Vunakanau was an increase of nearly 100 aircraft in the past two days. It might be an accident or it might be that the Japs suspected we were up to something in the Solomons.

I went up to General MacArthur's office, gave him the information, and told him that I still intended to attack Vunakanau with the maximum possible B-17 strength on the morning of the 7th as I believed there was a possibility that the Japs had read the Navy's mail and that this might be the reason for the recent heavy air reinforcement in the Rabaul area. I said I would like to strike earlier but the airplanes had to leave Australia on the 6th in order to refuel and take off from Port Moresby early on the morning of the 7th. In spite of the fact that the crews had been working day and night to get the planes in commission, if I backed up the schedule one day it would mean the difference between a decent-sized attack and one of four or five planes. The General agreed and again asked me how many B-17s I expected would take part in the raid. I told him I hoped that twenty would leave Mareeba and Townsville for Port Moresby on the 6th and I saw no reason why my original estimate of sixteen to eighteen over the target would not hold. He was very enthusiastic and wanted to know who was going to lead the show. I told him that I expected my group commanders to lead their groups in combat and that Colonel Carmichael was the group commander. He nodded approval and remarked that, if the operation went as I expected, it would be our heaviest bomber concentration flown so far in the Pacific war. He said not to forget to award as many decorations as I thought fit. He gave me authority to award all decorations except the Distinguished Service Cross, which he wanted to present himself as that was the highest decoration he was allowed to give. However, if a particularly deserving case came up and I wanted to decorate someone on the spot, he told me to just go ahead and he would confirm it.

During the afternoon, orders came down from GHQ covering our support of the South Pacific Guadalcanal operation. Instead of ordering the Allied Air Force to support the show by a maximum effort against the Jap airdromes in the Rabaul area and putting in the usual sentence about "maintaining reconnaissance," the order went into a page and a half of details prescribing the numbers and types of aircraft, designating units, giving times of take-off, sizes of bombs, and all the other things that the Air Force orders are supposed to take care of.

I immediately went up to see Sutherland and had a showdown with him on the matter. I told him that I was running the Air Force because I was the most competent airman in the Pacific and that, if that statement was not true, I recommended that he find somebody that was more competent and put him in charge. I wanted those orders rescinded and from that time on I expected GHQ would simply give me the mission and leave it to me to cover the technical and tactical details in my own orders to my own subordinate units. When Sutherland seemed to be getting a little antagonistic, I said, "Let's go in the next room, see General MacArthur, and get this thing straight. I want to find out who is supposed to run this Air Force."

Sutherland immediately calmed down and rescinded the orders that I had objected to. His alibi was that so far he had never been able to get the Air Force to write any orders so that he, Sutherland, had been forced to do it himself. I asked him if he prescribed for the Navy what their cruising speed should be and what guns they would fire if they got into an engagement. He didn't say anything. No wonder things hadn't been running so smoothly.

I spent the rest of the afternoon working out a reorganization of the Air Force. I decided to separate the Americans and the Australians and form the Americans into a numbered Air Force of their own which I would command, in addition to commanding the Allied show. The Australians would be organized into a command of their own and I'd put Bostock at the head of it. My Allied Air Force headquarters would remain a mixed organization. I would not only command it, but be my own chief of staff until I could get someone from the United States to take over. To keep the Australians in the picture I would leave Group Captain Walters in as Director of Operations and Air Commodore Hewitt as Director of Intelligence for the Allied Air Force headquarters. As American assistants, Lieutenant Colonel Beebe would work with Walters and a Major Benjamin Cain would be Hewitt's assistant director. Besides letting the Australians have something to say about the show, I really hadn't an American capable of doing either of these jobs unless I took Walker or Whitehead out of the combat end. I couldn't afford to do that.

It was quite noticeable in New Guinea and up at Mareeba that the combat units were short-handed. Back in Brisbane it was equally noticeable that officers were falling all over themselves. I decided to have a real housecleaning right away. Those who were not pulling their weight could go home and the rest would move north to take their turns eating canned food and living in grass huts on the edge of the jungle. Furthermore, there were a lot of those kids up in New Guinea who ought to come back to Australia for a couple of weeks' rest. Many of them told me that they had been in combat continuously for four or five months.

The next morning I flew to Mareeba, assembled Carmichael and his squadron commanders, and gave them instructions for the raid on Rabaul. The primary target was to be Vunakanau, the secondary target Lakunai, with the shipping in Rabaul Harbor on third priority. It looked as though twenty bombers would take off the next morning for Port Moresby. The group was quite elated over the fact that they were to finally put on a big show. I cautioned Carmichael and the squadron commanders about holding formation, as they would undoubtedly run into plenty of Jap fighters. They were a bit concerned about their formation work as they were not used to flying "in such large numbers together." I couldn't help but shudder at this evidence of a lack of formation training when twenty airplanes appealed to the kids as a large formation. I told them that from now on we would utilize every opportunity to practice formation flying as well as quicker take-offs and assemblies in the air. I said that the exhibition I had seen at Port Moresby of six B-17s taking an hour to get off the ground and never assembling into formation before arrival at the target was something that I did not want to see again. I gave them a long talk on the subject of what would happen to a bomb if it were hit by a machine-gun bullet. Hundreds of tests had shown that the bomb would not blow up as the crews believed, so there was no excuse for unloading the bombs just because an enemy airplane headed in their direction. From that time on, we wanted the bombs put on the target. They seemed glad to get the information which relieved them of one of their biggest worries.

In the afternoon I went over to Charters Towers and looked over the 3rd Light Bombardment (Dive) Group and the air depot on the other side of the field. The 3rd, which used to be the 3rd Attack Group back home, did low-altitude strafing and bombing work. They still wanted to be called an Attack Group, so I told them to go ahead and change their name. That organization had trained for years in low-altitude, hedge-hopping attack, sweeping in to their targets under cover of a grass cutting hail of machine-gun fire and dropping their delay-fuzed bombs with deadly precision. They were proud of their outfit and they liked the name "attack." Now the powers that be had changed their name to "Light Bombardment parenthesis Dive" and they didn't like it. I knew how they felt. I had been an attack man myself, had written textbooks on the subject and taught it for years in the Air Corps Tactical School. It seems like a little thing but it really isn't. Numbers, names, and insignia mean even more to a military organization than they do to Masons, Elks, or a college fraternity.

The 3rd Group was a snappy, good-looking outfit, but their combat operations had been at a low ebb for some time on account of a lack of equipment. One squadron was equipped with A-24 dive bombers which hadn't enough range to reach any of the targets in New Guinea. One squadron had A-20s, but these had arrived in Australia without any guns or bomb racks. It would be a couple of weeks before installations could be completed. A third squadron had B-25s with only one front gun. This squadron had seen quite a bit of action, but they had taken heavy losses from the Jap head-on fighter attacks so that the airplane was not very popular. The fourth squadron had no airplanes at all.

The Charters Towers air depot was not paying its way. It had a lot of hard-working earnest kids, officers and enlisted men, who were doing the best they could under poor living and eating conditions, but their hands were tied by the colonel in command whose passion for paper work effectually stopped the issuing of supplies and the functioning of the place as an air depot should. He had firmly embedded in his mind the idea that he could not issue supplies if he was taking inventory or if the requisition forms were not made out correctly. Requisitions coming in from New Guinea and even from the 3rd Group on the other side of the field were repeatedly being returned because notations were made on the wrong line or the depot was too busy sorting out articles that had just been shipped in. He told me that he thought "it was about time these combat units learned how to do their paper work properly." I decided that it would be a waste of time to fool with him so I told him to pack up to go home on the next plane back to the United States. The excuse would be "overwork and fatigue" through tropical service. I told Bill Benn to get his name and make arrangements by telephone with my office at Brisbane to ship him home. A good-looking major on the depot staff told me he knew what I wanted done. I put him in command and told him that beginning immediately I wanted all requisitions, verbal or written, filled at once regardless of the state of inventory or whether or not the forms were filled out properly and that whenever possible he was to fly the equipment or parts to points where they were needed in order to speed up maintenance and get airplanes back into commission.

The ingenuity of the kids at Charters Towers was an inspiration. In spite of canned Australian rations and hot, dirty, dry-season living conditions in tents in bush country, they were trying. The junior officers were helping all they could. The seniors had been the stumbling block. There were very few spare instruments, so the kids salvaged them from wrecks and repaired them. There was no aluminum-sheet stock for repair of shot-up or damaged airplanes, so they beat flat the engine cowlings of wrecked fighter planes to make ribs for a B-17 or patch up holes in the wing of a B-25 where a Jap 20-mm. shell had exploded. In the case of small bullet holes, they said, they couldn't afford to waste their good "sheet stock" of flattened pieces of aluminum from the wrecks, so they were patching the little holes with scraps cut from tin cans. The salvage pile was their supply source for stock, instruments, spark plugs—anything that could be used by any stretch of the imagination.

One youngster had built, out of scraps of junk, a set for testing electrical instruments. I asked him if he could assemble the stuff, which he had spread all over a table, into a box the size of a small suitcase and give me a drawing of it so I could send it back to the Air Force Materiel Division at Dayton and have it standardized as a field testing apparatus.

He grinned all over and said he could and would. I told him after he finished that job to make about six more of them as it was the best thing of its kind I had ever seen and I wanted to put them around the other places where repair work was going on.

A radio came in from Scanlon asking for ten days' leave as he was tired and needed a rest. I wired back to take his leave and to bring his things with him as he was not going back to New Guinea. I then sent a note to Whitehead telling him to take command in Scanlon's place, told him the details of the coming bomber mission on Rabaul and what I wanted done to the Jap airdromes in New Guinea.

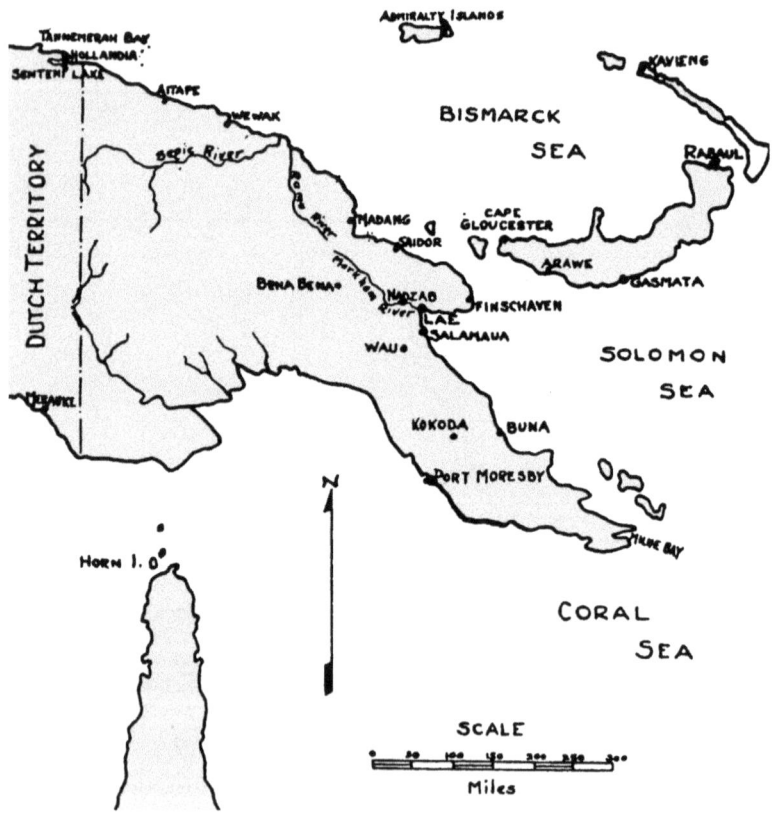

3. NEW GUINEA
August—September, 1942

M Y ESTIMATE turned out all right. Twenty B-17s flew to Port Moresby and eighteen took off on the big strike on the morning of the 7th of August. Twenty Jap fighters intercepted our bombers about twenty-five miles short of the target, but the kids closed up their formation and fought their way to Vunakanau where they dropped their bombs in a group pattern that was a real bull's-eye. The Japs still had the same 150 planes lined up wingtip to wingtip on both sides of the runway. The pictures looked as though we got at least seventy-five of them, besides setting fire to a lot of gasoline and blowing up a big bomb dump on the edge of the field. In the air combat we shot down eleven of the twenty Jap fighters that participated.

We lost a B-17 piloted by Captain Harl Pease, who really had no business in the show. He had been out on a reconnaissance mission on the 6th and came back in the afternoon to Mareeba with one of his engines out of commission. He and his crew decided they simply could not miss the big show, so after working for several hours on another plane which had been declared unserviceable, they got it running and arrived in Port Moresby about one o'clock on the morning of the 7th. One engine on this airplane was missing badly but Pease and the crew went to work on it and a couple of hours later took off with a load of bombs and joined Carmichael and the rest of the group. During the combat coming into Rabaul, Pease was on the wing which bore the brunt of the attack. By skillful handling of his airplane, in spite of the fact that the bad engine by this time had quit entirely, Pease held his place in the formation and his crew shot down at least three of the Jap fighters. Shortly after he had dropped his bombs on the target, another wave of enemy fighters concentrated their attack on his airplane which they evidently realized was crippled. The B-17 burst into flames and went down. No parachutes were seen to open.

I recommended Pease for a Medal of Honor; General MacArthur approved and forwarded the recommendation to Washington. The next day a wire came in from the War Department confirming the award. Carmichael was given a Distinguished Service Cross, while a number of lesser decorations were awarded outstanding members of the group.

The Marines landed at Guadalcanal with practically no opposition. Tulagi was also taken, but with a little more fighting. There was no Jap air interference. Admiral Ghormley wired General MacArthur a congratulatory message on the success of our attack on Vunakanau and the fact that it had broken up the possibility of Jap air interference with his landing in the Solomons. General MacArthur added his congratulations and I sent both messages to the 19th Group with my own.

During the day, we intercepted several Jap radios which were appeals by the Nips in the Solomons for help from the Rabaul air units. The Jap commander at Rabaul replied that he couldn't do anything for the boy in the Solomons on account of our "heavy air raid on his airdromes." The next day we intercepted another message which showed that only thirty bombers were serviceable at Vunakanau. Jap prisoners, taken at Guadalcanal airdrome (Lunga), confirmed our observation that the day before there were approximately 150 bombers at Vunakanau at the time of our attack. I issued instructions to an Australian squadron at Townsville, equipped with Catalina flying boats, to work on both Vunakanau and Lakunai for the next three nights. The squadron would put over only three or four planes each night but I hoped that would help the Japs to worry a little more.

Group Captain Walters finally brought me the report on aircraft strength that I had asked for a few days before.

On the books I had, in the United States part of the show, 245 fighters, 53 light bombers, 70 medium bombers, 62 heavy bombers, 36 transports, and 51 miscellaneous aircraft, or a total of 517.

This didn't look too bad, until I found what the real story was. Of the 245 fighters, 170 were awaiting salvage or being overhauled at Eagle Farms. None of the light bombers were ready for combat, and only 37 mediums were in shape or had guns and bomb racks to go to war with. Of the 62 heavy bombers, 19 were being overhauled and rebuilt. There were 19 different types among the 36 transports and less than half of them were in commission. The 51 miscellaneous turned out to be light commercial or training types which could not be used in combat.

The Australian Royal Air Force listed 22 squadrons, but most of these were equipped with training planes doing antisubmarine patrol off the coasts of Australia itself. Two fighter squadrons in New Guinea had a total of 40 planes, and four reconnaissance squadrons had a total of 30 aircraft.

There was also a Dutch squadron of B-25s, supposed to be training at Canberra, but Walters said they were a long way from being ready for combat.

All told I had about 150 American and 70 Australian aircraft, scattered from Darwin to Port Moresby and back to Mareeba and Townsville, with which to dispute the air with the Jap. He probably had at least five times that number facing me and could get plenty more in a matter of a few days by flying them in from the homeland. I issued orders that no more airplanes were to be salvaged. We would rebuild them, even if we had nothing left but a tail wheel to start with.

The 19th Group returned to Australia the night of the 7th and got ready for the next raid. The airplanes had been pretty badly shot up and a lot of repair work and engine changes had to be accomplished, but on the morning of the 9th Carmichael was off to clean up the Jap airdrome at Lakunai. This time, however, the fates were looking the wrong way. Heavy weather interfered with the mission so much that only seven B-17s made the raid. This time only fifteen Jap fighters intercepted. The bombers fought their way through to the target, put down a good pattern of bombs on Lakunai, and shot down five of the Jap fighters. The photographs were not too good but it looked as though we destroyed at least fifteen more Jap airplanes on the ground, besides building several good-sized fires and hitting a vessel in Rabaul Harbor. The vessel was an afterthought. On one of the bombers, the bomb-release mechanism did not function in time to bomb Lakunai, but it was fixed immediately after and on the withdrawal, as the formation was passing over the harbor, the bombardier pulled the string when he found himself lined up on a big Jap merchant vessel. Two of his eight bombs were direct hits and the ship was burning nicely as the lads left the area.

All of our planes were badly shot up. One of them had to land in the water off the east tip of New Guinea. We sent out a couple of Australian flying boats and picked up the crew. Another of the B-17s got back to Port Moresby on two engines but collapsed on landing. Whitehead reported that he would try to rebuild it but might have to use it for spare parts to fix up others.

Whitehead had been pounding away at the Jap fields in New Guinea. In the last three days fifty-five individual planeloads of bombs, totaling at least that number of tons, had been unloaded on the enemy airdromes at Lae and Salamaua. It was hard to tell how many planes the Nips had lost on the ground at these two fields, but the photographs showed a lot of wrecks that wouldn't fly for a long time if ever, some piles of ashes where airplanes used to be, and runways full of bomb craters which had not been filled up for the past two days. We hadn't seen a Jap airplane over New Guinea for five days, so our fighters had been busy shooting up everything that moved or looked like Jap property, all along the trail from Kokoda to Buna. During the past week, my estimate was that the Nip had lost over a hundred airplanes in the Rabaul area and between thirty and forty in New Guinea. Of course, he could replace his losses fairly easily, while if I lost that number, I would be out of business.

On the 7th I had sent a wire to Washington asking for authority to organize a numbered air force. I said that if they were not using the number five, I'd like to call it the 5th Air Force. On the 9th in came a wire from General Marshall telling us to go ahead and organize the 5th Air Force from our own personnel. In the message, General Marshall said, "Heartily concur in your recommendation." He had told me in Washington that he didn't think much of mixing nationalities in the same organization.

On the 10th I went to Townsville with Bill Benn and started talking skip-bombing with Walker and Carmichael. Walker was not for it. He was an old bombardment man himself and, as an instructor at the Air Corps Technical School, had written the book on bombardment tactics and was a great believer in high-altitude formation bombing. It was an excellent method for a big target like an airdrome or a town but not so good against a turning, twisting target like a moving vessel on the open sea. Carmichael was against skip-bombing, too, but for a different reason. He thought the morale in the 9th Group was a bit delicate already and that introducing a new method of attack would break them. They had taken a lot of punishment in the past few days and were beginning to talk about the virtues of night bombing. That, of course, would be much safer but only good for area targets. Carmichael was probably right. I would have to start the skip-bombing scheme with some other outfit.

While talking with Walker, I found out that there was supposed to be another heavy-bombardment group in the theater, the 43rd, but all they had left was the flag and a couple of guys to hold it up. The group had run out of airplanes about two months previously, so all the officers and men were scattered over Australia in little details running emergency landing fields, weather stations, doing guard duty over supply dumps, building camps, and everything but what they were trained for. I ordered the pieces picked up and assembled in the Townsville area, where I decided to equip them with B-17s as fast as I could get them out of overhaul and repair, or from the United States as replacements.

Lieutenant Colonel Paul B. Wurtsmith, the commander of the 49th Fighter Group over at Darwin, came in to see me. I had sent for him a couple of days previously after hearing about the excellent job his outfit had done in maintaining their airplanes and shooting down raiding Jap aircraft. For some reason his nickname was "Squeeze." He looked like a partially reformed bad boy. He believed in himself, was an excellent thief for his group, took care of his men, and they all followed him and liked him.

I told him I was going to put him in charge of the newly organized 5th Fighter Command. If he made good, I'd make a general out of him. If not, he would go home on a slow boat. I said for him to go to Sydney for ten days' leave and then move to Port Moresby and set up his headquarters. As soon as I could move the two RAAF fighter squadrons then in New Guinea to Darwin, I'd move the 49th from Darwin to New Guinea to help the 35th Fighter Group keep the air clear of Jap aircraft in New Guinea. Wurtsmith grinned and said, "General, I don't like boats to travel on, even when they are fast."

Some information came in that looked as though the Japs were going to run supplies and troops into Buna from Rabaul in a few days, so I told Walker to send all the heavy and medium bombers he could get in shape up to New Guinea, said goodbye to Royce who was to leave for home that night, and flew up to Port Moresby to talk with Whitehead and see what we could do to stop the Japs from getting ashore.

With all the shooting we had been doing along that trail from Buna to Kokoda and all the fires and explosions we had caused, the Japs might be low on food and ammunition or even troops.

On the morning of the 13th air reconnaissance picked up the Jap convoy of one light cruiser, two destroyers, two subchasers, and two large transports off the south coast of New Britain, headed for Buna and escorted by Jap fighters.

During the day we dispatched a total of twenty-two B-17s and six B-26s on four different raids against the Jap convoy. Each ran into heavy Jap fighter opposition and as a consequence the best our bombardiers could claim was that they had splashed water on a couple of the Jap ships. All of our planes returned, although one B-26 collapsed as it landed. We had a couple of crewmen wounded and all that the gunners claimed was one Jap fighter plane definitely, and two others probably, shot down.

I went down to Seven Mile Airdrome to meet the crews as they landed. They were all tired, sweaty, and mad. One tall cocky-looking bombardier, who looked about twenty, gave me the answer when I asked him if he made any hits. The response was, "No. General, we've got to have some new vitamins. Some new kind to make us tough." I remarked that he looked pretty tough to me already, but the kid was serious.

"No," he said, "I'm not tough enough yet. When I'm bending over that bombsight trying to get lined up on one of those Jap ships and the bullets start coming through the windows in front of me, they take my mind off my work. I want some kind of vitamin that will make me so tough I won't notice them. Then maybe I can concentrate on getting some hits." I know what he meant. He was right. We were not going to make hits until we could keep those bullets out of the bombardier's cockpit. That new vitamin he wanted was fighter cover. We didn't guess very well when we designed our fighters with insufficient range to do the job in the Pacific where distance was the main commodity. As soon as I could get those P-38s with their extra range and maybe add some more with droppable tanks hung under the wings, that kid and the rest of them would get their new vitamins.

The Japs unloaded their ships that night at Buna and were well on their way back by the time six B-17s made an attack on them the next morning. Once again the Jap fighters interfered with the bombing. Our planes all came back, another man was wounded, they shot down two Japs, but hit no ships. It was another bad day. We sure did need those P-38s. I went over to a first-aid tent by the side of the runway where a wounded sergeant bombardier was having his leg fixed after a Jap bullet had gone through the calf.

"Tough luck, Sergeant," I said. "How did it happen?"

The sergeant hadn't noticed me come in. He was puffing away contentedly on a cigarette and watching the bandaging job but not saying a word. He looked up, hesitated, and the tears started rolling down his cheeks.

"General," he said and began to choke up, "General, there I was the lead bombardier, with the other two planes dropping on my signal. I'm lined up swell and coming along just right, when about two hundred feet short of the bomb-release point that damn Zero dives in and shoots me in the leg and it startled me so I pulled the bomb-release handle. The damn bombs fall off and the two guys behind me release theirs when they see mine go and there we have a swell pattern of eggs two hundred damn feet short of that Nip boat." The tears kept coming and the sergeant kept talking. "I didn't mean to do it, General, but when that bullet hit me it made me jump and—"

"Sergeant," I interrupted, "for god's sake shut up and let me pin this Purple Heart on you. The first thing you know you'll have me bawling, too."

I had to shut him up as I was getting lumps in my throat, too. Somehow I'd have to get some more bombsights. We had enough to install in each third airplane, so the lead bombardier of each flight of three did the aiming and the other two planes flew in close and dropped when they saw the leader's bombs go. With a screwy system like that, if the lead bombardier missed, all the bombs missed. If the lead airplane was shot down or the lead bombardier was shot, we really were in a fix. If I could just get one bombsight for each bomber, we'd cut down on the misses and maybe stop that sergeant's tears.

We heard that General Eisenhower had been directed to go ahead on the North African invasion in November. That meant that probably we would be more than ever a forgotten theater, especially as far as airplanes and shipping were concerned.

The Australians lost the Kokoda airdrome on the 16th. It wasn't much as a landing field, but the light Jap fighters could and might operate from it. Kokoda was only ten minutes' flying time from Port Moresby. Just to be on the safe side, I told Whitehead to drop an occasional load of bombs on the runway to keep the Japs from using it.

I found that the morale was not all gone in the 19th Group. It was principally the veterans of the Philippine and Java fighting who wanted to go home. Most of the crews that had come out from the States during the last three months were still as cocky as the day they arrived. They were not at all convinced that the Nip was any superman.

A B-17 reconnaissance airplane landed that afternoon after its third mission in twenty-four hours. The plane was really shot up. The brakes and flaps were out and two engines ruined, after a hot fight with some Jap fighters up north of the Vitiaz Straits between New Guinea and New Britain. The pilot stalled down the best he could, but the plane rolled the whole length of the field and then, by ground-looping at the far end, the pilot brought it to a stop just before pitching over into a swamp. The kid, a red-headed, freckle-faced rascal, who probably shaved once a week, coolly slid back his side window, looked around, and then remarked to the world at large, "Huh! I'll bet I'm the first sonofabitch that ever used this much of this runway."

He taxied back along the runway to the hard standing assigned to this particular B-17, where the ground-maintenance crew was waiting to put it in shape for another mission. The top-turret gunner crawled out and stood there in a cocky pose —all five feet three inches of him—with his cap back on one ear, chewing gum and sporting a confident smile. A favorite with the gang, he had five Jap planes to his credit.

"How's she going, Joe?" called one of the admiring grease monkeys.

"One hundred per cent," retorted Joe the gunner. "Yesterday and today, three flights, three fights. Just right."

The gang laughed admiringly. Joe was enjoying himself, too. He was strutting a bit, but why not? Who had a better right?

"Boy, did I see a pretty sight this morning," he continued. "Just off Finschaven, on the way home, we get hopped. One of them Zeros peels off up ahead about five hundred yards, half rolled, and comes in head-on. I let him have a nice long burst at about two hundred yards. He never straightened out of his dive. I watched him all the way. Do you know, he hit that water so clean he hardly made a splash. Boy, was it a pretty sight!"

Except for a small seven-plane raid on Milne Bay out on the east end of New Guinea, where we were trying to put in a couple of landing fields, we hadn't seen a Jap plane over Papua since August 5th until on August 17th the Nips came over Port Moresby on a really good-sized raid. Our continuous air offensive had evidently annoyed them, as the raid was made by twenty-four bombers escorted by about the same number of fighters. We lost eleven aircraft on the ground at Seven Mile Airdrome. In addition, the operations building was burned down, several trucks destroyed, two hundred drums of gasoline went up in smoke, and the runway was hit in several places. The Jap left eight calling cards in the shape of long-delay time-fuzed bombs which exploded at intervals all the way up to forty-four hours. Several men were wounded by bomb fragments. Once again our warning service was inadequate so that, although we managed to get twenty fighters in the air prior to the attack, no interception was made on the Jap formations which came over at better than 20,000 feet. We remained on alert all day for a follow-up attack, but it did not come. If he had come back again with a low-level show about an hour after the first one, he would have just about cleaned me out. Maybe he was short of aircraft, but I didn't think so. He just did not know how to use his air decisively.

The next day Benn and I flew along the south coast of New Guinea from Port Moresby and spotted three locations for emergency landing fields for airplanes to land on in case the Port Moresby area was fogged in or if they ran out of gas on the way back from a mission. I arranged to put the natives to work cutting the kunai grass, removing trees, and leveling the places off.

We then flew across the mountains to the north coast and looked over Wanigela Mission, the place Bull Garing the Australian had told me about. It was a natural. We could land there anytime and occupy and supply the place by air. From here we flew along the coast to Buna and then along the trail to Kokoda and back to Port Moresby. The coast between Wanigela Mission and Buna was not too good for airdrome locations, but there were a couple of grass-covered flats along the way that with a little fixing could be used by DC-type transports. About ten miles west of Buna, however, the country was wide-open. The country was quite flat and covered with kunai grass that did not look to be over four feet high, so that the ground was probably dry and hard. There were several fires in the Buna area and along the trail to Kokoda where our fighters and bombers had been operating while I was flying around Wanigela, but I did not go close enough to make any detailed inspection. The Jap antiaircraft boys might have resented my presence and, besides, the photographs which would be taken immediately following the raid would furnish much better information than I could get.

Driving along the strip at Seven Mile after I landed, I noticed a soldier on a bulldozer pushing dirt along to surround something or build a circular wall around it. I stopped and asked him what he was doing. He stopped the bulldozer and said, "Oh, we got one of them delay-action bombs down there in the ground about ten feet that ain't gone off yet. The engineers have dehorned most of the others, but this one is a new breed that they don't know about, so I'm just pushing dirt around her so she'll just fizz straight up when she goes off." I thanked him for the information. He wiped the sweat off his forehead and calmly went on with his job.

The bomb went off that night about eight o'clock. As he had

predicted, "She just fizzed straight up." It made quite a lot of noise and a lot of dirt went up into the air, but no damage was done.

These kids adapted themselves fast. They were almost blase about bombing and you could hear them comment on or criticize a Jap formation or their own with equal calmness. It helped to keep me from worrying, when I saw them so unconcerned.

An old sergeant of one of the B-25 squadrons came to see me. He said, "General, we got five B-25s out of commission for lack of wheel bearings. If they sit around here, the Japs will come over some day and burn them up sitting on the ground."

"Sergeant," I said, "I know about it and there just are no wheel bearings in supply anywhere here or in Australia."

"But, General," he replied, "I know where there are some wheel bearings. Last month a B-25 was shot down near Bena Bena. There is a Lieutenant Hampton over with the troop carriers who knows about a field near there and if you will let him take me and three other guys, some rations, a kit of tools, and a couple of tommy guns and some ammunition with us, we'll salvage the bearings and some other stuff off that B-25 and do ourselves some good instead of waiting around here until we get taken out. That B-25 didn't burn and we might get a lot of good loot out of her."

Bena Bena was on a plateau up in the middle of New Guinea, inhabited by the partially reformed cannibals discovered there by explorers and gold seekers about twelve years before. We were not sure how many of the natives of Bena Bena were for us and how many were for the Japs. Lieutenant Hampton was the hot-shot troop-carrier pilot who was supposed to be able to land and take off a DC-3 transport out of a good-sized well. The field at Bena Bena at this time was not much, but it could be negotiated easily by Hampton and, with a little luck, by any other good transport pilot. The only complications were that we didn't know whether the Japs were patrolling the area or not and there was always the sporting proposition of flying an unarmed transport, with no rubber-covered, bullet-proof gasoline tanks, within easy range of Jap fighters at Lae and Salamaua. But the sergeant was right. We needed those wheel bearings.

Whitehead was with me. We talked it over, and Whitehead suggested that Hampton fly the sergeant and the three grease monkeys in under cover of the next strike we put down on Lae and Salamaua. I agreed.

The next morning the little expedition flew to Bena Bena, where Hampton left the four men and made a date to meet them four days later.

That was the time the sergeant estimated it would take him to get to the wrecked B-25 and finish his salvage job.

Four days later Hampton landed again at Bena Bena. He and the co-pilot sat in the plane, engines running, tommy guns stuck out the windows, and ready for a quick getaway if things began to look suspicious. There was no sergeant, none of his gang, and no natives. An hour went by. It looked as though the estimate of the situation had been bad. Maybe the Japs had the sergeant and his three men. Maybe the natives had decided that the white man had lost the war and had joined the Japs. Maybe there was a meat shortage at Bena Bena. Maybe—

Hampton had just about decided that something had gone wrong and it was time to think about getting out, when the sergeant came running all out of breath out of the jungle trail onto the field and yelled, "Hold everything, Lieutenant. The gang will be here in just a minute. We got delayed because we found a wrecked fighter plane about five miles east of here and we got a lot of good loot out of that, too."

A few minutes later the "gang," consisting of the other three men and a hundred natives, appeared, loaded to the limit with everything they could stagger under. They had brought back most of both wrecked airplanes. Wheels, landing-gear parts, ailerons, rudders, fins, pieces of aluminum sheet cut from the wrecks for future repair work, propeller parts, instruments—and the precious wheel bearings.

Three hours later the flying "air depot" landed at Port Moresby. Three days later five B-25s and three P-39 fighters joined the list of aircraft in combat commission.

Twenty-five P-38s, the first of the fifty promised me by General Arnold, arrived by boat at Brisbane. I sent word to Connell to come north from Melbourne, take charge of setting them up, and work twenty-four hours a day on the job. Also to give the Australian sheet-metal industry a contract to make about 10,000 150-gallon droppable gas tanks to hang under the wings so that we could extend the range. The pilots for all fifty airplanes had already arrived, among them Lieutenant Dick Bong, my bad boy from San Francisco. A Captain George Prentice, a veteran from "Squeeze" Wurtsmith's 49th Fighter Group, was assigned to command them. I wanted a veteran as a leader so that the new youngsters would have someone they could look up to and from whom they could get instruction in the art of shooting down Japs. They didn't have any back in the States to practice on. Prentice had one official victory to his credit and had flown a lot of combat missions up at Darwin.

It wasn't long before I began having engineer and quartermaster troubles. They belonged to the Service of Supply (SOS) and ran the base organizations at ports like Townsville and Port Moresby, were in charge of construction of docks, roads, and airdromes, and were responsible for getting supplies in by boat and for feeding us. There were not enough engineers at Port Moresby to do all the work in a hurry, so the tendency was to spread the men thin and skimp the job in order to make a showing. One of the results was that they were not paying enough attention to the drainage problems that would confront us when the rainy season arrived about two months later. Runways were being built without enough culverts to carry off the water and I was afraid that, with our airdromes washed out, we would get caught with our airplanes on the ground some day. The stock answer of the engineers was that they knew all about building airdromes and would appreciate it if we tended to our flying and let them do the construction work. I told the base commander that for his sake I hoped he was right when the rains did come.

The food situation was really bad. Australia had plenty of fresh meat and vegetables but the SOS had loaded up with a lot of canned stuff. It was not bad for the first three or four days, but after that the men threw most of it away, as it was simply unpalatable in that steamy climate after a hard day's work. The American kid wants fresh meat and gravy, once in a while. If he doesn't get it, especially when he knows there is plenty available, he goes stale and tires easily. If he loses his appetite, he doesn't eat, loses weight, and is soon a prey to all the ills and diseases of the jungle country. The alibi given me was that there were not enough refrigerator boats, but the cargo vessels that came in to Port Moresby seldom had their refrigerators full when they left Australia. The crews of these vessels insisted on fresh meat and they were not particularly interested in what the combat troops got. Some adjustments surely could have been made, but the SOS attitude was that they had provided food and we should eat it and shut up.

Outside of General MacArthur, most of the American ground people seemed to be looking for reasons to criticize the Air Force. I kept hearing that no favoritism should be shown the Air Force and that they deserved no special treatment and so on. As far as I could find out, we had not been getting any special favors, but, as long as we were the only ones doing any fighting in the American forces, I was going to see that if any gravy was passed around, we got first crack at it.

In order to cut down the malaria and dysentery rate, I said something about putting concrete floors in the mess buildings and screening them in, but was informed that this was unnecessary, it was not in accordance with the regulations on what constituted combat-zone construction, and anyhow there were no building materials, cement, or screen wire available.

I sent instructions to Connell to get together a stock of cement by hook or crook and to load every plane going to New Guinea with it so that we could concrete all our kitchen and mess-hall floors. I told him also to buy up all the screen wire he could find, even if he had to get it one roll at a time in the hardware stores. We needed refrigeration, so Connell was to buy any electric or oil-burning types that we could carry in a DC-3 transport and fly them to New Guinea, too, along with the motor generators to operate the electric equipment. To satisfy the paperwork advocates, I'd have to get our tables of equipment amended. When those tables had been drawn up in Washington several years before, no one thought about what we would need to fight the Japs in New Guinea, but still there were people who looked upon a table approved by the War Department as something as sacred and inviolate as the Constitution of the United States.

On August 21st, I returned to Brisbane and saw General MacArthur. I discussed the operations in New Guinea and, when I got to describing the terrain along the north coast from Milne Bay to Wanigela Mission and Buna, the General wanted to know where I had gotten my information. I told him, "From visual inspection from a hundred-feet altitude." I then spent the next ten minutes listening while General MacArthur bawled me out for flying around where there might be some Japs. He said he had decided that he needed me to look after the Air Force and that, from now on, I would stay south of the Owen Stanley Mountains unless I got direct permission from him otherwise. I decided it would be better to wait a few days before proposing that we seize Wanigela Mission with an airborne show.

Special intelligence indicated the Jap was up to something. It looked like a convoy run with a special landing force. It might be a reinforcement for Buna, but there were indications that it might be to capture Milne Bay where we had two brigades of Australian troops and some engineers fixing up a couple of airdromes. One was already in operation, with two Australian fighter squadrons installed there. Jap air reconnaissance had been quite active in the last day or two. So I sent word to Whitehead to watch for Jap bombing raids on Milne Bay to begin any time, but probably soon.

My 3000 parachute fragmentation bombs arrived but there were no racks in the A-20 light bombers that would take them, so I turned the job of making the racks and installing them over to a character named Major Paul I. (Pappy) Gunn. Pappy was a gadgeteer par excellence. He had already developed a package installation, of four fifty-caliber machine guns, that fitted beautifully into the nose of an A-20 with 500 rounds of ammunition per gun. He had learned how to fly with the Navy years ago. No one knew how old he was but he was probably well over forty, although he looked you straight in the eye and said thirty.

When the war broke out, Pappy was doing civilian flying in the Philippines, so the Army mobilized him and now I had him as a major. It was a private war with him, as he had a wife and four children in Manila. He didn't know whether they were still alive or not. Anyhow, he didn't like Japs.

People called Pappy eccentric and said he told some wonderful stories that were too good to be all true. I don't believe that Pappy ever took a chance on ruining a good story by worrying about the exactness of some of its details, but he was a godsend to me as a super-experimental gadgeteer and all-around fixer. There was absolutely nothing that feazed Pappy. If you asked him to mount a sixteen-inch coast-defense rifle in an airplane, Pappy would grin, figure out how to do it, work day and night until the job was finished, and then test the installation by flying it himself against the Japs to see how it worked.

His stories of his exploits were something that no one should miss. His actual accomplishments, however, which Pappy seldom mentioned, were just as good as the astounding stories he delighted to tell.

Right now he was breaking his back to fix up those A-20s to carry my parachute frag bombs. I told him I wanted sixteen of them ready in two weeks. Pappy said he would have them. No one else but me believed he could do it, but when I offered to make a small bet on the possibility, I found no takers.

The next day Bill Benn and I flew to Melbourne, looking over Tocumwal on the way. Tocumwal was evidently designed to rival our big supply establishment and procurement center at Dayton, Ohio, so that the boys would feel at home when the war moved down to Australia. Beautiful runways, nice big hangars, concrete and steel test stands, warehouses—everything was well built and dispersed all over the place so as not to be vulnerable to the Japanese bombers. Of course, the nearest Jap airdrome was still 3000 miles away, but it was explained to me that we should play safe in case the Japs came a little closer. About a mile from the field a little village of quarters was being built to house the Tocumwal

Depot officers. The base hospital was three miles from the field. There was plenty of transportation; the buildings were so widely dispersed that, if walking were necessary to get from one building to another, so much time would be consumed that the men would get very little work done.

The place was loaded with airplane and engine parts, propellers, lumber, furniture, lighting fixtures, plumbing supplies—almost anything you could think of. The depot commander's real pride and joy, however, was his filing system, with its stock records. The only difficulty he seemed to be having was that he could not get the inventory up to date; new stocks were coming in all the time and messing up their figures. That made it difficult to keep track of things and at the same time fill requisitions, but it wouldn't be long before the war moved down into Australia and then the supplies would be where they belonged.

I went on to Melbourne where Major General Lincoln and his staff met me and drove to his headquarters. I was introduced around and told about the organization and what they were doing. I told Lincoln to close up his whole shop, except for a small office to look after our contracts with Australian industry in the Melbourne area, and to move to Brisbane right away. All supplies in the Melbourne area were to be moved to Brisbane for the time being, until I could shove them still farther north. Lincoln had another big organization in Sydney which I said to close down to another small office like the one I proposed for Melbourne.

Everyone seemed to be taking on a stunned look at this point, so I brought up the subject of Tocumwal. No more supplies were to be sent there for any reason, the airplane repair work was to be finished as soon as possible, and everything except the runways and buildings was to be moved to the Brisbane area or farther north as fast as I could get storage space at Townsville. All construction work, including that of the pretty little village, was to be stopped immediately, regardless of the state of completion, and the building materials held for shipment on my orders. This stuff would probably go to Townsville, as I intended that to be my main supply and overhaul depot for Australia until we moved out of the country entirely.

Lincoln at this point began to look a little worried. He asked me what I was going to do with Tocumwal and reminded me that we had put a lot of money in the place. I told him to give it to the Australians, or back to the aborigines if the Aussies wouldn't take it. When I was invited to inspect the filing system, I'm afraid I was not very polite. I was not interested in his paper work. I had seen too much of it already. From now on, if he had the supplies, I wanted them issued on oral, written, or telephoned requests, and issued fast, with no more alibis about taking inventory or worry about whether the form was filled out properly. We could not sink Jap ships or shoot down Jap planes with papers and filing cabinets and I was far more interested in getting the planes flying than in having a beautiful set of records that we were not even going to take home when the war was over. Just ask the Australians what happened to their records and filing cabinets at Singapore. I understood they had some excellent ones there.

I spent the night with Lincoln and a few of his top staff out at the house which Billy Robinson had turned over to Brett. A beautiful place, full of good-looking furniture, rugs, and paintings. Excellent meals and nice comfortable beds. No wonder the boys didn't like to give it up. They probably thought I was crazy. They were probably right.

Everyone in Melbourne had gotten himself fixed up with as much rank as the job would allow. Up in New Guinea I noticed that squadrons were commanded by captains and lieutenants, whereas the jobs called for majors. I told Benn to remind me as soon as I got back to Brisbane to stop all promotion in the non-combat outfits until the squadrons were commanded by majors who had earned that rank by demonstrating their leadership in actual combat. I remembered that in New Guinea one of the big gripes was that they never received any promotions while the boys in Melbourne and Sydney lived "the life of Riley" and got increased rank besides.

The next morning I had Lincoln assemble his whole staff and tried to give them the picture of what we were up against in New Guinea. That was where the war was and it was not moving to Australia. Those youngsters up there were our customers and customers are always right. Our only excuse for living was to help them. The payoff would be Jap ships sunk and Jap planes shot down. As far as I was concerned, the ones accomplishing that job were going to get top priority on everything. We might work ourselves into having stomach ulcers or nervous breakdowns, but those things were not fatal. The work those kids up in New Guinea and at Darwin were doing, however, had a high fatality rate. They deserved all they could get. Most of the crowd appreciated what I was talking about. The others would go home.

Over at RAAF Headquarters I met Air Vice-Marshal George Jones, the Australian Chief of Air Staff. He wanted me to help him get American airplanes and equipment. I told him I would do everything I could to put both Australians and Americans into air combat. I explained the new set-up of an American Air Force and a separate Australian combat air organization. Jones agreed that he thought it would work much better but was not too keen when I said I was going to put Bostock in command. These two just didn't like each other. It dated from a few months before when Sir Charles Burnett, an Englishman who was acting as Australian Chief of Air Staff, was called back to England. Bostock, the ranking RAAF officer, was passed by and Jones, considerably his junior., was selected for the job. Why, I didn't know. As a matter of fact, except for the feud which sometimes was a nuisance, I liked the situation as it was. I considered Bostock the better combat leader and field commander and I preferred Jones as the RAAF administrative and supply head.

I also called on Mr. Drakeford, the Australian Air Minister. He was the head of the locomotive engineers' union and had no aviation background, but he was trying. I considered that he was sincere and honest and would help in every way possible. My call was just a courtesy affair, as I had no real business to discuss with him, but he appreciated it. One thing that I liked about him was that he didn't pretend to know anything about aviation or the strategy or tactics involved in the use of air power. He looked after the interests of the RAAF when it came to budget matters, allocation of manpower, resources, and industry, and left the operating end to the operators.

During the evening the news came in that Squeeze Wurtsmith's 49th Fighter Group at Darwin had intercepted a Jap force of twenty-seven bombers escorted by fifteen fighters and had definitely shot down six Jap bombers and eight fighters, with three more bombers probably destroyed. We had no losses. General MacArthur wired me congratulations. I added mine and passed them on to Wurtsmith.

Back in March 1942, on the fourth of March to be exact, Lieutenant Robert Morrisey had shot down the first Jap plane for the 49th Fighter Group. During the celebration which followed, Squeeze Wurtsmith had produced a magnum of brandy which he announced they would open when the 49th achieved its 500th official destruction of a Jap airplane in air combat. None of this shooting-them-sitting-on-the-ground stuff was to count. They had to be wing shots. They had to get them in the air. The group was still carrying that huge bottle around as if it were a household god. It was not even a very good grade of brandy and it was definitely green, but that made no difference to the kids. They were not thinking about how it would taste. It had become a symbol of a goal that they were shooting for and which they were certain they would reach ahead of any other group. The fight at Darwin brought the score to sixty. Only 440 to go.

On the morning of the 24th I returned to Brisbane. I asked Bill Benn to find out how many B-17s and crews had already been assigned to the 63rd Bombardment Squadron, which would be the first of the four to comprise the 43rd Bombardment Group which I was resurrecting. His report showed eleven B-17s and twelve complete crews had arrived at Torrens Creek Field, about a hundred miles southwest of Townsville. I told Bill he was fired as aide and to go up there and take command of the 63rd Squadron. Bill grinned and left to pack up. After a couple of weeks' training under Bill's supervision, I figured on attaching the squadron to the 19th Bombardment Group for operations and starting formation of some more squadrons, until I got enough to put the 43rd Group in business on its own.

It looked as though we were due for trouble in New Guinea soon and I was more certain than ever that it was Milne Bay. The Japs raided the place that morning. The two RAAF fighter squadrons there intercepted the Nips and chased them off, getting two definitely destroyed and two more probables. The Aussies had one P-40 fighter damaged out of the twenty-three they put in the air.

I called Walker at Townsville and told him the Jap air raid on Townsville looked like a pre-invasion show. I ordered two squadrons of B-17 bombers to be at Port Moresby ready for action at daybreak the next day and two more squadrons to be there that afternoon. Also all effective medium bombers (B-26s) of the 22nd Group were to go to Port Moresby immediately and go on alert for action on arrival. One of the B-17 squadrons was to hit Vunakanau and Lakunai airdromes at Rabaul that night with incendiary bombs.

I then sent word to Whitehead to knock out the Jap field at Buna and keep hammering it as long as the Japs tried to use it.

If I could keep the Jap air force whittled down so that it couldn't support their landing operation, that operation wouldn't go far and they would have trouble supporting any troops that they did put ashore.

Eight B-17s hit the Rabaul airdromes during the night. After the attack, fires visible for forty miles were reported by the returning crews. I figured that might help.

I had told the fighter pilots of the 35th Group when I was in New Guinea the preceding week that if anyone shot down a Jap airplane in air combat, I would give him a decoration. In the afternoon of the 25th, seven P-39s found a "hole" in the bad weather hanging over Port Moresby and raided the Jap airstrip at Buna. As the P-39s appeared over the edge of the airdrome, ten Jap fighter planes, which were taxiing out to take off, stopped. Over at Port Moresby at Fighter Command Headquarters where they were listening in on the fighter radio frequency, they heard one of our pilots say, "Get off the ground, you so-and-so's. Don't you know the General won't give us any ribbons for shooting you while you're sitting?"

It sounds screwy, but the Jap fighters and our own operated on the same frequency and actually hurled insults at each other in the air during combat. A lot of these Japs spoke fairly good English, but their habit of hissing some of their syllables generally identified them. At any rate, on this occasion following the insult, they started taking off, but after three of them were shot down before they could get the landing gears retracted, the rest stopped, got out of their cockpits, and ran for cover. So the boys finished the job by setting fire to the remaining seven Jap planes "while they were sitting."

Around noon, flying in wretched weather, a reconnaissance plane located a Jap convoy of 3 destroyers, 2 minesweepers, 2 transports estimated at 8000 tons each, and 2 unidentified smaller ships, about 120 miles north of Milne Bay and headed south. Our B-17s took off to attack but owing to the weather did not locate the Jap vessels. Just before dark, the Aussies from Milne Bay put twelve fighters carrying two 300-pound bombs each over the convoy. Although a 300-pound bomb was rather light for this type of work, it was all they had. They sank one of the minesweepers and then heavily strafed all the rest of the ships in the convoy. All the planes returned and the Aussies were swaggering wider than ever. Those rascals are fighters.

During the night, the Japs landed between 1500 and 2000 troops on the north side of Milne Bay about fifteen miles east of our Number Three airstrip which was under construction. The other two strips were about four miles west of Number Three.

All day long on the 26th, in the face of the worst weather I had seen in New Guinea so far, we kept taking off to attack the Jap vessels and the troops and supplies that had been landed. Most of the planes had to return without being able to get through. Some of them, however, found targets in spite of the weather. One flight of six B-17s reported sinking one of the Jap transports and damaging a minesweeper. The B-26s and the Aussie P-40s sank Jap landing barges, destroyed four light tanks that the Nips had managed to get ashore, and set fire to piles of supplies on the beach. We had one B-17 shot down in flames and another so badly damaged by antiaircraft gunfire that it folded up on landing back at Port Moresby. Another had the whole nose shot off, killing the navigator and badly wounding the bombardier. Still another had five antiaircraft shell holes in each wing, each large enough for a man to jump through. One pilot had his bombs stick in the racks. After making twelve runs on the Jap vessel, he was still unable to get the bombs off, so he brought them back home. With literally hundreds of shrapnel and shell holes, the skin of the wings and fuselage practically in ribbons, and two engines shot out of commission, I don't yet understand how he made it.

In spite of our attacks, the Japs pushed west along the trail, driving the Australian outposts ahead of them, and by nightfall were within three miles of our Number Three strip where the Australian resistance had stiffened.

Just before midnight, a B-17 returning from a reconnaissance mission to Rabaul, piloted by Lieutenant R. E. Holsey, landed on a narrow strip of sandy beach on the south coast of New Guinea about fifty miles east of Port Moresby. As a result of an argument with the Jap antiaircraft defenses at Rabaul, two engines had been knocked out and a third was not doing too well. It had suddenly died altogether. Hurriedly releasing a couple of flares, Holsey had done a beautiful job of putting the B-17 down and keeping it right side up on that carpet-sized strip of sand. The crew worked all night removing everything they could take out of the airplane and working on the engines to get them in shape. We sent down a thousand feet of steel mat to lay on the sand for a take-off strip, chopped down some trees, and waited for the tide to go down so that the wheels would be out of water for the run. Holsey and the sergeant crew chief, who had finally got all four engines running after a fashion, then took the B-17 off. Just as the plane had been literally jerked into the air by the pull of more horsepower than the engines had ever been designed for, one of the engines died, but the others held and twenty minutes later the plane was back at Seven Mile strip waiting for the barge to bring back the rest of the crew and the stuff they had stripped off the airplane and which now had to go back on again. Four days later Holsey announced that he and the plane were ready to go again.

Fighting like fanatics, with their supplies gone, their tanks destroyed, and with nothing but rifles and bayonets, the Japs drove the Australians to the edge of Number Three strip. General Clowes, the Australian commander, ordered preparations made for the evacuation of Milne Bay. We pulled the two RAAF fighter squadrons back to Port Moresby, keeping the Milne Bay Number One strip open for refueling purposes only.

As part of the preparations ordered by General Clowes, the Australian canteen at Milne Bay was mined so that at the last minute it could be blown up to prevent the Japs from getting hold of the supplies there. During the night a native cow, wandering around outside the canteen, alarmed one of the guards, who fired his rifle as a signal that the Japs were breaking through. Someone pushed the plunger and blew the canteen up. The next day, when it was discovered that the Jap advanced troops were still several miles away and that a large part of the beer had survived the explosion, the Australian troops availed themselves of the golden opportunity to get all the beer they could drink without the necessity of drawing their rations. That evening the Japs started withdrawing.

Major Victor Bertrandais, in civil life an executive of Douglas Aircraft, and Captain Eddie Rickenbacker's crew chief in World War 1, reported to me for a job. Vic was a driving go-getter. I'd known him for over twenty years. He said he had reported to General Joe McNarney in Washington about a week ago. Joe said, "Go on out to Australia and report to George Kenney. He'll either chase you home on the next plane or give you a job and work you to death." I told Vic I had a job for him. First, to go to Townsville and see where he was to build a depot to be manned by 3000 to 4000 men, overhaul 100 engines a month, erect or repair 200 airplanes a month, and do everything else I happened to think of, besides running a supply depot to keep Whitehead going in New Guinea and Walker in northeastern Australia. Then he was to go to Tocumwal, pack up everything there, and transfer it to Townsville, personnel and all. He had carte blanche. All I asked was speed and results. I introduced him to Connell and explained what I wanted Bertrandais to do. I told Connell to give Vic all the help he could and to push our stocks of supplies out of Brisbane and forward to Vic as fast as he got buildings erected in which to store them.

We wound up August with a low-altitude attack on Lae which caught the Nips by complete surprise. After the B-26s had blown huge craters in the runway, set fire to a big fuel dump, and wrecked a number of buildings, the A-20s swept in, strafing antiaircraft gun positions, aircraft on the ground, vehicles, and buildings. They left the place a mess. Just as the A-20s came over the edge of the airdrome, two of the pilots on the right side of the line-abreast formation, noticed that the porch of a large plantation house, on a little hill just north of the strip, was crowded with Japs sitting there having lunch. The two lads banked over and emptied their guns into the house, setting it on fire. They circled around for about five minutes to take care of any survivors who might come out, but with great glee they reported that there was no sign of life and that the house was blazing furiously when they left. The next night Tokio Rose said that our sneaky tactics were what might be expected of the gangster-murder type of people that the American air forces were recruiting.

General Rowell, the Australian commander in New Guinea, sent me a nice note of appreciation:

HQ N G F
31 Aug 42

Commanding General,
Allied Air Forces,
Port Moresby.

I desire to place on record my thanks for the assistance the Allied Air Forces under your command have given to the Army in the past few days, not only in the way of direct ground support, but also in keeping the enemy air force away from our troops by the attacks on Buna.

The fact that the weather was dead against securing really decisive results in attacks on shipping was beyond your control and does not detract from our admiration for the determined efforts made by your bomber pilots to find the enemy and sink his ships.

Major General Clowes has asked that a record be made of his appreciation for the work of the Fighter and G.R.Sqns R.A.A.F. stationed at Milne Bay. I am convinced that, when the story is complete, it will be found that their incessant attacks for three successive days proved the decisive factor in the enemy's decision to re-embark what was left of his force.

S. L. Rowell, Lieutenant General
General Officer Commanding New Guinea Force.

Now that the Jap attempt to seize Milne Bay had failed and the Australians were mopping up the survivors who had not been evacuated, an interesting sequel to the story had come to light:

When the Japs started raiding Milne Bay on August 24, we had flown a fifty-caliber machine-gun antiaircraft unit down there from Port Moresby. One of the pairs of guns was mounted in a revetment on the west edge of Number Three strip. Work was abandoned on the strip when the Japs got a little too close for comfort, and on the afternoon of the 28th a detail under a sergeant of the antiaircraft unit was getting ready to pull the guns out, when suddenly a Jap soldier moved out of the jungle opposite the gun position and stood there looking around. He didn't see the sergeant, who was behind the gun, or the others, who had ducked fast. The sergeant whispered, "Watch me get that s-o-b," slipped in the ammunition belt, swung the guns around, took aim, and pulled the trigger. Nothing happened, except that the Jap turned leisurely around and disappeared as the sergeant suddenly woke up to the fact that he had forgotten to unsafety the guns. The curses started rolling forth but stopped as about a hundred Nips pushed out of the jungle in a bunch and started to move out on the strip. The lone Jap had just been a scout to draw fire or see if the coast was clear for the others.

The fifty-calibers now really got into action. Of course, the sergeant and his little gang say they killed hundreds of Japs. They probably did pretty well by themselves, at that. Anyhow, that was as far west as the Japs ever got.

On September 3d I returned to Brisbane and saw General MacArthur. I gave him a report on how things were going in New Guinea, as they appeared to me. The fighter defense had improved with the installation of a radar station at Yule Island about fifty miles northwest up the coast from Port Moresby and another at Rorona, thirty miles from Port Moresby in the same direction. Whitehead, with Freddy Smith as his Chief of Staff and Bill Hipps as Assistant for Operations, was doing a swell job, although the mess he had established for his own staff had easily the worst food in New Guinea. Whitehead was so busy he didn't know or care what he ate. Pappy Gunn had fixed up six A-20s with racks for my parachute frags and, as soon as the crews got a little training in the use of them, we would try them out the first time we could catch the Jap airplanes on the ground. We were watching Buna as a good place for the initial test.

I told General MacArthur that I had no faith in the Australians' holding Kokoda Gap. The undergrowth at that altitude was sparse and the Nips would move through it and around the Aussies and work their same old infiltration tactics. We were bombing along a trail that we could only see in spots. Maybe we could scare away the native bearers and make the Jap carry his own supplies but the surer way was to sink his ships and shoot down his planes. Even the Jap couldn't fight long without food and there was nothing to eat along that trail. The General said to keep on with my program as he had nothing but praise for the way the Air Force was operating.

The final story of the Milne Bay expedition was given us at the noon briefing in the War Room. The initial landing was by 1600 Japs, who were reinforced later by 600 more. One thousand were killed, 450 evacuated, 50 reached Buna by walking along the coastal trail, and 9 were taken prisoner. Of the 700 still unaccounted for, some were probably dead, some had escaped to the islands off the tip of New Guinea, and the remainder were still roaming the jungle and being hunted down by the

Australian patrols. Many of the dead showed signs of starvation. Their clothes were torn, their shoes rotted off from the mud and rain, and they had become so weak that they could no longer carry their rifles. The prisoners all admitted that our planes had done them wrong from the start. They were quite annoyed because their own airplanes had not supported them and had not even contested the air with ours.

Brigadier General Al Sneed had been Brett's senior American Air officer at Darwin. With the coming move of the 49th Fighter Group to New Guinea, Sneed was out of a job. I wired Arnold to give him a job back home as I had nothing for him here. I needed operators. He was not an operator. Arnold wired back okay, so I arranged for Sneed to leave in a couple of days.

September 8th was a bad day. First the news came in that the Japs had worked around the Australian defenses in the Kokoda Gap and were already five miles beyond at a native village called Efogi, on the downgrade toward Port Moresby, with the Aussies still retreating.

Then I had to ground all my P-38s. The leak-proof tanks were improperly made and now they were falling apart at the seams. I sent for every rubber expert, Australian and American, in the country to come to Brisbane to see if the tanks could be repaired. The other twenty-five P-38s of the fifty that Hap Arnold had promised me, arrived but without feeds for the guns, and six B-25s came in as replacements but these had no gun mounts, no guns, and no bombsights. The communications people had a new message from me to Arnold about every hour all day.

During one of the lulls in bad news I called Lincoln at Melbourne to find out how soon he was moving to Brisbane. When I learned that they were still waiting for me to confirm in writing my verbal instructions of August 22, over two weeks ago when I was down there, I almost pulled the telephone out by the roots. It wasn't long, however, before he understood that he was to move north rather soon.

The rubber experts said that some of the P-38 tanks could be reworked. The experts were all out at Amberley Field with their coats off, working with my mechanics and changing good tanks for bad ones. It looked as though I'd have enough in a few days to equip about twelve of the P-38s so that they could go north. I wanted those new pilots to get some experience flying in New Guinea before I put them into action against the Japs.

In the evening I spent about three hours with General MacArthur in his flat at Lennon's Hotel. I described Wanigela Mission and outlined my plan for occupying it and using it as a base for movements along the coast to Buna and particularly to the flat Dobodura plain, a few miles west of

Buna, where I wanted to establish a big air base. If we got on that north coast, we would not only extend the range and bomb-load of our airplanes, but we wouldn't have to climb over the Owen Stanleys and buck the thunderheads which made a barrier almost as bad as the mountains themselves. We could go into Wanigela Mission with a light plane or two, get the natives to cut the grass, and then throw in troops at the rate of a thousand a day, feed them, and supply them with ammunition, even if a couple of divisions were put into the show. With the Japs strung out all the way from Buna to Kokoda and beyond, this move would be a threat to the Japs that might develop into a quick defeat for them if we could move fast. MacArthur liked the idea but wanted to see where the Australians were going to halt the Jap drive for Port Moresby. If by any chance our airdromes in the Port Moresby area should be threatened or occupied by the Nips, obviously we might have to modify our plans considerably. He said he had inspected the 32nd Division (U.S.) a couple of days ago and, while their training still left something to be desired, he was going to send them to New Guinea shortly. The 7th Australian Division was now moving up to New Guinea and would be rushed up the trail to halt the Jap advance. MacArthur was certain that he would hold Port Moresby but he was not yet ready for the Wanigela Mission thing. He said for me to go back to New Guinea, watch the situation develop, and give him a report when I got back on the 12th. I took off for Port Moresby the next morning. On arrival I talked with General Rowell and his staff. He was busy getting barbed wire strung all over the place around the town. Whitehead's house was practically surrounded by barbed wire and trenches, as that hill was destined to be a resistance center if the Jap came to Port Moresby. It was apparent to me that if someone didn't stop him he would be there soon. The Aussies were still being driven back along the trail.

Rowell was planning a withdrawal to his perimeter defense line, which would put most of our airdromes in the hands of the Japs. I told him that I couldn't see why 12,000 white men should have to let two or three thousand Japs chase them behind the barbwire, but that if he did retire that far, I would evacuate the air troops, and then he would really be in trouble as Jap planes would occupy our fields and dive-bomb every vessel in the harbor or approaching the New Guinea coast. Then it would only be a question of time before he was starved out. Clowes down at Milne Bay would be cut off and he, too, would starve. With our air bases back in Australia, there would be nothing to stop the Nips from landing anywhere and taking Rowell out. With the Jap based in Port Moresby, Australia would be next on their list. He agreed that the situation would be bad but thought that he could hold out. If we couldn't stop the Jap advance by use of our air, it looked to me like another Singapore. One vital thing Rowell would lose by withdrawing would be his water line to the Laloki River. Another parallel with Singapore, where the British position became hopeless when they lost the only reservoir that supplied the city.

At present, our air opposition was all that was even slowing the Jap advance down. I hoped, by cutting off his supply line and driving his native bearers off the trail, to weaken and halt the Nip drive. Even a Jap had to eat and had to have ammunition to shoot. At the same time, some new blood was needed in New Guinea. I decided to recommend to General MacArthur that he let me fly some Yanks up there. They would come in bragging about taking over after the Aussies had failed. Actually they probably would be no better fighters when the chips were down, but they didn't know that and the Aussies didn't know it, so a little competition might help.

On September 12, Captain "Don" Hall leading nine A-20s of the 89th Attack Squadron demonstrated my parafrag bomb for the first time in war. The Japs the day before had filled up enough holes in the Buna strip to land twenty-two airplanes just before dark. They were still on the ground when the A-20 sneaked in over the palm trees and, strafing as they flew over the field in two waves, dropped forty parafrags per plane, stringing them over the airdrome about fifty yards apart. Seventeen of the Nip planes were destroyed, nine of them by Hall himself. The leading wave of four planes got some ack-ack fire, but the second wave reported "everything quiet as a grave." When the first parachutes blossomed, some of the Nips evidently thought it was a paratroop landing for they rushed out with their rifles and began to shoot. As soon as the super-sensitive fuzes on the noses of the parafrags touched the ground, however, the Nips found out their mistake. The fragments from that bomb will cut a man's legs off below the knees a hundred feet from the point of impact.

Following the attack by the A-20s we put five B-26s and seven B-17s over Buna to complete the job. They unloaded some 1000-pounders along the runway and from the looks of the photos I didn't think the Japs would be using Buna airdrome for a long time. They had bled themselves out of a lot of airplanes, trying to use the place. Sooner or later even a dumb Nip should understand that we didn't want him that close to us. During the past two days Buna got 35 tons of bombs and 33,000 rounds of ammunition.

I wired Arnold for 125,000 more parafrag bombs and sent word to my ordnance officer at Brisbane to get together with the Australian ordnance people to convert our standard fragmentation bombs into parafrags as fast as possible.

I returned to Brisbane and that evening reported to General MacArthur my conversations with Rowell and my own conclusions about the situation in New Guinea. I told the General that I believed we would lose Port Moresby if something drastic did not happen soon. The Japanese were advancing so fast that the Australians had had no opportunity to organize any resistance and their withdrawal had been so fast that they had abandoned the food dumps at Kokoda, Efogi, and Myola Lake that we had put so much effort into building up by air drop during the previous two weeks.

I believed that Rowell's attitude had become defeatist and that this attitude had permeated the whole Australian force in New Guinea. In some quarters there was evidence of panic. The troops were worn out and the constant stream of sick and wounded coming into Port Moresby had visibly affected the morale of the troops there, who had no hesitancy in saying that they wanted no part of fighting up on the trail. There was a definite lack of inspiration and a "don't care" attitude that looked as though they were already reconciled to being forced out of New Guinea.

I advised getting some American troops up there as soon as possible to stop stories now going the rounds that the Yanks were taking it easy in Australia and letting the Aussies do all the fighting. The Jap propaganda leaflets all pictured the Americans taking good care of the Australian wives while the Aussie sweat it out in the jungle. It wouldn't do to have this propaganda take effect. The Germans, in appealing to the French in 1940, blamed the British troops for doing the same thing. I was in France at that time and I know it had an effect on the French, who gradually developed a definite coolness toward the British.

I asked General MacArthur to let me fly a regiment to New Guinea right away and to let General Blamey put them into the fighting. The General said he agreed that we must send some Americans to New Guinea right away and said to be in his office the next morning when we would meet with the staff to work out plans for the move.

4. THE BUNA CAMPAIGN: I
September—October, 1942

SEPTEMBER 13th turned out to be a pretty good day. I like that number. General MacArthur announced that he was ordering the American 32nd Division to New Guinea. I asked him to let me fly the first regiment in, as it would be two weeks before they could get loaded and transported there by boat. The General's staff didn't like the idea at all. They wanted to make the move "in an orderly way." I kept stressing the importance of the time element, as the Aussies up on the Kokoda trail were still withdrawing and they didn't have much farther to go before they would be behind the barbed wire at Port Moresby with no air support.

General MacArthur asked me how many men I would lose flying them from Australia to New Guinea. I told him that we hadn't lost a pound of freight yet on that route and that the airplanes didn't know the difference between 180 pounds of freight and 180 pounds of infantryman. The General said he hated to hear his doughboys compared to freight but for me to fly a company up there and we would see how long it took and how the scheme worked out. It was decided that we would start at daybreak on the morning of September 15th with a company from the 126th Infantry of the 32nd Division. I figured I'd have them all up there by that evening and then I'd see if the General would let me fly the rest in.

General MacArthur wanted to know all about the parachute fragbomb attack on the Buna airdrome. I told him of the success and that I was so sure that it had proved the value of the bomb, I had radioed Arnold for 125,000 of them. In the meantime, we had started converting our regular fragmentation bombs to parafrags. That afternoon the General awarded me a Purple Heart for meritorious service in developing the bomb and utilizing it successfully for the first time in warfare.

I called Walker and told him to get busy with parafrag attacks on the Jap fields at Lae and Salamaua if they should land any planes there or start filling up the existing bomb craters. The B-26s and the B-17s with heavy bombs were to follow up the low-level attacks and dig deeper and better holes in the airdromes for the Nips to work on.

Whitehead sent word to me the next day that he had abandoned work on a projected strip at Rorona and pulled the radar set out of there, as Jap patrols had been reported only a few miles away. The Nip advance had reached Ioribaiwa Ridge, which was on the Kokoda-Port Moresby trail, only thirty miles from Port Moresby and twenty-two miles from my new field at Laloki.

Admiral Ghormley of the South Pacific Theater called for help. He wanted us to beat down the Jap air strength at Rabaul on the 15th and 16th of September so that it wouldn't interfere with his sending a convoy of reinforcements and supplies into Guadalcanal at that time. I told General MacArthur I would put one squadron of B-17s on Rabaul each day and have the RAAF Catalinas work the Nip airdromes over at night. I called Walker and gave him the change in mission for the B-17s. They were already loaded with 2000-pound bombs to dig up the fields at Lae and Salamaua, so they had to unload and put in the smaller bombs for destruction of airplanes on the ground in the Rabaul area. These sudden calls for changes in the mission were not popular with the crews, but I told Ken to explain that Ghormley was in real trouble again.

On the 15th, at first light, 230 doughboys of the 126th Infantry, with small arms and packs, loaded on board a mixed collection of Douglas and Lockheed transports at Amberley Field. At six that evening, Whitehead radioed me that they had all landed at Port Moresby. I rushed upstairs to General MacArthur's office to give him the good news and asked him to let me haul the rest of the regiment. He congratulated me most enthusiastically but told me that he had already ordered the rest of the regiment shipped by boat and that the loading had already begun. I said, "All right, give me the next regiment to go, the 128th, and I'll have them in Port Moresby ahead of this gang that goes by boat."

The staff again opposed—a little more forcefully this time, as though they were afraid this foolishness might get out of hand. General MacArthur asked Sutherland when the 126th was scheduled to pull out of Brisbane Harbor. Sutherland replied, "The morning of the 18th." MacArthur said, "Tell General Eichelberger that George is flying the 128th Infantry to Port Moresby, beginning the morning of the 18th. George, you and Bob get together and make your arrangements." I said, "Yes, sir," and went to work.

I called Mr. Drakeford, the Australian Air Minister, and got twelve transports released from the civil airlines to report to me the afternoon of the 17th at Amberley Field for a week's duty. All bombers overhauled anywhere in Australia from then on until further notice would assist in hauling troops to New Guinea and any airplanes coming in from the United States would be commandeered, civilian ferry crews and all, to help out.

Making arrangements with Major General Bob Eichelberger presented no problem. We were friends of twenty years' standing. He liked this air-movement business, as it was fast and Bob liked to move fast once he was told to move. He wasn't going up to New Guinea right away, but he expected that if enough American infantry did go up there, the command would eventually be big enough to warrant sending him, too.

Brigadier General Donald Wilson arrived from Washington the next day. I had asked Hap Arnold for him some time ago. I appointed him Chief of Staff, Allied Air Forces, relieving Air Vice-Marshal Bostock, who took over the newly created RAAF Command. Wilson also had to double in brass as my Chief of Staff for the 5th (U.S.) Air Force.

I took off for Port Moresby to confer with Whitehead and tell him about the movement of the 32nd Division to New Guinea. All our troops were glad to see the American doughboys arriving. Even though only 230 had been flown in, our crewmen, who had been wearing pistols as they worked on airplanes for the past few weeks, felt that from now on things were going to be different. I hoped they were right. Those Yanks were eager-looking boys, but I guess the Aussies were, too, when they first came up to New Guinea. The Australians at Port Moresby were quite interested in the arrival of the American infantry. They were curious, wondering, but obviously glad. I believe that the move was made at just about the right time psychologically, for the Japs were still advancing, although not quite so pell-mell as a few weeks previously. The dead ones looked pretty thin.

The Buna airdrome appeared deserted. Wrecked airplanes were still where the bombs hit them and the bomb craters had not been filled in. It looked as if the Nip had decided to abandon trying to use it as an airdrome.

I told Whitehead to get me some low-altitude pictures of Wanigela Mission and to watch for any sign of Jap occupancy or activity down there.

On the 17th the Jap advanced patrols reached the edge of Imita Ridge where they were halted by elements of the Australian 25th Brigade. Imita Ridge is only eighteen miles from Laloki, and from there it is only another eight miles to Port Moresby itself. Our fighters and light bombers shuttled back and forth all day, strafing and bombing the whole length of the trail from Imita Ridge to Kokoda and on to Buna. I returned to Brisbane, bringing Whitehead with me as far as Townsville. Whitehead looked tired, so I decided to let him and Walker exchange jobs for a while to let Whitey eat some of that good Australian food while Walker got used to M-and-V rations and gained operating experience running the advanced echelon in New Guinea. I made the exchange and then went on to Brisbane.

During the evening I saw General MacArthur in his apartment at Lennon's Hotel, talked with him about the situation in New Guinea and my impression of the effect of the arrival of American infantry upon both Australians and Americans up there. I said I thought he ought to look over the show in New Guinea himself at the earliest opportunity. I believed that his presence in the Port Moresby area at this time would be a real boost to morale. He said, "All right. Let's leave tomorrow. I'll be your guest."

We took off about noon the next day and landed at Townsville where we spent the night. The first airborne contingent of the 128th Infantry had passed through without mishap and Walker radioed that the planes had all landed at Port Moresby and were on their way back to pick up another load. General MacArthur was happy as a kid over the way the movement was going and quite enthusiastic about the way Bertrandais was getting along with the establishment of the big air depot I had told him to build. Vic had already started a production-line type of airplane repair work and had made plumbers, carpenters, electricians, and any other specialists he needed out of just plain GIs, which was all I had to give him. His mess probably put on a special effort for General MacArthur, but it was not much better than it had been when I had eaten there before on my return trips from New Guinea. The General said to Bertrandais, "Well, now I know who to steal from if I ever need a good cook." Vic pretended to look a bit worried and said, "General, if you need a good carpenter or steam-fitter or plumber I wouldn't mind so much, but cooks are hard to get." MacArthur laughed and said, "Don't worry, I've already got a good one." He asked Bertrandais where he got all the carpenters who were busy erecting hangars, warehouses, shops, and camp buildings all over the place. Vic said he gave each of the men a hammer and some nails. Anyone who hit his thumb more than once out of five times trying to drive a nail was eliminated. The rest became carpenters. General MacArthur "bought" Bertrandais.

During the evening I talked with him about occupying Wanigela Mission by air and then moving northwest along the coastal trail to Buna, with air supply keeping the show going the same way that we were supplying the Aussies on the Kokoda trail. As soon as we could take out Buna, there was a similar set-up to capture Lae by landing troops on the flat grass plains at Nadzab on the Markham River about twenty miles west of Lae. General MacArthur listened, asked several questions, and said, "We'll see." If I could prove our capabilities of transporting troops and supplying them by air during the next week or so, I figured he would give me a go-ahead on Wanigela. It was too bad that his staff hadn't his vision and intelligence. They were the stumbling blocks, not MacArthur.

The next morning we flew to Port Moresby. The General got quite a kick over the fact that we made a turn over the Jap positions on Imita Ridge, coming in to land at Seven Mile strip, where the air movement of the 128th Infantry was in full swing. Ten troop carriers that had cleared Townsville before we took off were just beginning to land at Seven Mile, while just off the edge of the strip the bulldozers were busy filling in three deep craters made by the Japs the previous night when they had raided us again.

The General watched the orderly way the transports were coming in, being unloaded and refueled, and taking off for Townsville while the troops were quickly formed up and marched off to their camping area. He liked the show and said so, chatting for some time with doughboys, mechanics, and airplane crewmen. I showed him around the town and the construction work on the docks and the roads. He didn't think much of the speed being made or of the quality of work and told the engineers about it in no uncertain terms.

After lunch with Walker at Whitehead's mess, which was still the worst in New Guinea, we drove out to Laloki where sixteen P-38s which had arrived a few days before were being worked on and flight-tested. I introduced the P-38 pilots. General MacArthur was glad to meet and talk to them and they could feel it. As usual, anyone that saw him and talked with him or listened to him for fifteen minutes was sold on Douglas MacArthur. This gathering was no exception.

All the newspapermen in New Guinea were at Laloki. There were Pat Robinson of INS, Dean Schedler and Ed Widdis of the AP, Harold Guard of UP, Byron Darnton of the New York *Times*, Lon Sebring of the *Herald Tribune*, Jack Turcott of the New York *Daily News*, Al Noderer of the Chicago *Tribune*, Ed Angly of the Chicago *Sun*, Frank Prist of Acme Pictures, Martin Banett of Paramount News, and Bob Navarro of March of Time. Geoffrey Hutton and George Johnston represented Australia and William Courtney and Dixon Brown were covering the war for the British public. General MacArthur gave them an off-the-record talk as background for future releases. He emphasized that we were not going to be run out of New Guinea and that he would fight it out with the Jap here and not in Australia. He paid a real tribute to me and the Allied Air Force, giving us credit for breaking up the Jap show at Milne Bay and for gaining control of the air over New Guinea. He predicted that the Jap drive on Port Moresby had reached its limit and that we would see the Jap withdrawing across the Owen Stanley Mountains soon, as he could not keep going without a supply line. That the air attacks had ruined his supply line and therefore his capability of continuing the attack.

General Blamey came to Whitehead's house to see General MacArthur in the evening. They made a date to go out along the trail the next morning to inspect the troops. They could go all the way to Imita Ridge by jeep. The trail was a bit rough, but passable that far. Mr. Drakeford, the Australian Air Minister, also was in. I thanked him for his cooperation in turning over the civil air transports to me for the move of the troops to New Guinea. He promised to do anything he could to help. I promised to return the transports by the 25th of September.

General Blamey did not want to use the Americans for the present on the Kokoda trail. He intended to use them to cover the Laloki River line from infiltration. General MacArthur offered no objection. Blamey was sending General George Vasey with the 7th Australian Imperial Forces Division units up the trail to chase the Japs back to Buna. He had a lot of faith in Vasey and these troops. Also I believe that as a matter of pride he wanted the Australians to get themselves out of the mess they were in.

General MacArthur and Blamey spent nearly all the next day "jeeping" along the trail all the way to Imita Ridge where the fighting was going on. The Japs were trying to capture the ridge but were still being held. Replacement troops coming up were now fresh and the supply situation was quite good, as stuff could be brought in by jeep and trailer.

The Buna Campaign: 1

General MacArthur made quite a hit with the Aussies. They were glad to see him up on the trail. Good psychology.

I didn't go along. I hate to walk, anyhow, so I spent the morning with Bill Benn, playing with skip-bombing on the old wreck on the reef outside Port Moresby Harbor. The lads were doing quite well. A nice-looking lad named Captain Ken McCullar was especially good. He tested ten shots and put six of them up against the wreck. At 200 mph, altitude 200 feet, and releasing about 300 yards away, the bomb skipped along like a stone and bumped nicely into the side of the ship. Sometimes, if the airplane was too low or flying too fast at the time of release, the bombs would bounce clean over the vessel. Our delay fuzes were coming along. We couldn't get anything out of the United States for some time, so we were modifying the Australian eleven-second delay fuzes into four- to five-second delay. So far they worked pretty well. Sometimes they went off in three seconds, sometimes in seven, but that was good enough. I knew I would have to prove the scheme before I could really put much heat on the gang back home for an American fuze.

The troops of the 128th Infantry were still coming in on schedule at the rate of about 600 men every twenty-four hours. When I gave him the figures that evening, General MacArthur was really enthusiastic about the air show. He said that the Aussies told him we were saving the situation up on the trail. He said the AIF troops were real soldiers and that they would fight. Blamey stopped all conversation about withdrawing any farther and said that in a few days he was passing to the offensive to drive the Nips back where they came from. It looked to me as though it wouldn't be long now before Rowell went south. He and Blamey didn't act like bosom friends, from where I sat.

During the afternoon General MacArthur drove over to the hospital on the edge of Port Moresby and pinned a Silver Star on Vern Haugland, an AP correspondent who had parachuted from a B-26 over the jungle on August 7th and had finally been picked up by the natives twenty-five days later. The plane had encountered bad weather coming to Port Moresby from Australia, the pilot had become lost, and finally, as the fuel was exhausted, had ordered everyone on board to take to their parachutes. Haugland had landed in the Owen Stanley Mountains without food or matches. By the time he was rescued and hospitalized his weight had dropped from around 160 pounds to 90. He had nearly starved to death and was desperately ill and delirious when he stumbled onto a native trail which led to a remote Papuan village. He was actually only half conscious of what was going on as the General spoke to him and pinned the medal on his pajama top. The little ceremony seemed to act like a tonic to Haugland, however. From that day he improved rapidly and three weeks later was flown back to Australia to recuperate. In another month he was back at work.

General MacArthur also inspected the American infantry being flown in. He said they were fresh and full of pep and ready to go, although they were far from being seasoned troops. Blamey's decision to use battle-tried Australians was probably a good one.

During the day I saw some pretty experiments on some 300- and 500-pound bombs which we had been fixing up to annoy the Nips.

To cut up aircraft on the ground we had wrapped these bombs with heavy steel wire, and we dropped them with instantaneous fuzes on the end of a six-inch pipe extension in the nose. They looked good. The wire, which was nearly one-quarter inch in diameter, broke up into pieces from six inches to a couple of feet long, and in the demonstration it cut limbs off trees a hundred feet away which were two inches thick.

The noise was quite terrifying. The pieces of wire whirling through the air whistled and sang all the notes on the scale and wailed and screamed like a whole tribe of disconsolate banshees.

The Jap built pretty good dirt revetments around his airplanes, so, to get at them from above, we were experimenting with the delay fuze that came with the photo-flash bomb which could be set for night photography to fall two or more thousand feet before it went off. We set this fuze so that it would explode after dropping 10,000 feet. Then, with the plane dropping it from just above that altitude, we hoped to get an air burst to get at the revetted planes from above or maybe sometimes at Nips in slit trenches. If the bomb hit the ground first, it had an instantaneous fuze in addition to the other so it wouldn't be wasted. In the test I witnessed, one exploded on contact and acted about like any other bomb with an instantaneous fuze. The second went off about 200 feet in the air and showered fragments all over the map but probably would have not done much good as the fragments were dispersed too thinly over too great an area. The third exploded at about fifty feet and really cleaned house. The bushes and jungle growth were cut up and blown away over an area big enough for a baseball diamond. We decided to fix up some 1000- and 2000-pounders from air bursts, and maybe wire-wrap them, too, and some day give the Nips a surprise.

On the next test we scheduled one airplane to fly at sea level and radio the barometric pressure up to the one dropping the bomb, to see if we could get the altitude determination as true as possible. I hoped this would give us more air bursts around fifty feet, which was the altitude we wanted.

I told "Ken" Walker to try using instantaneous fuzes on shipping, as I thought we had overestimated the effect of "near misses." The reconnaissance reports following our bombing attacks with 1/10 second-delay fuzes had shown that the convoys were still proceeding at the same speed after the bombing attacks, so evidently the near misses had been going so deep before the explosion took place that all we got was a big wave, which didn't even slow the Nip vessels down. I believed, however, that an instantaneous explosion near a vessel, up to perhaps fifty yards away, would push fragments into it, killing Japs, cutting steam and fuel lines, and maybe setting fires. If we got a deck hit on one of these unarmored ships with a big bomb, it would sink anyhow, whether the fuze was 1/10-second-delay or instantaneous.

Ken didn't like the idea and his Naval liaison officer didn't think much of it, either, but I told them to try it for a while and see what results we got.

General MacArthur and I returned to Brisbane on the 21st.

Just as the General left for his office, two B-17s, replacements from the United States flown by civilian crews who belonged to the Boeing Aircraft Company, landed at Amberley Field. They got out of the planes and asked how they could get into town. I told them the next town they saw would be Port Moresby, New Guinea, as I was loading thirty doughboys in each of those B-17s and they were to fly them to the war at the rate of one load every twenty-four hours. They were tickled to death at the idea of taking part in the show and said, "Okay. Give us something to eat and we'll push off whenever you are ready."

If anything happened to these civilian ferry crews, I'd catch hell, but if we didn't get troops into New Guinea I'd catch hell, too.

Evidently our attacks on the Jap airdromes around Rabaul to support the South Pacific operation were as successful as the kids claimed they were. Just as I was leaving for another visit to Port Moresby, General MacArthur sent me a letter of commendation which I transmitted to Walker and asked him to bring to the attention of the crews that did the job.

GENERAL HEADQUARTERS
SOUTHWEST PACIFIC AREA
OFFICE OF THE COMMANDER-IN-CHIEF

September 23, 1942

SUBJECT: Commendation
TO: Commander, Allied Air Forces

1. Admiral Ghormley has informed me that he and Admiral Turner are convinced that the attack by the element of the 19th Bombardment Group upon Vunakanau on the 16th contributed materially to the successful accomplishment of Admiral Turner's mission in Guadalcanal. Admiral Ghormley asks that there be conveyed to the crews of the six B-17's his commendation, "Well done."

2. It gives me great pleasure to convey to you this message.

DOUGLAS MacARTHUR,
General, U.S. Army

On arrival at Port Moresby I went out to General Blamey's field headquarters just outside of town and had a long conference with him, General Rowell, and members of the Australian staff on the possibilities of a land advance along the Kokoda trail toward Buna, plus one along the Kapa Kapa to Jaure trail, on which we had a reconnaissance party, then from Jaure down the Kumusi River to Wairopi on the trail, halfway between Kokoda and Buna. These two movements would be combined with an airborne occupation of Wanigela Mission, followed by an advance along the coast to Buna itself. I had discussed this three-pronged movement with General MacArthur on our trip back to Brisbane and he had asked me to sound out Blamey and get his reaction to the plan. I discussed with Blamey the amount of help the Air Force could give the troops and told him that we could take care of all the air lift and air-dropping requirements of all three forces, assuming that the Australian 7th Division would go over the Kokoda trail, one American battalion over the Kapa Kapa-Jaure trail, and that the rest of the American 32nd Division would be flown to Wanigela Mission for the advance up the coast. The way we were flying in troops from Australia ahead of schedule had impressed Blamey that we had more airborne capability than he had dreamed of. He told me that as soon as he could get moving forward along the Kokoda trail and felt that he was in the clear there, he would go for the Wanigela Mission thing and give me an Australian battalion from Milne Bay as the occupation force. The Kapa Kapa-Jaure trail idea he was not sure of and he preferred to wait until our reconnaissance party, then out prospecting the route, returned and gave us an idea whether or not the route was feasible.

Blamey had taken over personal command of the Allied Land Forces in New Guinea. Rowell was not even consulted any more.

All Jap attempts in the last two days to capture Imita Ridge had failed. The Aussies were getting ready to move forward and do some attacking of their own. The tired militia troops had been replaced by General George Vasey's veteran 7th Division fighters, who were fresh and eager to go. For the ninth successive day we pounded the Japs along the trail with every airplane we could put in the air.

Don Wilson radioed me that General Hap Arnold was due in Brisbane late the 24th or early on the morning of the 25th, so I took off in one of the "troop-carrying" B-17s which was returning to Brisbane. It was their third straight round trip in three days.

When I got to Brisbane I got a message from Walker saying that the last of the 128th Infantry had arrived at Port Moresby. The 126th, which was still at sea, was not due to arrive there for another two days and then they would have to unload. I thanked the civilian ferry crews who had brought out the B-17s and told them they could go home now and wired Drakeford that he could have his civil airline crews and planes back. After writing a lot of letters of appreciation to the two batches of civilians who had really made it possible for me to do the job on time, I went up to see General MacArthur, give him the news, and crow a little over getting the airborne 128th Infantry in ahead of the waterborne 126th. He grinned like a kid and was quite complimentary.

I showed him pictures of Wanigela Mission, which looked quite good for landing purposes, and told him of my conference with Blamey. I said Blamey had agreed to put a battalion of Aussies in there just as soon as he got moving on the Kokoda trail and that I was able to fly them in, feed them, and then fly in and supply the whole 32nd Division as it moved up the coast to Buna. I reiterated the need to build a big air base on that flat plain area west of Buna, from which we could hit Rabaul and Lae without first having to climb over the 13,000-foot Owen Stanley range.

General MacArthur approved the air seizure of Wanigela Mission and told me to deal with Blamey on it directly. I was almost certain that General MacArthur would soon give me a go-ahead on the ferrying of the 32nd Division across, too. Sutherland and the rest of the GHQ staff were definitely opposed to the whole scheme, but if we didn't run into any trouble at Wanigela Mission, I felt sure the General would give me a green light in spite of his staff. I didn't expect any trouble. The Japs hadn't a hundred men east of Buna.

Hap Arnold's whole time schedule, from the minute he was due to arrive in Brisbane until he returned to Noumea via Townsville, Port Moresby, and Guadalcanal, came in that evening by the "punch card" code. This code, which we used when the message needed to be secret for only a short time, could be broken easily in an hour. The worst thing about his schedule was that, in addition to broadcasting it to the whole world, it called for his flying from Townsville to Port Moresby and from there to Guadalcanal in daylight in an unarmed airplane. I rearranged his schedule so that these two flights would be at night. His staff didn't use their heads very well, unless they wanted the Nips to pull an interception on the Old Man. I anticipated having a bit of fun explaining to Hap that I had rearranged his schedule and his route.

About noon on the 25th, Arnold, Major General Streett and Colonel Bill Ritchie, who looked after SWPA affairs in the War Department Plans Division, arrived at Amberley Field. I explained to Hap that I had rearranged his time schedule and why. He looked a little sheepish and said, "Okay, let's go in town and eat, and then when can I see General MacArthur?" I told him I had arranged the meeting for that afternoon after lunch. After conferring with General MacArthur and hearing a lot of nice things about myself and chuckling at General MacArthur practically ordering Hap to give me anything I wanted, Hap and I spent the rest of the day going over my problems and needs.

He said he could not give me any more groups but would do the best he could on giving me replacement aircraft, although he warned me that every time an airplane came out of the factory about ten people yelled for it, with the European Theater getting the first call.

He promised to take care of crew replacements for combat losses and combat fatigue. He was glad I wanted to send home the Philippine and Java veterans, as he needed them to put more experience in his training establishments back home. I told him he certainly needed it, as the kids coming here from the States were green as grass. They were not getting enough gunnery, acrobatics, formation flying, or night flying. He promised to do something about that as soon as he got back. I told him that the 19th Bombardment Group would have to be pulled out, both combat and ground crews. They were so beaten down that psychologically they were not worth fooling with. I asked him to exchange them for the 90th Bombardment Group, then in Hawaii, equipped with B-24s (Liberators). General Emmons, in command there, needed an experienced group and I needed a fresh one. Arnold agreed and said he would arrange the swap. He asked about having two types of bombers to maintain, but I told him I didn't care as long as he would send me replacements. The B-17 was preferred in Europe, but I was not particular. I'd take anything.

He wanted to know all about the parafrag bombs and how they were turning out. I told him the story and he promised to push a program for big-quantity output right away.

He was quite complimentary about the way the show was going in this theater but seemed worried about my building a depot as far forward as Townsville and stocking our supplies up there. When I told him that Townsville would really be the rear-area depot, as I was going to start one in Port Moresby right away, he said he thought I was crazy, that it would be bombed out, and from the way the New Guinea situation looked to him when he left Washington, the Japs would probably capture the place soon. I told him that the Japs were all through as far as Port Moresby was concerned and he would soon see them back across the Owen Stanleys and that, furthermore, the gain in being able to repair shot-up airplanes in New Guinea, instead of shipping them by boat back to Townsville, was worth losing my depot from bombing once a month.

He laughed about my scheme to get troops on the north coast of New Guinea by way of Wanigela Mission and said, "More power to you, if you can put it across. I'll get you some more transports somehow, if I have to steal them."

I asked him to stop his staff from trying to tell me how to run my show out here. He wanted to know what I meant. I reminded him of a radio signed "Arnold," which came in a few days before, wanting to know why I was bombing airdromes when experience had shown that they were not profitable targets and that Rabaul was full of shipping within range of my B-17s and B-26s. I told him that bombing airdromes had given me control of the air over New Guinea, allowed me to mess up the Jap advance over the Kokoda trail, and air-drop supplies to feed the Australian troops there, had deprived the Japs of air support at Milne Bay, and had made Port Moresby almost a rear-area rest camp. In regard to the B-26s going to Rabaul, while when his staff laid their rules on the map it might show that Rabaul was in range from Port Moresby, the fact remained that after the B-26 had climbed over the 13,000-foot Owen Stanley range with a load of bombs it just did not have enough gas left to make the trip and get back home.

Hap said he'd put a stop to such foolishness. He promised to read the riot act to his staff as soon as he got back, but I knew that sooner or later some more radios would slip by and get me all mad again. One thing that helped, however, was that a very important guy named Douglas MacArthur believed in me. He would not let me down and I would not let him down. I was quite sure that he knew that, too.

Big news came in from New Guinea that night. The tide had turned. The Aussies attacked, drove the Nips out of their advanced positions at the foot of Imita Ridge, and were pushing them back on Ioribaiwa Ridge.

The next morning I took Arnold to Townsville, inspected the depot with him, and then sent him on to Port Moresby, with Whitehead accompanying him. I told Whitey to come back to Townsville as soon as Arnold left and finish getting some rest. I returned to Brisbane that night after Hap took off for New Guinea. I was to hop over to Noumea on the 28th to meet Arnold again, in a conference with General Miff Harmon and Admiral Ghormley on cooperation between SWPA and SOPAC. Sutherland was to go along to represent MacArthur.

General MacArthur asked me to fly the headquarters of the 32nd Division to Port Moresby, starting the next day. That made about 4500 troops of that division flown to New Guinea from Australia. The remaining regiment, the 127th Infantry, was to go up by boat.

On the 28th I flew to Noumea, taking Sutherland along. At lunch with General Arnold I brought up the question of what to do with Lincoln. I had abolished the Melbourne rear echelon and consolidated its functions in my Brisbane headquarters, so that Lincoln was surplus. Miff Harmon said he needed someone to be "King" of Fiji and would give Lincoln the job. Arnold later agreed and said he would issue the orders for the transfer as soon as he returned to Washington.

In the afternoon I attended a conference on board the Argonne, Admiral Ghormley's flagship in Noumea Harbor. Besides Ghormley, Admirals Nimitz, Turner, and Callahan were there and Generals Arnold, Harmon, Streett, Sutherland, and myself.

The Navy wanted me to make mass raids on Rabaul airdromes and shipping as a primary mission. I told them that we were doing everything we could and were just as anxious as they were about the Jap strength at Rabaul. I said we were then planning and getting ready for a large-scale attack on the airdromes there and would also try to burn the place down but that knocking off Jap convoys to Buna, maintaining air control over New Guinea, and helping out our ground forces were all missions that had to be attended to, so that I could not set any definite dates. Ghormley said he appreciated everything we had done to help their tough situation at Guadalcanal and hoped we would have all the luck in the world when we hit Rabaul in force. I told him that he could rest assured that we would do all we could, as General MacArthur scrupulously responded to his directive from the Joint Chiefs of Staff in Washington which called for cooperation between the two theaters. I hoped that some day when they got out of trouble they could come over and help us out. I liked Ghormley but he looked tired and really was tired. I don't believe his health was any too good and I thought, while we were talking, that it wouldn't be long before he was relieved.

That evening the news came in that the Aussies had captured Ioribaiwa Ridge. The Japs were really dug in and fought to the last man but they were out of food and short on ammunition.

I spent the night with Miff Harmon and the next morning said goodbye to Hap Arnold and flew back to Brisbane.

They told me there that Lieutenant General Rowell had left New Guinea. The Australians said his new job was to be in Melbourne. They didn't know what it would be but assured me that it would not be very important. It is a bad idea to be a loser in a war.

The last of the headquarters of the 32nd Division, with Major General Forrest Harding in command, were flown to New Guinea the 29th.

General MacArthur, in spite of adverse recommendations of his entire staff, who insisted I was reckless and irresponsible, decided to take the north coast of New Guinea by an airborne, air-supplied movement of the 32nd Division from Port Moresby to Wanigela Mission and on up the coast to Buna. The Aussies were to make the initial occupation of Wanigela Mission by flying in a battalion from Milne Bay. This would all be coordinated with the drive of the Australians over the Kokoda trail to Wairopi, where they would be joined by one battalion of the 32nd Division which would move from Kapa Kapa to Jaure and then down the Kumusi River to Wairopi.

The next morning a note came down to me from General MacArthur. It was a copy of a message he had just sent to Washington.

Chief of Staff
War Department, Washington, D.C.
Recommend the promotion to Lieutenant General of Major General George C. Kenney, 0-8940. This officer commands the Allied Air Force, composed of the Fifth Air Force and the Royal Australian Air Force, South West Pacific Area. His position justifies the rank of Lieutenant General. Allied Land Forces and Allied Naval Forces, the latter of far less strength than Air Forces, are commanded by men of corresponding or higher rank. General Kenney has demonstrated superior qualities of leadership and professional ability.
                                                                                                    MacArthur

Quite a nice way to end the month.

Major William Benn's 63rd Bombardment Squadron had really gone to town, justifying everything I'd ever said about Bill and his qualities of leadership.

Their first combat mission was on the night of September 19th, when five B-17s of the squadron attacked Jap shipping in Rabaul Harbor. They admitted that no hits were scored. The crews were evidently peeved at missing the targets, so they dropped down to 200-feet altitude and machine-gunned three vessels in the harbor with all the ammunition they had. Fires were reported breaking out on all three. In spite of heavy antiaircraft fire, all five B-17s returned. There were no casualties and surprisingly few holes in the airplanes.

Benn promptly put the squadron back on bombing training, with individual crews taking their turns on reconnaissance missions. At dawn on October 2nd, six planes of the squadron repeated the attack at Rabaul. The results of training paid off this time. Two large Jap transports were set on fire by direct hits, a cruiser heavily damaged by a direct hit, and an ammunition dump near Lakunai airdrome was blown up. That lad Captain Ken McCullar, whom I saw doing some excellent skip-bombing a week ago, hit one of the Jap ships on this attack. A couple of hot boys named Murphy and Sogaard got the other two.

All day on October 4th, the natives around Wanigela Mission worked like beavers cutting a strip through the kunai grass big enough for us to land transports. The next morning we turned the corner in the New Guinea war when we started flying a battalion of Australians to Wanigela from Milne Bay. By nightfall we had landed the whole battalion of approximately 1000 men and, in addition, about a hundred American engineer troops from Port Moresby, to put the strip in shape for bigger operations later on. No Jap troops were encountered and no Jap aircraft flew over the area all day. As far as we could find out from the natives, the Nips had no idea that we were on the north coast of New Guinea.

General Blamey was quite enthusiastic about this modern warfare but told me he had really had to order General Clowes to provide the troops from Milne Bay. Clowes didn't approve of this method at all.

While the Aussies were being flown to Wanigela Mission, we covered the operation by attacks on Buna and Rabaul to give the Japs something to look at besides our airborne troop movement. Benn's squadron put eight B-17s over Vunakanau airdrome and dropped twenty tons of bombs with excellent effect, setting fires, destroying aircraft on the ground, and digging huge holes in the runway. The squadron was intercepted by thirty Jap fighters, but the B-17s shot down eight of them and all eight B-17s returned without even having a man wounded or an engine shot out. In a few weeks that 63rd Squadron had developed into a cocky, swaggering outfit if I ever saw one.

I went back to Brisbane and told General MacArthur that we had started on the road to Tokio and that, as soon as Buna was out of the way, a similar show would put us into Nadzab and let us capture Lae from the back door. I said I wanted to do some reconnaissance and start planning soon. The General said to go ahead on the planning but that no final decision would be made until Buna was occupied. If the Buna job could go through quickly, I was sure he would be ready to roll on up the coast. I'd need some more transports to handle the Nadzab show, but Arnold had promised to take care of that.

Some time previously we had started shipping a lot of cement and other building materials, as well as a few refrigerators, to New Guinea by boat, but as soon as the stuff was unloaded the engineer colonel commanding the base issued instructions for the engineers to take custody. Of course, the regulations said that the engineers were charged with the purchase, storage, and issue of such things, but as they wouldn't get them for me I'd been getting them myself. I told Connell to ship as much of the stuff as he could by air direct to the airdromes, where Whitehead would see that it got where it belonged, and to put things like refrigerators in airplane engine boxes. The engineers didn't touch boxes and crates labeled Air Force supplies.

One of Wurtsmith's sergeants one day went down to the dock to pick up a shipment of engines. One of the boxes was not the right size and shape for an Allison engine, although it was so labeled. The sergeant said, "That's no engine crate. Open her up and let's see what kind of a shenanigan is being worked. I'm not going to load up this truck with someone else's junk."

The box was opened, exposing to view a nice new refrigerator which was supposed to go to the very squadron to which the sergeant belonged.

"You're right, Sergeant," said the engineer officer in charge of the unloading crew. "I guess they must have made a mistake in the marking of that box."

We not only lost the refrigerator but, from then on, any suspicious-looking box or crate would probably be opened before we could cart it away. That dumb sergeant was not very popular in his squadron for several days.

Our P-38s were grounded again, with leaks in the air intakes of the cooling system. We were now practically rebuilding the wings to plug the leaks on ten of the airplanes at Port Moresby, eighteen at Townsville, and twenty others at Brisbane. It looked as though I wouldn't have enough ready for operations until around the middle of November.

General MacArthur sent for me on the 6th and told me that the Navy over in the South Pacific Area was planning another operation for October 11th and wanted us to support it by attacking Rabaul. I had already planned to do something about that place. For the past couple of weeks the Japs had been unloading supplies and troops there. They might be building up for a reinforcement of their forces in the Solomons or maybe around Buna. The shipping had not been a good target, as the Jap vessels had been coming in just before dark, unloading at night, and leaving before daybreak. The airdromes in the Rabaul area were well beaten up and the Jap air strength was low. I told General MacArthur that my recommendation to help SOPAC was to burn out the town of Rabaul the night of the 8th-9th and repeat the attack the next night. At the same time I'd put the Australian Catalinas to work on the Jap airdromes on Bougainville Island. In case there should be a Jap reaction to our bombing of Rabaul, I would put down a big attack on the morning of the 9th on the Lae airdrome to stop the Nips from Rabaul from using it as a refueling point on the way to or from Port Moresby for their fighters which did not have the range for the full round trip. When my B-17s got back to Port Moresby after their attack on Rabaul, I'd keep a standing fighter patrol in the air to cover the refueling of the bombers and their take-off for their home bases in Australia. The General approved my plan and I told Whitehead and Walker the dates to put it into effect. I told Walker I expected that we would have nearly thirty planes on the opening night and maybe twenty on the second attack.

The Australians were still advancing along the Kokoda trail. The Japs were withdrawing so fast the Aussies had lost contact with them. The trail was littered with dead Japs and graves where others had been buried. The bombs and bullets took out a lot of them but most of them just starved to death.

During the night of the 8th-9th, the weather over Bougainville was so bad the Catalinas could not get through, so four of them dropped their incendiary bombs on Rabaul, setting fires that lit the town up nicely for thirty B-17s, a record number to date, which showered the place all night with instantaneous-fuzed 500- and 300-pound bombs and incendiaries. The whole town was left a sea of flames, which were still visible eighty miles away an hour after the last airplane had left the area. The next morning the photo-reconnaissance plane found the place still covered with smoke and could not get any pictures. All of our planes returned.

The next night seven Catalinas lighted up Rabaul again for us and eighteen B-17s repeated the attack. Five of these bombed Lakunai airdrome, building many fires. The town was still burning when the bombing started and blazed up all over again when the incendiaries began to rain down. Bill Benn's 63rd squadron had ten B-17s in each attack. That had become the hottest outfit in the whole air force.

General MacArthur sent me a message stating: "Please tell the Bomber Command how gratified I feel at their fine performance over Rabaul." I did.

I got word that day that General Arnold, on his return to Washington the previous month, had agreed with the Navy that in an emergency Admiral Ghormley could divert airplanes on the way to me and use them himself, "provided that they can be used more effectively in the SOPAC area." They should have merged these two theaters a long while before and let one man decide where the equipment was most needed for the campaign as a whole. I had already lost fifteen P-38s, some B-17s and DC-3s (C-47s) on account of these "emergencies."

During the night of the 11th-12th, the Navy pulled off their show, sinking a Jap light cruiser and three destroyers and damaging a heavy cruiser and a destroyer. Our Navy lost a destroyer and had a heavy cruiser, a light cruiser, and a destroyer damaged. The operation covered the landing at Guadalcanal, beginning on the 12th, of the American Division under Major General "Sandy" Patch, relieving the First Marine Division which originally landed there over two months before on August 7th.

On the 14th, 670 troops of the 128th Infantry of the 32nd Division were flown to Wanigela Mission. Those were the lads who were headed for Buna along the coastal trail.

On the 15th, on the way to Port Moresby, I stopped at Townsville and Mareeba and pinned over 250 decorations on that many proud chests. The ceremony at Townsville took nearly an hour and the one at Mareeba over two hours. By the time I got through, I had worn most of the skin off the thumb and forefinger of my right hand. It was a great show.

The 19th was all pepped up about going home. The rumor was out. I told Walker that, as soon as the actual orders came in, no one scheduled to go home would go on any more combat missions. I never saw a flyer yet who didn't worry about this "one last mission" business. I don't like it myself.

At Whitehead's, for dinner that evening, I talked with Major General Forrest Harding, the commander of the 32nd Division. He was quite enthusiastic about the way things were going on the air movement of his troops to the north coast. He said the Air Force had really opened his eyes as to what could be done with airplanes and that from then on he was on our side of all the arguments. Harding was a nice guy. He was anxious to move and I believed with luck he would make good. Some of the GHQ staff said he was inexperienced, but so was everyone else fighting Japs in New Guinea. If the ground forces had tried to fight a slow-moving, walking war the way the textbooks taught, the combination of malaria, dysentery, Japs, and jungle would have ruined a lot of reputations. Harding wanted to move fast and was willing to gamble on the planes keeping him supplied with food and ammunition as he moved along the trails, instead of building up big supply dumps and depots along a line of advance which was also intended to be a line of retreat if things should go wrong.

The troops that we flew to Wanigela Mission were moving west along a trail that, according to the map and the natives, led to Buna. The battalion of the 126th Infantry moving across the mountains from Kapa Kapa was having trouble. The trail was extremely rough and so grown up in places that it had to be cut out to get through. The pass over the Owen Stanleys was nearly 9000 feet, a lot higher than the 6500-foot elevation in the Kokoda Gap. They hadn't run into any Japs so far and the Jap had shown no sign that he knew anything about our build-up at Wanigela Mission.

Harding said the morale of his troops was good and he hoped no one would stop him from moving. He remarked, "General Kenney, I'm drinking the same stuff you guys are these days. I don't know what it is or what it is going to do to me, but it sure tastes good."

I hoped with Harding that they would turn him loose, but this mixed-command organization might trip him up. General Sir Thomas Blamey was the Allied Ground Forces commander in New Guinea. The Australians under Vasey were heading for Buna along the Kokoda trail and Blamey probably hoped they would get through in a hurry and retrieve the reputation of the Australians and incidentally himself. Vasey, a grand fighter, leading a grand fighting division, could go only so fast up that trail and the Australians would be pretty tired and a lot thinner by the time they got to Buna. It was not unreasonable to think that Blamey would not relish the idea of the Aussies getting to Buna all worn out and finding the Americans already in there, having been flown right up to the edge of the war and then walking in and taking all the prizes. If they got talking about "coordinating" the efforts of Vasey and Harding and they delayed too long, this could be a difficult show. It was dry season right then on the north coast of New Guinea, the trails were mostly dry and the swamps were shrunk to their minimum sizes, but about the middle of November the rainy season would hit us. The records indicated that the normal annual rainfall was around 150 inches. That meant that in a short time the trails would be muddy ditches and the flat coastal plain would be mostly swamp. In turn, that would mean more delay and more time for the troops to get malaria and dysentery. About that time, those eager-beaver but inexperienced troops of ours would be liable to lose their eagerness and high morale and bog down. If they bogged down, Harding would not come out of this war a big shot.

Sutherland complicated the picture. He had no confidence in the Australians. Every time I talked to him he tried to impress on me the terrible consequences if we should have a reverse. He said if anything went wrong, General MacArthur would be sent home. He didn't mention it, but in that case Major General Sutherland would also be out of a job. There wasn't any need for a reverse, if we kept moving. We had more men than the Jap. We owned the air over New Guinea. We were bombing and machine-gunning his troops and burning up his supplies. We could supply by air, while the Nip had to run an air blockade with his vessels every time he wanted another bag of rice, another round of ammunition, or another Jap soldier to replace his losses.

Admiral Ghormley called for help again. On the seventeenth we put five Catalinas on Buka, six B-17s on Buin, seven B-17s on Jap shipping in the Faisi anchorage, and seven B-17s on Vunakanau airdrome. The next night sixteen B-17s hit shipping off Buin. By this time I had to go to General MacArthur and tell him that my bomber crews were worn out and, regardless of anyone's needs, they needed a rest and the airplanes needed maintenance. Just then a wire came in from General Emmons in Hawaii, saying that the first squadron of twelve B-24s of the 90th Bombardment Group would arrive in Australia on October 23rd. I recommended that we send home the twelve crews of the 19th Group most in need of rest, in twelve of the old-model B-17s immediately and follow with the rest of the 19th Group as fast as the crews and planes of the 90th Group arrived from Hawaii. The General approved the plan.

Major General Rush B. Lincoln left on the 19th for Noumea to report to Miff Harmon.

I put out an order requiring everyone in the Fifth Air Force to wear long trousers and long sleeves. The funny thing about it was that it not only made sense but I knew it would be enthusiastically obeyed. A month before, however, it would have been exceedingly unpopular.

When the Americans first came to New Guinea and saw the Aussies wearing shorts and shirtsleeves cut off above the elbow, it appealed to them as a smart idea for that hot, humid, jungle service. Just as an experiment, I put long trousers and long-sleeved shirts on one squadron of a fighter group and shorts and short-sleeved shirts on another squadron for a month. At the end of the trial period, I had two cases of malaria in the long-trousered, long-sleeved squadron and sixty-two cases in the squadron wearing shorts. The evidence was good enough for the kids as well as for me, so I issued the order. Later, in New Guinea, I noticed that Vasey's 7th Australian Division were wearing "longies."

On the 22nd, an Australian detachment occupied Goodenough Island, about fifty miles northeast of Milne Bay. They found a few Japs left over from the Milne Bay invasion, but the opposition did not amount to much. We planned to put a radar out there to help out our warning service and, as soon as I could spare some engineer effort, to build a fighter strip there, too.

General MacArthur asked me when I was going to decorate Bill Benn. His 63rd Squadron had sunk or damaged more Jap shipping during the past month than all the rest of the Air Force put together, without losing a man or an airplane. I told the General that I'd like to decorate Bill but hesitated to hurry the thing as he had so recently been my aide. I didn't want the rest of the gang to think there might be a bit of favoritism. The General smiled and said, "Yes, that's right, you can't be too careful about such things, so I'll decorate him myself. Go write me up a nice citation for the Distinguished Service Cross." I had it in his office in an hour.

The first squadron of B-24 bombers arrived that day from Hawaii and at the same time a radio came in from Arnold, telling me to check the anti-shimmy collars in the nose-wheel gears for cracks and to ground all airplanes that showed cracks. We checked them. They all showed cracks. I wired the information to Arnold and asked for replacements to be flown out immediately. We tried welding the cracks. That just made more cracks. I told Connell to get all the tool shops in Brisbane busy making some out of steel. The cracked ones were not steel, but we didn't know what they were made of so we played safe and made them strong enough. Now the B-24s would be no good to me for another couple of weeks. In the meantime, I'd sent twelve B-17s back home, so I was just out both ways.

We put over another trick on the Nips on the night of the 23rd. Twelve B-17s went over Rabaul Harbor after air reconnaissance the day before had reported a concentration of shipping that looked worth while. The first six bombers were from the newly organized 64th Squadron of the 43rd Group. They bombed from 10,000 feet and, while the Jap searchlights lit up the sky and the antiaircraft guns blazed away, the other six bombers from the 63rd Squadron came in at 100-feet altitude and introduced skip-bombing to the Nips. Captain Ken McCullar, whose airplane already had been credited with sinking or damaging four Jap vessels, sank a Jap destroyer with two direct hits amidships. Captain Green scored direct hits on a light cruiser or large destroyer, a small cargo vessel, and a medium-sized one. The crew reported that the cruiser had her stern under water and was on fire all over when they left. The cargo vessels were both sunk. A lieutenant named Hustad hit another cargo vessel, estimated at 10,000 tons, setting fire to it. Hustad reported that she was blazing nicely and listing a little when he left but he could not claim the vessel as definitely destroyed. The six planes from the 64th claimed damaging four other vessels with hits or near misses.

I sent word to Benn to come down to Brisbane. He arrived about five that afternoon. I immediately took him upstairs to General MacArthur's office where the "Old Man" congratulated him and pinned on his Distinguished Service Cross. I told Benn to stick to moonlight nights and bombing with flares for this skip-bombing business. I didn't want the gang to do it in daylight, as they hadn't enough forward firepower to beat down the Jap deck guns as they came in for the kill.

The next morning a courier came in with a letter from Whitehead giving us news of the progress of the 32nd Division troops after they left Wanigela Mission. After walking west two days and covering about fifteen miles on a very poor trail, they came to the Musa River, the end of the trail and nothing but swamps on the other side. After spending another day without locating any trail on the other side of the river, they got some native canoes, paddled down to the mouth of the river, and turned northwest along the coast to a native village called Pongani. Four hundred men were ferried to Pongani in this way and with the help of the natives were now clearing out a few bushes and cutting the grass for a strip big enough for a DC-3. The troop carriers would then fly to Pongani the rest of the troops, who had left the Musa River and gone back to Wanigela Mission. So far no Japs had been encountered on the north coast or by the battalion which crossed the mountains from Kapa Kapa and had now reached Jaure at the headwaters of the Kumusi River.

The Japs were so strung out along the trails from the Buna area to Kokoda that they couldn't have had much strength left at Buna. We should have been able to take the place in another couple of weeks if we could have just kept moving before the rainy season arrived.

I wrote to Arnold and asked him to take back that permission he gave the Navy to hijack airplanes on the way to me. A couple of weeks previously they had decided they had an emergency and stopped ten of my transports at Noumea and I couldn't get them to give them back to me. In the meantime, Whitehead had sold the Aussies on the scheme of an airborne show at Nadzab to take Lae out from the back the way we were going to take Buna. General MacArthur would have bought the idea as a follow-up to the Buna campaign, but without those ten transports I simply could not show the airlift necessary to do the job. In addition, due to the lack of these transports which I had been promised and which I had figured on, the capture of Buna itself had probably been delayed. The Army was not sending any more men to Wanigela right then, as they were afraid I couldn't supply them and at the same time keep on dropping food to the Australians on the Kokoda trail and the troops on the Kapa Kapa-Jaure trail. My job would be easier after the Aussies recaptured Kokoda strip. We could land supplies there instead of dropping them along the trail, where only about half the stuff was ever recovered by the troops, but I hated to wait and watch that rainy season getting closer and closer all the time.

On the night of the 25th, Benn's 63rd Squadron, which now constituted about all the heavy-bomber strength I had, put on another skip-bombing party at Rabaul. Eight bombers took part, demolishing a gunboat, probably sinking a 5000-ton cargo vessel, and badly damaging two other large cargo vessels. McCullar was again one of the lads getting hits. Lieutenants Hustad, Anderson, and Wilson were the others.

I was really in a bad way for heavy bombers. The 19th Group was out of the picture, going home. The twelve planes of the 90th Group were on the ground with cracked nose-wheel collars. Incidentally, the replacement collars shipped out by air were also cracked when they arrived and my production of steel substitutes would not solve the problem for another week or so. The new 64th Squadron had nine B-17s, but the crews were still pretty green. The third new squadron of the 43rd Group, the 403rd, had eight airplanes, but those crews were still greener. That was the complete picture on heavy bombardment in the Southwest Pacific Area at that time and I had lots of work to do. The reconnaissance pictures showed Rabaul crowded with ships.

The Navy had just won another battle over in the Solomons, which they called the Battle of Santa Cruz. The Japs lost 250 planes and a light cruiser. Two of their aircraft carriers, a cruiser, and three destroyers were damaged. SOPAC lost the aircraft carrier Hornet and a destroyer and had a battleship, a carrier, and a destroyer damaged. The Japs turned back. The action was all by carrier and land-based aircraft.

The Buna Campaign: 1

Just before the Battle of Santa Cruz, SOPAC must have asked Washington for help and sold their case, for on the 27th we got a copy of a radio from the War Department to Emmons telling him to expedite the departure of the next squadron of the 90th Group to Miff Harmon in the South Pacific. The message also said to hurry up the departure of the rest of the group and have it, too, report to Harmon for temporary duty if Admiral Nimitz wanted it. I hoped the Santa Cruz victory would cheer Ghormley up so that he would let those B-24s keep on past Noumea. I had an emergency myself.

On the 28th Ghormley called on MacArthur again for help, asking that we hit Jap shipping at Buin-Faisi and lend them a squadron of P-38s at Guadalcanal as soon as they got gasoline up there. I told General MacArthur that the shipping at Buin-Faisi was well dispersed and transitory, while Rabaul was congested with vessels at anchor, and that I would have only eight P-38s in shape in about another ten days, as we were still reworking the wings to plug those leaks in the intercooling system. He directed that I use the bombers on Buin-Faisi and wire Harmon that we would lend him eight P-38s in ten days.

On the 29th nine B-17s hit the Buin-Faisi anchorage, setting on fire an 8000-ton cargo vessel and damaging a cruiser and a destroyer. On the 30th, the 43rd Group commanded by Colonel Roger Ramey, just in from Hawaii and a real leader, put twenty-four of the group's thirty B-17s over Buin-Faisi and Rabaul. At Buin-Faisi fourteen bombers got directs on a battleship and two other unidentified vessels and damaged a light cruiser and an aircraft carrier. The other ten hit a large cargo vessel and two destroyers in Rabaul Harbor. Photos taken the next day showed all three vessels half under water and aground. That night, eight Australian Catalinas also attacked shipping in the Buin-Faisi anchorage. We wound up the month of October with a daylight attack by nine B-17s on Buin which blew up an unidentified ship and damaged a light cruiser, a destroyer, and a large cargo vessel. That night, nine Catalinas continued the attacks on Buin-Faisi.

On November 1st, as the Australians recaptured Kokoda and reported that they had counted over 2000 dead Japs on the trail since they captured Ioribaiwa Ridge on September 28, twelve B-17s again hit Buin-Faisi, damaging a destroyer and a couple of cargo vessels. We were keeping our fingers crossed, for we had not lost a bomber during the raids of the past four days, but at this point I found my bomber effort again about shot until the crews could get some rest and the airplanes get some much-needed maintenance.

For five weeks we had not seen a Jap airplane over New Guinea, except for an occasional night bomber. On the afternoon of November 1st, six of our light bombers making their customary raid on Lae were intercepted by from nine to twelve Jap fighters. The lads shot down one Jap and came on home. A reconnaissance plane scouting over Buna reported that he had seen seven Jap fighters as he ducked into a cloud to avoid them. This looked like the preliminary to another Jap convoy movement to resupply or reinforce Buna, and there I was caught with my only heavy group worn out. I sent word to Ramey to send what he had in commission up to Port Moresby prepared to attack shipping at daybreak tomorrow.

The next day, November 2nd, as we had guessed, two Jap vessels, escorted by two destroyers and with fighters escorting them, tried to run into Buna. I wired Whitehead to tell the kids that those ships were the targets and that I had plenty of decorations for any crew that sank a ship. Captain Ed Scott, leading seven B-17s of the 63rd Squadron, all that the group had in shape, made an attack late in the morning, scoring some damaging near misses, slowing one of the vessels down to a walk and then strafing its decks with machine-gun fire. They were intercepted by nine Jap fighters, who interfered considerably with the bombing. Scott got three of the Jap fighters and brought all his planes back. In the afternoon he led his outfit out again and repeated the attack. This time they sank the already crippled vessel and set fire to the other. I made good on the decorations I had promised.

The Japs must have got the fire out, however, and probably landed their cargoes during the night, as the next morning's reconnaissance picked up a Jap vessel escorted by two destroyers north of Buna headed back toward Rabaul. The weather closed down, preventing further attacks, although I don't know what we would have made them with, as Scott's outfit had only three bombers in commission that morning.

General MacArthur wired congratulations to General Blamey for the recapture of Kokoda and sent me the following:

To—Commander Allied Air Forces
From—GHQ SWPA

Please express to all ranks of the Air Corps concerned my admiration for the magnificent part they have played in the campaign which has resulted in the capture of Kokoda.

MacArthur

A few days previously I had received a message from General Miff Harmon asking me to locate and send back to him a B-17 and crew which was supposed to go to him but for some reason had left New Caledonia and gone to Australia. I checked up and found that the pilot was a classmate of a lot of my 43rd Group gang and that originally he had been ordered to come to me. By the time he got to New Caledonia, his orders were changed, assigning him to SOPAC. On arrival at Plaines des Gaiac field in New Caledonia the kid got gas, in order to "test his engines which had been giving him trouble," and headed for Brisbane. On arrival he demanded gas in a hurry so he could get to the war. With such commendable enthusiasm, he got service, flew to Townsville, reported in, and the next night was over Rabaul on an attack.

He had had three combat missions to date and I hated to let him go, but after all he had broken too many orders, so I ordered him to go back to Miff. I wrote a nice letter to Harmon telling him what a grand kid the boy was, all about his combat record and spirit, asked him to take care of him and above all not to have him suffer for wanting to carry out his original orders, which meant serving with the gang he knew while he was learning to fly. I also said that the lad was not trying to shirk combat but had exhibited commendable courage and ability and that I was confident that he, Miff, would give him a break or I would not be sending him back to New Caledonia.

A couple of days later Miff wired me thanks and said he would take care of the kid.

I finally got an airplane of my own. A few weeks previously I had told Bertrandais to fix up an old wrecked B-17 for me with a table, a bunk, and a few chairs in it, so that I could use it as a flying headquarters, taking along my new aide, a bright lad named Captain "Kip'" Chase, and a couple of staff officers or guests on my travels around the theater. The crew was one of the hot ones of the 19th Bombardment Group that had a lot of combat time and needed a rest but didn't want to go home. Captain Wilbur Beezley was the first pilot, one of the best handlers of a big airplane I had ever seen. Now I wouldn't have to thumb rides back and forth on repaired airplanes going back to New Guinea or take up cargo space on one of the transports.

I wired Arnold that sooner or later I had to have some bombers. The second squadron of the 90th Group had come in, as Ghormley decided he did not need them, after all, but I had received no B-17 replacements for over a month. I decided to keep the rest of the B-17s of the 19th Group and send the crews back home by the Ferry Command, which could take care of about eleven men a day. If Emmons or Arnold complained, I'd tell them that I had an "emergency," too.

I asked General MacArthur if he agreed to that action. He laughed and said he believed he'd have me shot if I didn't keep those B-17s. That was good enough backing for anyone.

Colonel Art Meehan, who a few years before used to do a good job for West Point during the football season, reported in from Hawaii on November 4th. I sent him up to Walker to take command of the 90th Bombardment Group and start operating as soon as I got those nose-wheel collars fixed.

The Australians fixed up the Kokoda strip and we started flying in supplies and taking out the sick and wounded. The field was nothing but a strip, one hundred feet wide and half a mile long, on a slope so steep that the planes landed uphill, turned around, and took off downhill. At that time in early November, with the continuous rain at the altitude of 6500 feet, the mud on the field itself was from a foot to a foot and a half deep. How these kids got in and out of the place with a heavily loaded DC-3 I don't know, but they did.

5. THE BUNA CAMPAIGN: II
November—December, 1942

On November 6th I flew to Port Moresby, taking with me General MacArthur, his aide, Lieutenant Colonel Morehouse, and my aide, Captain Chase. The General had established an advanced headquarters at Government House, the former residence of the Australian Governor of Papua, which was about a mile west of Port Moresby. MacArthur said he was going to stay there until the Buna show was over. He wanted me to be there with him in personal command of the Air Forces in New Guinea, except for short trips to the mainland that I would have to make to keep things moving forward and to look after the rest of the operations in Darwin and Townsville.

Sutherland and several members of the GHQ staff had come up a few days before with a small detachment to open the place, put in communications, establish the offices and the mess, and get it ready for the General to occupy.

The house was quite comfortable. It was a big rambling one-story affair on a little knoll overlooking the harbor. The palm trees and shrubbery were quite decorative and wide screened-in porches all around the house kept it cool even in the middle of the day.

The raising and lowering of the flags, United States and Australian, were done every day with appropriate and sometimes curiously played bugle music by a detachment of Papuan native infantry. They were good-looking specimens, coal-black, fuzzy-haired, barefoot, and wearing a white wraparound knee-length skirt with a red sash. With their cartridge belts and rifles, to us they were quite picturesque and to the rest of the natives quite impressive.

I saw General Harding at Whitehead's, where I had dinner. Harding said he had received instructions to hold up his advance and not hurry troops over until a lot of supplies were built up on the north coast at Wanigela and Pongani. He wanted to move and said that in a couple of days he would be ready to advance on Buna. From Pongani west, the going had been much easier and he had the natives starting to cut the grass so that we could land troops and supplies there. An engineer colonel by the name of Jack Sverdrup, a tall blond reincarnation of Leif Ericson, who led the prospecting party over the Kapa Kapa-Jaure trail, was in charge of finding and preparing landing fields between Wanigela and Buna. He had already located and prepared six strips along that route and this had immensely facilitated the airborne movement. For some reason or other, Sverdrup had worked miracles with the natives, who seemed to be willing to work harder and longer hours for him than for anyone else.

Over at Government House I told General MacArthur that the weather forecast for the next few days looked good and that I wanted to get all the 32nd Division possible, with their food and ammunition, across the mountains. Also that, with Dobodura coming into the picture, the maintenance problem was fairly easy.

General MacArthur said Harding was to keep moving and for us to keep on ferrying troops and supplies across as fast as possible. He wanted Harding to get into his assault position as soon as he could and to keep us informed as to the situation.

The headquarters of the 128th Infantry was now at a little native village called Mendaropu, about a mile north of Pongani and only twenty miles from Buna.

I was sure that General MacArthur wanted to move fast but he had a lot of factors to consider. In the first place, while I knew he had a lot of faith in me and the ability of the Air Force, we were operating on a very thin shoestring. The constant requirement to support the SOPAC effort, the troubles I was having getting enough heavy bombers to do the job with, the delay in putting the P-38s in action, and the loss of those ten transports to SOPAC were enough to cause anyone to hesitate at what his staff kept insisting was a rash commitment of his forces. That staff had not approved the airborne idea from the start and they kept saying that we were already out on a limb and should go slow.

General Blamey would not order Harding to jump, for several reasons. He naturally wanted to see his Australians get in on the kill. He probably thought they had earned something besides watching the Americans rush in and reap the glory. He was not sure of the fighting ability of the American troops, while he had complete and justifiable confidence in Vasey and his veteran 7th Division. Blamey, moreover, was not particularly impressed by the 32nd Division staff work as shown by the way they continued to mess up the planning of the daily haul of troops and supplies across the "Hump," as the kids referred to the Owen Stanley range. I had to admit the Australian planning was much better. I hoped Vasey could move faster along the trail to Buna, for if Harding had to wait much longer, so that the Aussies could coordinate their final attack on the Jap beachhead with his, I was afraid the rainy season would have started.

Bill Benn came in to see me. He had been made Chief of Operations for Walker's 5th Bomber Command. I was glad Walker did it, for Benn had been pushing his luck a little hard for the past month. Benn, of course, didn't like the idea of being pulled out of combat, but he had more than done his share already. Captain Ed Scott took over command of the 63rd Squadron that Benn invented.

On the 8th we started landing troops at Dobodura on the strips that Jack Sverdrup had cut. Dobodura was only ten miles from Buna and we had two complete regiments of the 32nd Division over the Hump. Harding was ready to go. I talked with General MacArthur that evening about the situation. Pretty soon the malaria and dysentery would begin to hit that 32nd Division and I wanted to see them go while they were still full of fight. Harding could jump off by the 10th easily, but if he had to wait for the Aussies to join hands with him it would be another ten days, the Nips would all be back in Buna and dug in for a last-ditch suicide defense. With the rains coming on to add to our troubles, we might not take Buna before Christmas and then the troops would be too worn out for any operation against Lae for a long time. General MacArthur told me had a conference with Blamey scheduled for the next morning.

The conference was held and the decision announced. Harding's attack was to be coordinated with that of the Australians when Vasey got down to Popondetta. The Aussies were then to drive for Gona while Harding launched a two-pronged drive for Buna, one along the coast and the other from Dobodura. Lieutenant General Herring, an exceedingly capable Australian, was to be the field commander of the two Allied divisions. Blamey wanted it that way. Sutherland and the rest of the GHQ staff supported his recommendations. The way General MacArthur had treated me left me nothing to complain of but I still wished he had let Harding drive on Buna without waiting for the Aussies to come up.

We kept plastering Buna and the Jap installations at Soputa and Popondetta, two villages to the west. I found I had another fireball, in the 3rd Attack Group, named Lieutenant Ed Larner. While he was leading his squadron in an attack on Jap artillery and machine-gun positions at Soputa, an antiaircraft burst under the tail tipped the nose of Larner's airplane over so that he hit the trees at the end of his strafing run. After tearing through the tree tops for a hundred yards, he brought the plane back to Port Moresby with the wings dented in, one engine full of leaves and branches, and the whole length of the bottom of the fuselage grooved where the top of a palm tree had been in the way. The lifting surface of the wings was so badly damaged that he had a landing speed of nearly 175 miles an hour, but Larner got away with it, landed, and taxied up to the line. As he got out I gave him a Silver Star and made him a Captain.

That lad was good. He had fire, leadership, and guts. That plane was one of the worst-looking wrecks to be still flying that I'd ever seen. Ninety-nine out of a hundred pilots would have bailed out before trying to land it. Larner turned in his report reading: "... following this accident I was able to make only two more strafing passes before the plane became so unmanageable that I thought it best to return to base where repairs could be made." The lad was a bit cocky, bragged some, and swaggered, too, but it was all right with me. He had a right to.

On the 12th the Aussies and the battalion of the United States 126th Infantry, which had moved down the Kumusi River from Jaure, joined at Wairopi. At this point there used to be a wire-rope suspension bridge across the Kumusi. The natives heard the white man talk about "wire rope," so they built a village at the end of the bridge and called it Wairopi.

Harding's artillery commander wanted some artillery to knock out the Jap bunkers. He said he could take a 105-mm. howitzer apart and put the pieces in an airplane. I promised to do what I could, so asked General MacArthur to let me try it. He sent word to General Fuller, the American 41st Division commander in Australia, to put a gun at Amberley Field for us to play with. My aide, Kip Chase, and an artillery officer left immediately to see if they could stow the pieces in a B-17 and bring it to Port Moresby, where we would reload it in DC-3s and fly the gun over the Hump to Dobodura.

The 126th Infantry completed its air movement to Pongani on the 13th.

On November 14th we got word that Eddie Rickenbacker had been rescued on the 12th with all the crew except one, who had died of exposure. They had been missing since October 21 on a flight from Hawaii to Canton Island. This was good news. Eddie and I had been close friends, dating back to World War I, when we were on the front together. He was scheduled to rest up for a while and then come out here to see us.

Bill Benn came over for a chat. He said the 43rd Group wanted to call themselves the Kens Men and did I have any objection. I told him that I had none and to tell the gang that I felt highly honored. The next time I inspected the group they had painted out the cute, scantily clothed girls and substituted the words "Kens Men" in block letters a foot high. I was flattered, of course, but I sort of missed the pretty gals.

The next day a B-17 landed at Seven Mile Airdrome carrying a 105-mm. howitzer, a tractor to pull it with, the gun crew of eight men, fifty rounds of ammunition, a tool kit, and the camouflage net to shield it from the eyes of Jap aviators. General MacArthur went down with me to see the gun unloaded. His grin was worth the work of getting it up here. How they ever stuffed that 10,000 pounds of gun, ammunition, crew, and miscellaneous equipment into that B-17 I don't know, but they went back to Brisbane that night to get the other three guns of the battery. The next morning two DC-3s took the pieces of gun Number One to Dobodura, where they were reassembled and put into action. By this time the ground troops and General MacArthur decided we could haul anything in an airplane. I wished his staff would get that way.

On the 15th the General cleared a press release stating that I had taken over personal command of air operations in Papua. This was the first time my name had been mentioned as even being present in the theater.

I finally got enough nose-wheel collars produced to equip the B-24s of the 90th Group and on the 16th they made their first mission. Eight B-24s took off to bomb the shipping in the Buin-Faisi anchorage at the south end of Bougainville Island. They got there but made no hits. On the way back they got separated in a rainstorm. Two landed in the water off New Guinea. Both crews were saved. The others landed at four different airdromes in northeast Australia.

The next day ten B-24s of the second squadron took off for Rabaul. The eleventh airplane swerved, hooked a wing on the trees at the edge of the runway, and caught on fire, exploding the bombs. The entire crew was killed and two other B-24s on the ground were completely wrecked.

Two got to Rabaul and set fire to a Jap cargo vessel. Two were missing. Colonel Art Meehan, the 90th Group commander, was on one of them. The other airplanes came back, but the crews were not sure where they had been.

I ordered Walker to take the whole 90th Group out of combat and put them on training status until they had learned more about night flying and navigation and had done some practice bombing and gunnery.

I needed them badly but their training was not what I had been led to believe. If I used them as they were, my losses would be beyond all reason and pretty soon I'd have another broken-down outfit like the 19th. Now I was back to the 43rd again. That's all I had, there wasn't any more.

The Australian 7th Division was in position that afternoon for the drive on Gona which was to start at daybreak the next morning. The 32nd Division was also in position and was scheduled to jump off on the drive to Buna the following day, the 19th. The rains had been hitting us hard for the past three days. A little more of that and the ground was going to get soft. The American troops were reporting a lot of dysentery. Something more to worry about.

The next day I went over to 5th Bomber Command headquarters. Walker brought up the subject of fuze settings to be used in attacks on shipping. He wanted to go back to 1/10 second-delay instead of instantaneous. The statistical boys had just given me their analysis of bombing for the past two months, which showed conclusively that the instantaneous-fuze setting was doing the job I expected. Convoys were slowing down after being attacked, more ships were being sunk, and many more were reported damaged and on fire.

I told Ken to have somebody go out and drop about four bombs at the old wreck on the reef outside Moresby and that we would then go out and inspect it.

After lunch we took a motorboat out as close as we could get to the wreck and a corporal rowed Ken and me about a mile the rest of the way. The evidence was there. The bombs had missed the vessel by twenty-five to seventy-five yards and yet fragments had torn holes all through it. Some of them were two to four square feet in area. I showed Ken the nice clean edges as compared with the rusty edges of holes and gashes made by previous practice bombings. Ken finally said, "Okay, you win. I'm convinced." I turned to the corporal and said, "Corporal, come back here and sit in the stern with me. General Walker is rowing us back to the motorboat."

Ken didn't say a word (except for a few of three or four letters when his oarsmanship went wrong) all the way back. I didn't kid him any more, so after a couple of drinks up at Whitehead's before dinner, Ken thawed out. Ken was okay. He was stubborn, oversensitive, and a prima donna, but he worked like a dog all the time. His gang liked him a lot, but he tended to get a staff of "yes-men." He did not like to delegate authority. I was afraid that Ken was not durable enough to last very long under the high tension of this show. His personal problem was tough because he kept himself keyed up all the time and he just couldn't seem to relax a minute.

Ken was the serious, studious type. In his early forties, of medium height, slight build, with dark hair and eyes and an intelligent face, he was a likable, hard-working asset to a command, but the combination of tropical service and tension was wearing him down. I hated to think of having to send him home but I had already started speculating on the problem of finding a replacement for him. He wouldn't like it but he would be of tremendous help to Arnold back in Washington, where his war experience could be used in the planning section.

The Australians started their final drive on Buna [ed. note: the island was actually Gona] in the rain at daybreak on the 18th.

In the evening the old reliable 63rd Squadron took on a Jap convoy of a light cruiser and two destroyers about fifty miles north of Buna and stopped the Nips from bringing in any supplies or troops. Ken McCullar thought the destroyer he hit was destroyed. He said it was dead in the water and burning from stem to stern when he left. Lieutenant O'Brien skip-bombed the other destroyer and reported that he saw it break in half and sink. Lieutenant Anderson got three direct hits on the light cruiser and said he didn't believe that the Nips would ever put out the fire that was burning when he left. The next morning the reconnaissance planes were halted by the weather so I couldn't get the final answer.

On the 19th the rain stopped all flying over the Hump. The Aussies were having heavy going in their move on Gona and the 32nd Division also reported slow progress on both of their drives toward Buna. No Jap resistance had been encountered yet. The mud was the main enemy so far.

I sent word to Major Pappy Gunn at Brisbane to pull the bombardier and everything else out of the nose of a B-25 medium bomber and fill it full of fifty-caliber guns, with 500 rounds of ammunition per gun. I told him I wanted him then to strap some more on the sides of the fuselage to give all the forward firepower possible. I suggested four guns in the nose, two on each side of the fuselage, and three underneath. If, when he had made the installation, the airplane still flew and the guns would shoot, I figured I'd have a skip-bomber that could overwhelm the deck defenses of a Jap vessel as the plane came in for the kill with its bombs. With a commerce destroyer as effective as I believed this would be, I'd be able to maintain an air blockade on the Japs anywhere within the radius of action of the airplane.

Continuing rain for the past two days had me worried about the supply situation on the north coast. We had lost a DC-3 when it hit a hill while trying to get down through the fog over Dobodura strip. With the help of Major Hampton, the crack transport pilot who pioneered on making the first landings on all our jungle fields, and the radio experts, we worked out a procedure of dropping supplies in bundles of 300 pounds each with a parachute attached. The pilot, flying on instruments, just as he would in the fog, flew toward the radio station on the airdrome, using his radio compass for direction. When the needle of the compass flipped showing that the plane was passing directly over the radio station, the pilot called to the sergeant in the rear, who pushed the bundle with its attached parachute out the door. The plane circled and repeated the performance until all the bundles were delivered. Sixteen trials with the pilot hooded and flying entirely on instruments at 2500-feet altitude put all the bundles in a one-hundred-yard-diameter circle. We had a radio station at the field at Dobodura. The next day, fog or sunshine, we would deliver supplies to Dobodura. Hampton was so enthusiastic, he said, "I hope it's foggy tomorrow." However, I hoped the weather man was right this time. He predicted good weather, clearing that night.

After watching the dropping tests and feeling pretty good, I went over to Government House to see General MacArthur. He was just finishing a session with his staff, who had told him that the troops on the other side of the Hump were low on food and that this day-to-day feeding business by air was unsound. They advised withdrawal and an early extrication from the whole show.

General MacArthur had just thanked them all very courteously and dismissed them. As I came in whistling, the General said, "Hello, George, let's go out for a walk." We walked out to a bench on the grounds and sat down. "George," said the General, "you know there are a lot of men over there eating their last meal tonight."

I said, "Yes, but tomorrow we serve breakfast at six-thirty and by noon I'll have five days' rations over to them." The General replied that he hoped so but that nothing had been delivered for the last three days, "I know," I answered. "On account of this damn rainy season, but our weather guesser says tomorrow will be fine. However, I don't care what the weather is any more. We have the thing licked now," and I described what we had been working on.

"George," said the General, "the Fifth Air Force hasn't failed me yet and I believe they can work themselves out of any trouble they run into. I'm not worried about it any more."

He chuckled as he told me of the staff conference, the advice he had been given, and the dire forebodings as to the future of the Buna campaign. He certainly stuck by me that day when the chips were down. I liked to work for Douglas MacArthur and I think he knew it.

The next day we built up the ration reserve on the north coast to a week's supply. The Americans needed most of it as they had been trying to build up a big ammunition supply and had been running too close to the line on food. The Aussie planning had been much better. They had made sure that they were going to eat, no matter what else happened.

On the 22nd we started moving the 127th Infantry, the last of the 32nd Division, over to Dobodura to act as a reserve to the other two regiments, which were having heavy going in the mud and had begun to run into some Jap opposition.

The boy Captain Ed Larner was at it again. He came back from a strafing attack around Buna with his tail bumper all scratched up where he had dragged it through the sand making a "low" pass at a Jap machine-gun position which had a heavy coconut-log overhead covering. Larner said he had to "look in the windows of the bunker to see what to shoot at." I was over at the airdrome looking at Larner's plane and saw his sergeant gunner. The sergeant sighed as he examined the tail bumper and said, "I guess I'll have to quit this pilot of mine. He's gone nuts. He runs into trees and tears 'em down and now he thinks he's a farmer and he's started plowing up the ground with his tail bumper."

I said, "Sergeant, I guess we can fix it up for you, all right. I've got a chauffeur over at my headquarters who wants to shoot a pair of fifty-caliber guns. How about swapping jobs?" The sergeant hesitated just a little before he answered, "No, General, I guess I'd better stick. You see, Captain Larner is so crazy he really needs me to look after him."

You couldn't separate that pair. Anyhow, I was not going to try.

On the 24th the Nips sent five destroyers down from Rabaul to run supplies and reinforcements to Buna. We picked them up coming through the Vitiaz Straits just before dark. With our troops already in trouble around Buna we didn't want the Nips to get any more men or supplies ashore, so before the B-17 crews took off to go on instruments over the 13,000-foot Hump, feel their way in the dark down through the clouds to a 2000-foot ceiling off the north coast, locate the Jap destroyers, and skip-bomb them by the light of flares, I told the gang there was a Silver Star for each member of a crew that sank a destroyer. One crew got their medals right after breakfast the next morning. Two other destroyers were hit heavily and left on fire and dead in the water.

About eleven o'clock that night I called General MacArthur, who was quite concerned about this latest Jap expedition, and told him I hated to wake him up but it was "one down, four to go." He said, "Don't apologize for news like that. Call me any time you can tell me that you are making some more Japs walk the plank."

At midnight I called again. "Two down, three to go," I reported. "Nice work, buccaneer," he laughed. "Keep me posted."

A little after one o'clock the morning of the 25th I called for the third time, "Another one hit but I can't claim a sinking. The weather has shut down to absolute-zero ceiling. I'm afraid some of those Nips will get ashore, but the kids did the best they could."

"It was magnificent," he called back. "Tell them I said so and I hope you pass out some decorations for the work they did tonight in this weather."

I told the kids the story and gave them his message. They decided they were for Douglas MacArthur, too.

Whitehead did a great job supporting the troops on the north coast. On the 25th, besides continuing to plaster the Jap positions with everything he had, he delivered 500,000 pounds of supplies to Dobodura and Popondetta. That meant over one hundred plane loads, which in turn meant that every transport flew three or four trips over the Hump. Blamey and Herring sent their congratulations dated November 25th as follows:

> "C in C Allied Land Forces and G O C New Guinea Force desire convey to General Kenney and General Whitehead their deep appreciation of the magnificent efforts being made and results achieved by Allied Air Force in supporting the Land Forces in offensive operations and in transporting men and materiel. The splendid effort today in carrying such large quantities to Dobodura and Popondetta is giving troops in the forward areas great heart for their continued onslaughts."

Bill Benn brought me a copy of Ken McCullar's report of his mission against the last Jap convoy. Here it is:

"On Nov. 24 we were ordered to attack a convoy of 5 destroyers, coming presumably to Lae. Part of our bombs were fuzed with 4 second delay and others with Y10 second delay fuzes. We spotted the convoy and climbed to about 3500 feet, cut our throttles back, and made the first skip-bombing run at 200 feet and 255 mph. The bombs hit just off the end or the boat and the AA [anti-aircraft] hit in the tail gunner's ammunition can, exploding about 70 shells and starting quite a fire. Sgt. Reser smothered this with a blanket and winter flying equipment until extinguishers were rushed back and the fire put out. On the next run, skip-bombing, our bombs hit directly on or very near the boat, starting a fire on the right front of the ship. By this time the radio operator and 2 more of the men were injured but not seriously. On run 3, the No. 1 motor was hit and all the controls shot away. The engine could not be feathered as the switch did not work. We climbed to 1500 feet and made a run from 1200 feet this time, the bombs hitting close and us getting hit again. We then climbed to 4000 feet as No. 1 motor was gone. Another run was made, dropping our last bombs, and No. 3 Engine cut out—having received a hit in some part of the fuel system. We feathered this engine but found we could not keep altitude on two engines. We flew for a while and tried to bring No. 3 in again. In the meantime, sending in our position and condition and course. No. 1 engine got red hot from the windmilling of the prop and it looked like any minute the whole thing would catch fire and blow up. We placed the navigator and bombardier in the back compartment of the ship, in case the prop flew off or we had to set it down. Evidently the prop ground loose from the engine at the reduction gear, for after a while the engine cooled off. Still losing altitude we began to work on No. 3. We got it to where it would pull 20 inches and after a while 25 inches. We threw out all the ammunition and excess weight and the airplane started to climb. Two and a half hours later we were at 10,000 feet, our ceiling, and luckily we found a pass to sneak through, landed O.K., and forgot about it."

> Kenneth D. McCullar
> Capt., Air Corps
> Pilot.

Hitler might have had a secret weapon as he claimed, but I'd bet it wasn't as good as Ken McCullar.

There had been a lot of rivalry among the P-38 pilots as to who would get the first official victory over a Japanese airplane in combat. The P-38s had been patrolling over Lae for several days, calling the Japs on the fighter radio frequency and insulting them by saying: "We are the P-38s. We are taking over the patrol of the Lae airdrome from you and if any of you bowlegged, slant-eyed so-and-so's don't like it and have guts enough to argue about it, come on up and we'll accommodate you." This is a polite version of what they said.

Until this particular date, the P-38s hadn't been able to get a rise out of the Nips, so they were carrying a couple of 500-pound bombs under each wing to dig a few holes in the runway with before they left.

One Jap pilot evidently got sufficiently insulted and taxied out to take off. One of the P-38 pilots, a big, good-natured New Orleans Cajun named Ferrault, dived down to take him. At about 2000 feet he realized that he was carrying a couple of bombs, so he hurriedly released them and pulled up out of the blast effect to be ready to tip over again and shoot down the Jap as soon as his wheels left the runway. At the top of his climbing turn he looked down and saw the two bombs hit the water at the end of the runway, which ran all the way to the beach, and splash the water up just in time to catch the Jap plane as it was pulling off the end of the runway and spill the Jap into the water for a total loss.

When the kids returned, I asked Ferrault if he had nerve enough to claim "the first Nip brought down in air combat in this theater by a P-38." He grinned and asked if I was going to give him an Air Medal. I had promised one to anyone that got an official victory. I said, "Hell, no. I want you to shoot them down, not splash water on them." I then asked the rest of the kids if they thought Ferrault was really entitled to anything more than credit for about three hours' combat time. They were fairly sure I was kidding so they all agreed with me. Ferrault was a bit uncertain himself when I left. Later on that evening I went over, got the gang together, and gave Ferrault his medal but told him that he'd better keep the whole thing quiet. The outfit kidded him for weeks about getting a Nip by splashing water on him.

The 32nd Division was not doing too well. Harding was getting the blame, as he had not weeded out incompetent subordinate commanders who didn't know what to do. The troops were shot full of dysentery and the malaria was starting to show up. We were flying back a lot of sick every day as well as a few wounded. The troops were green and the officers were not controlling them. They threw away their steel helmets and then wouldn't go forward because they didn't have them. They were scared to death of snipers and were beginning to imagine that every coconut tree was full of them.

General Blamey and General Herring came in to Government House for a conference during the afternoon of the 25th. General MacArthur suggested bringing the 41st Division up from Australia. Blamey frankly said he would rather put in more Australians, as he knew they would fight. Herring, who had just flown back from his headquarters at Popondetta, agreed with Blamey. They didn't think much of the fighting qualities or the leadership of our 32nd Division. I think it was a bitter pill for General MacArthur to swallow but he agreed that we would fly in the Australian 21st Brigade.

The trouble with the 32nd Division was that, in addition to being a green outfit, they sat around in the jungle for about ten days doing nothing except worry about the rain and the mud and listening to strange noises at night. They had been careless about their drinking water and as a result nearly everyone got dysentery. Their initial eagerness had gone and they simply were not as good combat troops at this time as the seasoned veterans that Vasey had in his 7th Division.

The fourth 105-mm. howitzer arrived from Brisbane, completing the battery I promised to fly to New Guinea.

At dinner that evening with General MacArthur and Sutherland, Dick was talking about the shortcomings of a democracy in time of war. He thought that a stronger, more centralized government would be much better and that there was too much debating by Congress on many issues that could be settled by the President himself. In time of war he thought it might even be advisable to stop having elections.

General MacArthur listened for a while and then told Sutherland he was wrong; that democracy works and will always work because the people are allowed to think, to talk, and keep their minds free, open, and flexible. He said that while the dictator state may plan a war, get everything worked out down to the last detail, launch the attack, and do pretty well at the beginning, eventually something goes wrong with the plan. Something interrupts the schedule. Now, the regimented minds of the dictator command are not flexible enough to handle quickly the changed situation. They have tried to make war a science when actually it is an art. He went on to say that a democracy, on the other hand, produces hundreds and thousands of flexible-minded, free-thinking leaders who will take advantage of the dictator's troubles and mistakes and think of a dozen ways to outthink and defeat him. As long as a democracy can withstand the initial onslaught, it will find ways of striking back and eventually it will win. It costs money and at times does look inefficient but, in the final analysis, democracy as we have it in the United States is the best form of government that man has ever evolved. He paused and said, "The trouble with you, Dick, I am afraid, is you are a natural-born autocrat."

Sutherland shrugged his shoulders, laughed, and then, trying to turn attention from himself, said, "What about George here?"

The General's eyes twinkled and he said, "Oh, George was born three hundred years too late. He's just a natural-born pirate."

Eddie Rickenbacker arrived in Port Moresby. He told me he was on a special inspection tour for the Secretary of War. Eddie, Walker and Whitehead, and myself spent a couple of hours in the afternoon at Whitehead's house, renewing old times and present troubles, and I then took him over to Government House where he was to stay as General MacArthur's guest. General MacArthur, Eddie, Sutherland, and I had Thanksgiving dinner that evening and we gave Eddie the story of the situation in New Guinea and of the Southwest Pacific Area in general. Eddie's story of the landing in the ocean, when the B-17 in which he was riding got lost and ran out of fuel southwest of Canton Island, was really dramatic. It was practically a miracle the way they survived for three weeks in those rubber boats. Eddie stuck to the same story but he told it better every time. I had already heard it up at Whitehead's but I was willing to listen again any time.

Rick got a great kick out of a story on Whitehead. Some of the staff were planning a ceremony on Thanksgiving at which an impersonator of the Jap Emperor was to present Whitehead with the Order of the Rising Sun for his excellent job of air supply of the Japanese troops while they were advancing along the trail from Kokoda toward Port Moresby. The Australians had retreated so fast that they didn't destroy the food dumps that Whitey had worked so hard to keep built up behind them. News leaked out about the staff's plan and Whitehead heard about it. He hit the ceiling at the idea of such levity at his expense and scared the plotters so badly that they all denied to a man that they ever thought of such a thing. I even swore to Whitehead myself that I knew nothing about it.

A great leader and aviator, that man Whitehead, and a driving operating genius, who planned every operation down to the last detail to insure success. He was no yes-man to anyone, but no one ever had a more loyal right-hand man than I had in Ennis Whitehead. He was durable, too. In spite of the way he punished himself, I instinctively knew he would last through the war. Perhaps it was because he had been quite an athlete in his younger days. Among other things, before he got into the Air Corps during World War I, Whitey had played professional baseball for a while. That was where he had gotten a badly smashed nose from colliding with a baseball. He was not the handsomest man in the world to begin with, but the baseball accident had not helped matters either. Neither had the extensive thinning of an already sparse crop of blond hair during the few years since he had passed his fortieth birthday. Just about average height, he had the figure of an athlete, the quick step of an active man, and a pair of the coldest blue eyes that ever bored through you. I certainly would have had a tough time if the Japs had had a Whitehead.

The next day Eddie and I toured airdromes all day and talked with the kids. Eddie was still a hero to the fighter pilots. After a lot of discussion about combat tactics, one of the youngsters said: "Colonel Rickenbacker, how many victories did you have in the last war?" Eddie told him the number was twenty-six. Quite a few of the youngsters said something about having only two or three or maybe only one and that they guessed that number twenty-six would stand for a long while to come. I said, "Eddie, I'm going to offer a case of Scotch to the first one to beat your old record." Eddie immediately chimed in, "Put me down for another case." We left amid a lot of grins, but twenty-six was still a long way off. Buzz Wagner's eight, that he had when he left to go back home a couple of months before, still headed the list.

The next day the story was all over New Guinea.

Harding was in bad. General MacArthur sent for General Bob Eichelberger to come up from Australia. Stories of inaction and even cowardice of our troops were filtering back. The officers didn't know their jobs. The commanders were too far to the rear. Instead of fighting, there seemed to be an idea that if they waited long enough the Japs would starve to death or quit. We were bringing back planeloads of shell-shocked and sick boys every day. The number of men who had actually been wounded was small.

On the 28th I took Rickenbacker back to Brisbane. We stopped in Townsville, where we paid a surprise visit to Bertrandais, Rick's crew chief in the old 94th Squadron in World War I. Bert with great pride showed Rick over his depot and then fed us the usual excellent meal. Rick was much impressed with the work of the Air Force in the Southwest Pacific and, like everyone else that ever had much to do with him, was impressed with General Douglas MacArthur. I heard Eddie tell his story again that evening at dinner. It was better than ever. Eddie's sincerity and an underlying religious streak really put it across. He expected to go back home in about a week. A great guy. I wished he could have stayed a little longer to talk some more to the kids.

The next morning, before taking off for Port Moresby, I looked over Pappy Gunn's job on the B-25. He was really going to town. He had a good-looking package of four fifty-caliber guns and 500 rounds of ammunition for each one, tucked away in the nose where the bombardier and the bombsight used to be. He was mounting a pair of guns on each side of the fuselage just under the pilot's window and three more underneath the fuselage. The ammunition feed for the three underneath guns was a tough problem. It looked as though we might have to leave them off and be satisfied with eight forward guns for a while. Pappy said that firing the guns on the ground at a target butt had knocked the rivets out of the fuselage skin, but he figured he could cure that trouble with longer blast tubes on the guns and by stiffening the gun mounts with steel plates. The airplane looked to me as though it might be getting a little nose-heavy. I asked Pappy how about the center of gravity. Pappy came right back with, "Oh, the C.G. Hell, we threw that away to lighten up the ship." I told him not to forget to do a little checking on the balance before he tried to take it off the ground and, if necessary, to put some lead in the tail to make it balance.

On the way to Port Moresby I stopped off in Townsville long enough to promote Bertrandais to the grade of colonel. He was on his way to a star the way he was getting results.

About noon a reconnaissance plane picked up four Jap destroyers southwest of Gasmata, New Britain, heading for Buna. The old reliable 63rd Squadron had gone back to Australia the day before for a well-earned rest and to do some maintenance after a hard week of combat operations. Walker sent the B-17s of the 64th Squadron after the Jap destroyers and radioed the 63rd to come back to New Guinea. The 64th ran into the usual Jap fighter cover and got no hits.

The 63rd with five B-17s, all they had in commission, arrived at Seven Mile strip about six o'clock that evening. The crews were tired, but they grinned as they asked me if I was in trouble again. I said yes and told them the story. They loaded up and started single-plane attacks. The weather was even worse than usual. The tops of the clouds over the mountains were so high they had to go through them and then break through to about a 2500-foot ceiling over the Solomon Sea. During the night all five B-17s of the squadron made two trips across the Hump without finding the Jap destroyers. I was down at the airdrome when the last B-17 came back with the news that they hadn't found the targets. Bill Benn said, "General, this is a tired gang." I said, "Bill, there are 20,000 American and Australian troops over there in trouble. If the Japs land another fresh batch of troops, they may lick us. In any event it will cost a lot of lives to take care of those extra Japs. You have fifty Americans aboard these five B-17s. There are at least 10,000 Americans over there depending on us. I don't want to lose one of those kids any more than you do, but if we have to lose someone to save those other 10,000 we will have to do it, that's all."

The whole gang was standing around listening. Before I could say anything more or Bill could reply, we heard some sergeant back there in the dark say, "Come on, what the hell are we waiting for? Let's put a load of bombs aboard and get going. There's 10,000 Yanks over there we gotta look after." All five B-17s made another mission, found the Jap destroyers, and nailed two of them. Just before the weather shut down the next morning at daybreak, preventing further operations, a reconnaissance plane reported two Jap destroyers headed north, fifty miles north of where they were attacked. The other two were either sunk or damaged so that they couldn't keep up. When the kids got back I told them to go to bed and stay there for twenty-four hours. No matter what happened I wouldn't use them before that, as they were really worn down. But what a spirit! You couldn't lose with a gang like that. I decided that as soon as they woke up I'd have a little ceremony and pin some medals on them.

General MacArthur really began to be worried about the caliber of his infantry. The news from Buna was bad. The 32nd Division attacks of the last two or three days had gotten nowhere. The troops just did not go. They acted scared to death of Jap snipers. There were cases of men throwing away their rifles, abandoning their machine guns, and running in panic. Their officers didn't seem to know what to do. We would fly food and ammunition to the fields around Dobodura, but if they already had enough to eat we couldn't get anyone to unload the planes. Our crews then would unload the planes themselves, but the supplies would be left there on the side of the runway. Harding should have purged that bunch of officers long ago. The Aussie 7th Division was plodding along but not moving very fast either. They were worn out from crossing the mountains and were taking it easy. It began to look as though we would have to fly in some fresh troops. There weren't many Nips in front of the Allied troops but they were on hard ground. They covered the approaches well with machine-gun fire and were fanatical fighters who could certainly take punishment. They gave no quarter and asked none. They refused to surrender, no matter how desperate their situation. The rainy season was on, and off the trails a man sank to his waist in the mud. Flanking action was fantastically difficult.

On the afternoon of November 30, Lieutenant General Bob Eichelberger reported to General MacArthur at Government House for instructions.

General MacArthur calmly gave Bob a resume of the situation. He said he couldn't believe that those troops represented the American fighting man of this war. He would be discouraged if he thought so. He believed that they needed leadership to galvanize them, to give them back the fighting, aggressive spirit of the American soldier. He said he knew they were not trained for this type of warfare, the climate was wearing them down, and they were sick, but that a real leader could take those same men and capture Buna. "Bob," he said, "that is the job I'm giving you—go get Buna." He went on to tell him that this was Bob's opportunity of his career, that he believed in him and was confident that he would come through, but then he said, "Now, Bob, I have no illusions about your personal courage, but remember that you are no use to me—dead." It was an inspiring set of instructions. MacArthur is a real leader. He knows how to inspire people to go out and work their heads off for him.

Bob is an old friend of mine. I'm fond of him. So as he left I wished him luck and promised to give him any air support he wanted, any time. I told him to send me a note back by air transport and in three hours I could give him air help if he would just tell me where he needed the bombs or bullets.

The Nips opened the month of December with another "Tokio Express" run of four destroyers with air escort, which we picked up about halfway between Gasmata and Buna late in the afternoon. Five B-17s and six B-25s attacked just before dark. The Jap fighters swarmed all over our bombers and, although six Japs were definitely destroyed and five more listed as "probables," the bombing was again interfered with. One of the Jap destroyers received a direct hit and was left on fire and dead in the water. The others received some damage from near misses and all four were heavily strafed by the B-25s. The Japs evidently decided that Buna would be a bad spot in which to try to unload and, shortly after dark, suddenly changed their course and landed their troops on the beach at the mouths of the Kumusi and Mambare Rivers, about forty miles northwest of Buna. All through the night and the next morning our planes were over the landing area strafing boats and landing parties ashore and sinking rafts on which the Japs were trying to float their supplies ashore. How many troops, out of the 1000 to 1200 men probably carried by the destroyers, managed to get ashore is difficult to estimate, but it is certain that they suffered terrific casualties. Very little of their equipment or supplies got ashore, so that I do not believe this expedition contributed very much to the capability of the Nip to hold his Buna position.

During the Jap retreat toward Buna, after the Aussies drove them out of Kokoda, we kept the bridge across the Kumusi River at Wairopi under daily bombardment. The Japs rebuilt and repaired the bridge several times, but finally they gave up and crossed the river on rafts, which also became strafing targets. Later we found that one of the casualties during an attempted crossing was Lieutenant General Tomatore Horii, the commander of the Jap forces in the Buna-Gona area and recognized as one of the foremost Japanese experts on amphibious operations.

General Harding came in to see General MacArthur the next day and get instructions. Harding was broken-hearted over being deprived of his command. I was really sorry for him. He might have made good and become a "big wheel," but —no luck. General MacArthur did a nice considerate job of letting him down. He told him that, after all Harding's combat activity for the past two months, anyone would be exhausted and it would be a good thing for him to take a leave. General MacArthur said for him to go to Australia, get rested, and then come back, and he would be given a job.

As we hadn't had anyone over Rabaul for several days on account of the weather, I asked for a photo reconnaissance of the place. Captain Ken McCullar was given the job.

On the morning of the 5th, Ken was at 28,000-feet altitude seventy-five miles from Rabaul when a turbo-supercharger disintegrated, wrecking both engines on the left side of his B-17. A normal person would have turned around and headed for home as the two-engined ceiling, as Ken had already found out, was not very high.

McCullar turned to the co-pilot, who told me the story that afternoon, and said, "There's no sense in going back now, after we've come this far, and besides it's a swell day for pictures.

We'll make it on the glide. Anyhow, didn't you see how anxious the Old Man was to get these pictures?"

They arrived over Rabaul at 19,000 feet on a two-motored power glide, took the pictures, turned toward home still losing altitude, shot at and scared off a couple of Jap planes that followed them for a while out of Rabaul, finally stabilized the altitude at 6500 feet and came on back to Seven Mile Airdrome.

As McCullar got out of the plane he said to me, "General, I think we got some good shots for you. I hope so, but if they don't turn out good I'll borrow another ship and go back this afternoon and get 'em while the weather is still good."

The pictures were the best ones of Rabaul I've ever seen.

That evening a good story came in from the Buna front. A couple of miles west of Buna Mission our advance had reached a point a few hundred yards from the coast and had stopped. After sweating it out in the mud for four days, a Sergeant Herman Bottcher got fed up and, disregarding his officers, called for volunteers to go through to the beach. Thirteen men followed him and they went through, dug in, and for two days drove off all attempts to dislodge them. The sergeant then worked his way back to get some more food and ammunition and report. There was considerable argument with his officers and a few harsh words from the sergeant, but he got what he was after and returned to his position on the beach. General MacArthur, when he heard the story, awarded the sergeant a Distinguished Service Cross and ordered him commissioned a captain.

One interesting angle was that Bottcher was not even an American citizen. He was a German soldier of fortune who fought against Franco in the Spanish Civil War and rose to the rank of captain. After the Spanish Civil War was over, he came to the United States and enlisted.

The Nips celebrated Pearl Harbor day on December 7th by raiding the Dobodura area with about fifty mixed fighters and bombers. They dropped some bombs on the strip and bombed our field hospital, killing two Australians and five Americans and wounding thirty others. Our P-39s and P-40s intercepted. Final score, fifteen Jap planes shot down. We had one P-39 slightly damaged—period.

Blamey sent me a message of congratulation for our work that day, which included not only the fighter combat but the support of the ground effort by nineteen bombers and the ferrying in of a big tonnage of food and ammunition. The message read: "To you, your squadron leaders and all of those associated with the magnificent success today, I wish to extend my heartiest congratulations."

Eddie Rickenbacker sent me a message from New Caledonia: "Hearty congratulations your swell day but why in hell did you wait until I left."

I flew down to Brisbane to see how Pappy Gunn was doing on the modification of the B-25 into a "commerce destroyer." As I got out of the airplane, Pappy was just landing after a test flight in the remodeled B-25. He didn't come anywhere near getting his tail down as he came in. The ship was quite evidently nose-heavy, but Pappy is an excellent pilot. He taxied up to the line and got out sweating like a horse. I said, "Pappy, how does she handle?" "Like a dream," Pappy replied, beaming all over. "And shoot," he went on. "Say, General—" I interrupted. "Pappy, don't you think a hundred or so pounds of lead in the tail would make it easier for you to hold that nose up and keep you from sweating so much?" "Oh, General," said Pappy, "I always sweat like that. I'm the nervous type. Still it might be a good idea to make it a little easier for some of these kids who haven't had too much experience."

In spite of Pappy's enthusiasm, the job was not done by any means. The ship was too nose-heavy to be entirely cured by putting lead in the tail, so we decided to move the two gun installations on each side of the fuselage back about three feet and put the ammunition for them back in the bomb-bay. It also looked as though we might have to omit the three guns mounted underneath the pilot, on account of the weight and on account of the blast knocking the rivets out of the skin. The package of four guns in the nose was working fine. We had stiffened the fuselage where the side guns were mounted, with steel plates, but the rivets still popped out. Pappy said he was going to put some sheets of felt between the steel plates and the skin to see if that would absorb the shock.

The Jap air attacks on the 7th looked like the usual tip-off that another Tokio Express run was coming, so we were not surprised when the next morning a reconnaissance plane spotted six Jap destroyers about fifty miles out of Rabaul Harbor and headed our way. That afternoon, eight of our B-17s attacked, scoring direct hits on three of the Nip vessels. The Japs turned around and went back to Rabaul. Three of them were quite low in the water and still burning as the bombers left. Seven of the Jap fighter cover were shot down. All our planes returned.

Sergeant Reser, the tail gunner in Ken McCullar's plane, was wounded in the thigh and buttocks during the combat but continued to man his guns until the party was all over. When McCullar got back we counted 109 bullet holes from machine-gun fire and two holes in the elevators from 20 mm. shells.

Tokio Radio on their evening program said, "Yesterday, the Beast, General George Kenney, Commander of the Allied Air Forces, returned to Australia from a conference in New Guinea with General Whitehead, the Murderer of Moresby. Undoubtedly some new and fiendish methods of warfare were decided upon by this pair of gangster leaders of a gang of gangsters from a gangster-ridden country."

It was a good line, but I would have liked to know how they found out that I had returned to Australia.

Vasey's Australians captured Gona on the 9th. Following close behind a strafing and bombing attack by three A-20s, seven B-26s, and twelve P-40s, the Aussies went in with their bayonets and hand grenades and mopped up. Over six hundred Japs were buried around Gona by the next day.

I wired Arnold that I'd simply have to have some more airplanes pretty soon because those I had were getting worn out fast and, in spite of everything we did about it, we did lose airplanes once in a while and others got so badly shot up that they were only good for spare parts to repair others. The continuous wear and tear of operating in and out of those rough makeshift fields and losses due to the weather had reduced my transport strength to a point where I began to worry about meeting my commitments over the Hump. Insufficient replacements were getting through to me, so I asked the Australian Air Minister Drakeford to help me out. He came through immediately. I got a message from him saying that he was sending me from the Australian airlines three DC-3s, one DC-2, two Stinson Trimotors, two Lockheed 10s, one Lockheed 14, and two British DH-86s. Air Vice-Marshal Jones, the Chief of Air Staff, RAAF, said he would send me fifteen Hudsons from his training command, manned by youngsters who hadn't had much time but would get experience by hauling loads in New Guinea. The Hudsons could only carry 1200 pounds of freight, but everything helped.

I recommended Squeeze Wurtsmith for commission as a brigadier general and took the papers in to General MacArthur as soon as I got back to New Guinea. I had just started the sales talk when he said, "I'll approve it and send it to Washington right away." Later I heard that one of his staff remarked, when Wurtsmith's name was mentioned as up for promotion, "That kid. Well, I hope he's twenty-one." My informant said General MacArthur turned and snapped, "We promote them out here for efficiency, not for age." The story is that a dead silence reigned in the room for several minutes. I think Squeeze was thirty-two at the time. It hadn't occurred to me even to look it up. He had what it took. That was enough for me and it seemed to be enough for General Douglas MacArthur.

The Japs were hanging on to a narrow strip of coast which included Buna village and the government mission and extended east for about three miles. Our troops were not making any progress but were so close to the Nip positions that we could not do any more bombing without endangering our own men. We would strafe, but the Japs were well dug in and their machine-gun and mortar positions were protected by coconut-log bunkers that bullets would not go through. Blamey decided to ferry some tanks up the coast from Milne Bay on rafts and see if he couldn't run the Nips out without sacrificing too many men. The tanks were to be manned by Australians from Major General Wooten's 9th Division, another outfit just back from Africa. We were to convoy the movement with plenty of fighters to make sure they got to Buna.

Over at Blamey's headquarters they told a story about a Jap prisoner captured at Gona. He was wounded and so emaciated from starvation that he could hardly stand, but when they brought him in for interrogation, he bit his tongue off so that he would not be able to talk and give away any information. He was so far gone already that this was the last straw. He died a few hours later. They were tough fanatics, with a psychology almost incomprehensible to us.

Pappy Gunn flew the remodeled B-25 up to Port Moresby. The felt pads took up the shock of firing the guns nicely, until they got wet and the pads dried out as hard as iron. Pappy then tried sponge rubber and the problem was solved. Putting the side guns back farther aft helped a lot and the ammunition feed problem was much simpler, too. Every time the bottom guns were fired the door that folded up behind the nosewheel fell off, so I told Pappy to just give me the airplane with the four guns in the nose, two on each side of the fuselage, and fix the top turret guns so that they could be locked to fire forward.

That would give us ten forward-firing guns, which I thought was enough to start with. I told Pappy to shoot 20,000 rounds of ammunition through the installation right away. If the ship held together, we would then go ahead on enough to outfit a squadron and let Captain Ed Larner start training it.

Taking advantage of bad weather, five Jap destroyers made the run down from Rabaul and at three o'clock on the morning of the 14th dropped their deck-loads of drums of food and ammunition to be washed ashore at the mouths of the Mambare and Kumusi Rivers for the troops they landed there on December 1st. They then pulled out and headed back to Rabaul. About noon I put twenty-three B-24s of the 90th Group to work on them. This was the first mission for the group since I had put them back on training status. They didn't hit any of the Jap ships, but they shot down eight of twenty Nip fighters that interfered with their bombing, and all the B-24s came back without any personnel casualties. The morale was now up so high that I decided they were ready to work and take some of the load off the 43rd Group.

Miff Harmon found out that we were adding guns to our B-25s, so he wired asking me to fix up 12 of his B-25s with extra guns and racks for parafrag bombs. He said he was sending one plane over the next day. I told him I'd fix that one up as a sample but that he would have to do his own work on the rest. It was going to take all the resources I could get to take care of my own outfit.

On the morning of the 16th General MacArthur told me that he had a complaint from the 32nd Division that some of our A-20s had shot up one of their barges moving along the coast, bringing supplies from Wanigela to Cape Endiaidere. No one had been hurt but they had been badly scared. The report said also that the two A-20s had dropped four bombs, all of which luckily had missed. General MacArthur said he told them it couldn't have been my A-20s, but I'd better look into it.

I did and found that the culprits were two Australian crews from a grand fighting outfit, Number 30 Squadron RAAF, flying A-20s, or Bostons as they called them, who had become separated from the rest of the squadron in poor visibility and thought that Cape Endiaidere was

Cape Ward Hunt where their targets really were. Everything east of Gona had been out of bounds for bombing and strafing for some time.

I went down to the 30 Squadron Area with Group Captain Bull Garing. We lined up all the squadron officers and bawled out the two pilots who had made the mistake. I told them that Australians should not bomb American barges any more than American pilots should bomb Australian boats. These international incidents caused bad feelings and were not good. Furthermore, I didn't like the lack of knowledge of geography that the error indicated and which was the cause of their attack on their own allies. However, I explained that what pained me more than anything else was the fact that they had missed the boat with all their bombs. I therefore considered that they were not fit to belong to an outfit in the Allied Air Forces and sentenced them to daily, morning and afternoon, bombing practice until Group Captain Garing could report to me that they could hit something with a bomb besides the Solomon Sea. I also wanted a certificate that each of them knew something about the geography of Papua, particularly since it was a possession of Australia.

The squadron looked a little puzzled for a while at my method of admonishment. Then a few eyes began to twinkle here and there and I knew that the Aussie sense of humor had got the idea. When I went back to Government House I told General MacArthur what I had done about it. He got a great kick out of the story.

I gave Walker and Whitehead a lecture that day and ordered them to stop flying combat missions. Walker had come back the night before from a reconnaissance mission with three feet gone from the left wingtip of the B-17 was flying in.

Flying under the low clouds in the dark looking for Jap barges along the coast, they had just got too low passing over a point of land and hooked the wing on a tree. They were lucky to get the plane back home. Walker had been over Rabaul several times already on the excuse that he should go along once in a while to see how his crews were doing. I told him that from then on I wanted him to run his command from his headquarters. In the airplane he was just extra baggage. He was probably not as good in any job on the plane as the man already assigned to it. In fact, in case of trouble, he was in the way. On the other hand, he was the best bombardment commander I had and I wanted to keep him so that the planning and direction would be good and his outfit take minimum losses in the performance of their missions. One of the big reasons for keeping him home was that I would hate to have him taken prisoner by the Japs. They would have known that a general was bound to have access to a lot of information and there was no limit to the lengths they would go to extract that knowledge from him. We had plenty of evidence that the Nips had tortured their prisoners until they either died or talked. After the prisoners talked they were beheaded, anyhow, but most of them had broken under the strain. I told Walker that frankly I didn't believe he could take it without telling everything he knew, so I was not going to let him go on any more combat missions.

Whitehead had been "inspecting" the Jap defenses and incidentally shooting at them in the Buna area and around Lae and Salamaua. A couple of days before he had come back from a mission with an antiaircraft shell hole in the wing of his B-25 big enough for him to jump through without touching the sides. I told him he was through, too, for the same reasons.

On the afternoon of the following day, the Australians landed eight General Stuart tanks just west of Cape Endiaidere and at daybreak the next morning they went into action, driving west along the quarter-mile-wide coastal strip held by the Japs and honeycombed with machine-gun and mortar defenses. The Nips were taken completely by surprise and for a while the Aussies made good speed, crushing the log bunkers and wiping out the Jap positions so that the ground forces could follow behind and complete mopping up the area.

The tanks made an advance of about a mile but then had to halt on the edge of an overflowed stream which was too deep for them to negotiate. By this time only four of the tanks were still in action. Two had been burned, one knocked completely out of action, and the other was hopelessly mired down and being used as a machine-gun position by the crew, which was waiting for darkness to cover their withdrawal. Blamey sent for four more tanks that he had down at Milne Bay. While waiting for them to come up, he said he intended to fight his way far enough west to put in some bridges for his tanks to cross on when they made the next attack, which he thought would about wind up the show.

The Japs at this time held a section of the coast from Buna Mission east, about a mile long and from one half to three quarters of a mile in depth. They probably had around 3000 men in that pocket. In the Sanananda Point area there were another 2000 or so isolated by the Australians at Gona, the Americans between Sanananda Point and Buna, and by both Americans and Australians about three miles from the coast in their rear. The Jap had organized his position quite well. He was on high ground, with almost impassable swamps between him and our troops. These swamps were crossed by a few trails but the Nip had them well covered by mortar and machine-gun fire and a host of snipers up in the trees. He didn't starve out as easily as we thought he would and he seemed to have plenty of ammunition. Blamey said he would have his tanks ready to go again in about four days. If that attack did not clean the Nips out, the infantry would have to go in and do the job before everyone got so sick that I had to ferry them back over the Hump. We were taking them across then at the rate of 1500 a week. Most of them were sick, not wounded.

Pappy Gunn came up with the commerce-destroying B-25. He had cured most of the troubles and the guns all worked, except for the three that Pappy wanted to have on the bottom of the fuselage. I told him to forget them and go ahead fixing up some more B-25s. I sent for Ed Larner, the hot attack boy, and told him to go to Australia with Pappy, help Pappy with the testing, and learn to like the airplane.

Captain Ken McCullar had put on a show the night before that will go down in the records. Over at Lae, with Lieutenant O'Brien dropping the flares for him to see by, McCullar had an engine hit and set on fire while coming in on a skip-bombing run which missed a Jap destroyer he was aiming at. With the flames streaming back past the tail surfaces, the crew thought it might be a good idea to head down the coast where, if they had to bail out, there was more chance of getting picked up by Allied forces or by friendly natives. McCullar said, "Oh, that fire will die out. Don't worry about that. I'm circling that damned destroyer at 100 feet while you gunners shoot out those deck guns. I'll teach those so-and-so ack-ack so-and-so's to mess up my airplane." He did, and the gunners did their stuff. Ken was right, the fire did die out, and, although another engine got hit and quit, they finally silenced the destroyer's deck guns, O'Brien dropped a couple of flares just right, and McCullar sank the ship with a couple of skip-bombing hits amidships at the waterline. Ken called O'Brien on the radio and told him to go home and tell the gang that he'd be "a little late on account of because he was on two engines." Coming up to the Kokoda Gap, it was evident that the B-17 couldn't make the 6500-feet altitude necessary to get through. McCullar passed the word to the crew to throw everything overboard that they could get loose. Guns, ammunition, food, life-rafts, clothing—everything loose, as Ken ordered, went, and the plane again headed for the gap. McCullar passed the word back to the crew, "I'm taking her through. When the pass looms up, if it looks too tough and I need to lighten ship some more, I'll pass the signal and you guys start bailing out, beginning with the tail gunner and working forward."

About fifteen minutes later, down on Seven Mile strip, we saw McCullar bring her in. I talked with the navigator who was in the nose.

He said, "We didn't have to bail anybody out but when we came to the pass it looked so bad I closed my eyes and counted fifty before I opened 'em. Sure was a pretty sight to see that slope falling away to the south."

I tried to get the General to go back to Brisbane for Christmas with Jean MacArthur and the youngster. He would have liked to go but he decided to stay until Buna was cleaned up. Blamey's last eight tanks were scheduled to push the next morning.

The Aussies drove their tanks as scheduled the next morning and got another half mile, when every one of them was knocked out of action or destroyed. There wasn't much left of the Jap holding now, but it was the strongest part of their old position. The four 105-mm. howitzers and eight Australian 25-pounders were placed in position for the final drive in a few days. We were to fly in plenty of ammunition so that the guns could blast the Japs out before the infantry jumped off.

The day before Christmas thirteen bombers attacked a couple of Jap transports in Gasmata Harbor, sinking one and damaging the other. A reconnaissance plane that evening, carrying four 500-pound bombs, flew along the coast from Gasmata toward Rabaul, looking for the cripple. Just about midnight we got this radio from the B-17: "Found ship twenty miles east of Gasmata. Dropped four bombs. All missed. Peace on earth, good will toward men."

Christmas was a quiet day. The weather stopped all flying. I had Christmas dinner at noon with General MacArthur at Government House and then spent the rest of the day wandering around visiting the parties at Whitehead's headquarters, at Walker's Bomber Command, and over at Squeeze Wurtsmith's. A couple of girls and a six-man orchestra that I had flown up from Australia put on a traveling show all over Port Moresby that went over big. Two Red Cross girls, who came up with them, trimmed up a grass hut with colored paper and passed out candy, cigarettes, and trinkets all wrapped up just like Christmas back home. They even fixed up some scrub brush that the kids dragged out of the jungle for a Christmas tree and trimmed it with tinsel and candles. It was good to see all the youngsters hanging around getting buttons sewed on or using any old excuse to look at and talk to an American girl again.

The day after Christmas eleven Jap "Zeke" fighter planes appeared over Buna, where they were engaged by twelve of our P-40s from the 49th Fighter Group. Four of the Japs were shot down. We had no losses. It was not a big fight but the third Jap to be shot down made the score 100 official victories for the 49th Group. Lieutenant Jack Landers was the one who made it, "only 400 to go" for that magnum of brandy. All over the theater, people were asking, "What is the latest score of the 49th

Group?"

On the 27th, the P-38s finally got what they had been looking for—a good fight, in fact, a Number One fight. Twelve of them were on alert at Laloki strip about noon when we got the warning. Captain Tommy Lynch led the show off the ground, and twenty minutes from take-off time the twelve P-38s were at 18,000 feet, 125 miles away, over Dobodura and made their interception with twenty-five Jap bombers and fighters, knocking down fifteen of them. Just as they crossed the mountains, some P-40s returning to Port Moresby at about 7000 feet, escorting the transports, called over the radio, "Hey, P-38s, bandits coming in to Dobo 18,000 feet up." Lynch called back, "Okay, P-40s, thanks. We'll drive a few down to your level."

Lynch shot down two, a little tow-headed kid named Lieutenant Richard I. Bong, my ex-bad boy from San Francisco, got two, and a Lieutenant Ken Sparks got a couple. Sparks had an engine shot out and some other damage while getting his first one and started to make a landing at Dobodura, but just as he was getting squared away to come into the field he saw a Jap dive-bomber off to one side so, although already flying on only one engine, he postponed his forced landing, shot down the Nip bomber, and then landed. His nose-wheel was badly shot up and would not come down, so Sparks' plane was banged up a little. It was repaired and flown back a day or two later. None of the other P-38s received any damage except a few bullet holes. Sparks crawled out of his damaged airplane, ran over to a DC-3 that was about to take off to return to Port Moresby, and got home almost as soon as the rest of his squadron.

I went over to the squadron headquarters to congratulate the gang and hear the stories. They were priceless. Sparks, the bubbling-over type, had a report a mile long. I told him he still owed me two Nips before he should look for a decoration. I had promised an Air Medal for each confirmed victory over a Jap in air combat, but, in the case of the P-38s, I said that the airplanes were so scarce that if a pilot got his airplane badly shot up I'd charge him two Nips which he would have to pay back before he got even with the board. Sparks grinned as though he was fairly sure that I didn't mean it. He was probably right. Dick Bong's report simply gave his time of take-off, the time of arrival over Dobodura, the fact that he had shot down two Japanese planes, and the time of arrival back at Laloki. I said to Whitehead, "Watch that boy Bong. There is the top American ace of aces of this war. He just started to work today." Whitehead liked Lynch. Lynch was a wonderful youngster and a sweet combat leader, but I guessed he would tire long before that cool little Norwegian boy Bong.

From their descriptions of the fight, most of them did everything wrong. They opened fire too far away, they tried maneuver combat with the Zeke which can outmaneuver them, and just got excited in their first show. I pretended to get mad and gave them the devil. Then I turned to George Prentice, the squadron commander, and said, "Huh, I'll bet you haven't even got any liquor to celebrate the first combat." Prentice said no he hadn't, so I said. "There you are, robbing me of my only three bottles of Scotch." I turned over to them three bottles that Mac Laddon of Consolidated Aircraft had sent me as a Christmas present and had just arrived that morning. I sent word to Bob Gross of the Lockheed Company, which made the P-38, that he owed Laddon three bottles and me three bottles for turning mine in to celebrate a Lockheed victory.

With their first big fight behind them, that P-38 squadron had the highest morale of any fighter squadron in the world. I wired the results to Hap Arnold and got back a nice message of congratulation which I sent over to the kids.

I think General MacArthur got as much kick out of the show as anyone in New Guinea. He had met about all the pilots in the squadron and I suspected that if I didn't put out some decorations soon the general would be on my neck.

The Jap ground troops probably didn't appreciate the party. Their own bombers dropped their bombs on the Jap ground positions when our P-38s hopped them.

In addition to the P-38 victory we had a pretty good day at Rabaul, where eighteen of our bombers attacked the shipping, sinking two cargo vessels, setting two others on fire, and badly damaging a destroyer with a direct hit on the deck. All our planes returned.

On the 29th Captain Ed Larner flew the B-25 eight-gun job to Port Moresby. I made him a major and put him in command of the 90th Squadron of the 3rd Attack Group, which I had designated to specialize in low-altitude work, including skip-bombing. I told Larner I wanted him to sell the airplane and the strafing tactics to his squadron. I wanted him to like the plane, make his squadron like it, practice shooting and skip-bombing on the old wreck on the reef outside Port Moresby until he didn't miss, and then we would find him a convoy to work on. I promised that if he sank that convoy I'd change the name of the Solomon Sea to Larner's Lake. The rascal grinned and said, "General, she's in the bag."

On the 30th the Allied attack sliced the Buna area up into isolated pockets which were cleared out in the next day or two. Blamey had brought eleven more tanks up and all the artillery we had was in position with plenty of ammunition to support the attack. I was glad those tanks were in there, as the mopping-up process would not be so costly in lives. Mopping up a soldier with the suicidal instincts of the Jap is no cinch. The Japs still held the spot called Buna Mission, but it wouldn't be for long, as they were completely surrounded and artillery and mortar fire were systematically blasting the place apart. The position was surrounded by swamp which stopped the tanks from getting at it or the show would have been over before. The Nips were invited to surrender. Their position was hopeless, they were outnumbered at least 20 to 1, they hadn't had a decent meal for two months, and they were down to their bayonets and a few hand grenades, but the answer was, "What's the matter with you Yanks? Are you yellow? If you so-and-so's are as tough as you say you are, why don't you come in and get us?" The artillery and the mortars started. The flamethrowers and the hand grenades finished the job.

The P-38s wound up the year with their second show. Eleven of them were escorting our bombers in an attack on Lae, where they engaged twelve Zeke fighters. The final score was: nine Japs definite, one probable, and two damaged. Sparks got another Nip by ramming him. Both Sparks and the Jap tried to pull out of a head-on attack at the last minute, but they had waited too long. The Jap had all of his right wing torn off and crashed and burned. Sparks lost two feet of his left wing but flew the plane back to his airdrome. In the two combats they had had so far, the P-38s had shot down twenty-four of the thirty-seven Japs they had encountered. The two airplanes that Sparks flew were the only ones seriously damaged. I liked the P-38 more every day.

With the capture of Buna Mission on January 2nd, 1943, all that remained of the Jap force was in the Sanananda Point area, where between 1000 and 2000 half-starved Nips were waiting to be mopped up. The Australians moving from Gona on the north and the Americans from the Buna area on the south were to take care of that. The 32nd Division was being pulled out and the clean-up of Sanananda, as far as the Americans were concerned, was to be the job of the 163rd Infantry of the 41st Division, which had arrived in Port Moresby a few days previously and which I had flown over to Dobodura. The regiment was commanded by Colonel Jens Doe, an old friend of mine. We were classmates at the Army War College in 1933. I didn't worry about that regiment. Jens was a real fighting leader.

We got some information that looked as though the Japs were about to try a reinforcement of Lae. I expected a convoy from Rabaul to make the attempt about the 6th, so I told Walker to intensify his reconnaissance along both the north and south coasts of New Britain and to put on a full-scale bomber attack on the shipping in Rabaul Harbor at dawn on the 5th to see if we could break the movement up at the source.

Colonel Jack Sverdrup, the engineer, came back from a trip to Bena-Bena and said that everything was arranged for a big native "sing-sing" or celebration in my honor on the 5th. Ten thousand natives were expected to attend and put on their dances and stuff for the "Number One Baloose Man." Baloose is pidgin dialect for airplane. A P-38 is a "one-man baloose, him two tails behind belong." They thought I owned all the airplanes, so they named me the Number One Baloose Man.

Walker wanted to hit the Rabaul shipping around noon, instead of at dawn as I ordered. He was worried about the bombers making their rendezvous if they left Port Moresby at night. I told him that I still wanted a dawn attack. The Nip fighters were never up at dawn but at noon they would not only shoot up our bombers but would ruin our bombing accuracy. I would rather have the bombers not in formation for a dawn attack than in formation for a show at noon which was certain to be intercepted. Ken had not been sleeping well and was getting tired and jumpy. The strain and the tropics were wearing him down. I decided that at the end of this month, if a couple of weeks' leave didn't put him back in shape, I'd have to send him home.

Six B-17s and six B-24s struck the shipping in Rabaul Harbor at noon on the 5th. Hits were scored on ten Jap ships. One vessel was sunk and six others left burning. Our bombers were intercepted by fifteen enemy fighters, three of which were shot down. Two of our bombers were missing. One of them carried Brigadier General Kenneth Walker.

Ken, for some unknown reason, had suddenly changed the take-off time early in the morning without notifying me, and then went along on the mission, in spite of the fact that I had told him not to. The returning bombers reported that his plane was last seen headed south about twenty-five miles south of Rabaul, losing altitude, with an engine on fire and two Jap fighters on its tail. I ordered all the Australian Catalina flying boats and all the reconnaissance planes to search the Trobriand Islands off the eastern end of New Guinea and the route to Rabaul. A report came in during the evening that Walker's airplane was down on a coral reef in the Trobriands. I told General MacArthur that as soon as Walker showed up I was going to give him a reprimand and send him to Australia on leave for a couple of weeks. Then General said, "All right, George, but if he doesn't come back, I'm going to send his name in to Washington recommending him for a Congressional Medal of Honor."

The next morning a Catalina rescued the B-17 crew off a reef in the Trobriands. It was not Walker's crew. I ordered the search kept up, but I had no hope for Ken. I was certain that his airplane had been shot down in flames and unless the crew bailed out they were gone. If they did bail out and the Japs had them, that was bad.

The expected Jap convoy was picked up about ten o'clock in the morning about fifty miles east of Gasmata heading for Lae. It was heavily convoyed by fighters, who intercepted our eight bombers and fifteen P-38s who made an attack. The P-38s shot down nine Japs and the bombers got six, but no hits were recorded on the Nip vessels. The convoy consisted of five transports escorted by five destroyers.

During the following day we sent a total of twenty-four bombers and seven P-38s after the convoy. Every attack was intercepted by the Jap fighter cover. We sank one of the transports and scored hits on two others and one of the destroyers, and during the day destroyed twenty-seven Jap aircraft in combat, but the convoy arrived in Lae Harbor during the night and started to unload.

On the morning of the 8th we continued the attack on the convoy which was still in Lae Harbor. Another transport was set on fire, beached, and finally destroyed, and three other Jap vessels badly damaged. Supplies on the beach and Jap troops were heavily strafed and bombed. The Jap fighters were very much in evidence all day, but we shot down twenty-eight more of them and destroyed another fifteen to twenty on the ground at the Lae airdrome. Altogether we used twenty-eight heavy bombers, twenty-two mediums, and thirty-two light bombers, which included nineteen Australian Beaufighters. Thirty-five of our fighters got into action, among them Lieutenant Richard Bong, who got his fifth victory over Lae. Captain Lynch got his sixth and Sparks his fourth.

The convoy pulled out during the evening and headed back toward Rabaul. I estimated that they landed another 5000 to 6000 Jap troops that we would now have to dispose of sooner or later. From the looks of the pictures they must have lost most of the supplies they unloaded, but our experience at Buna showed that the Jap could still fight even if he didn't have any supplies.

The GI craze for souvenirs almost got a doughboy over at Buna. Everything was supposed to have been mopped up there several days previously. This particular souvenir hunter went down into a good-looking dugout, but a few seconds later came bursting out with a Jap captain after him swinging a long Samurai sword and emitting shrill Nip four-letter words. The Jap had been calmly sleeping in his own private dugout and simply got sore at being disturbed. Some third party shot the

The Buna Campaign: II

Nip or, it is said, he would have caught the GI, who was unarmed.

On account of Walker's loss and the Jap convoy I did not go to the Bena Bena "sing-sing," but sent Whitehead's aide, Captain Beck, up to represent me and take the honors. Beck came back the next day and told me that I had missed a great show. He said that the chief had a present for me—two fifteen-year-old virgins. I asked him what he did about it. Beck said, "Oh, I accepted them for you and then turned them over to the Australian commissioner, who put them to work around the Government House. They are up there waiting for you." Beck probably did the right thing, so as not to insult the chief, but I decided that I'd better cross Bena-Bena off the list of places that I expected to visit.

That evening at dinner Sutherland and I were arguing about a separate Air Department. He opposed it. I argued for a Department of National Defense with three subdivisions, Army, Navy, and Air. Much to my surprise, General MacArthur broke into the conversation and said that a single department was the proper organization and that the Air should be separated and have the same autonomy as the land and sea forces. I replied that he hadn't thought that way in 1932 when such a bill was proposed. He said, "No, I didn't. At that time I opposed it with every resource at my command. It was the greatest mistake of my career."

It was a courageous statement from a big man. The big man can change his mind and admit his mistakes. It is only the small man who is afraid to admit that he was ever wrong and who looks upon such an admission as a confession of weakness. General MacArthur had changed his mind, decided he was wrong in 1932, and had the courage and the brains to admit it.

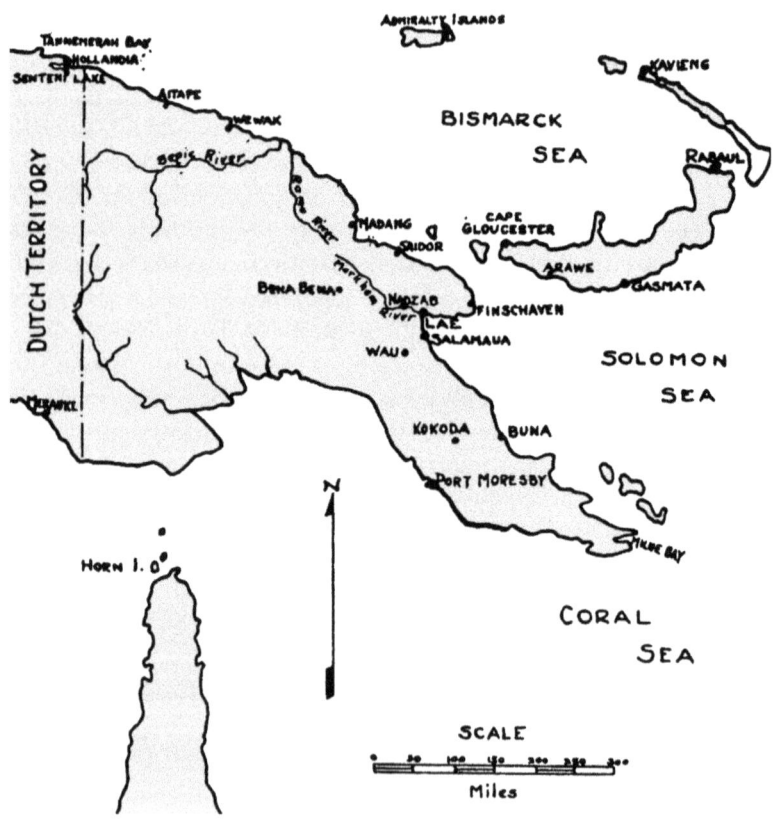

6. "OUR LOSSES WERE LIGHT"
January—February, 1943

GENERAL MACARTHUR and I with our aides returned to Brisbane on January 9th. The General's communique of that day announced the practical end of the Buna Campaign.

With no sign of Walker, I told Whitehead to take over the Fifth Bomber Command for the time being, in addition to his other duties, and wired Arnold to send me Brigadier General Howard Ramey, who was working for General Emmons in Hawaii. I had had Ramey as my Chief of Operations in the Fourth Air Force at San Francisco and had known him for years. I had a lot of confidence in him and believed he would do a real job for me. Arnold replied the next day that I could have him.

The forward gun protection in the B-24 turned out to be quite unsatisfactory. The Jap fighters found out about it and we were getting the airplanes shot up badly every time we got intercepted. There were four fifty-caliber guns in the nose of the B-24, but, as they shot through individual "Eyeball" sockets, only one could be fired at a time. It was a clumsy arrangement and didn't give the protection we needed, so I started Lieutenant Colonel Art Rogers of the 90th Bombardment Group installing a turret, which we took off the tail of a wrecked B-24, in the nose. This would give us a pair of power-operated fifty-calibers and should surprise the Nip the next time he tried a head-on attack against a B-24 so equipped. Rogers said the Air Depot in Hawaii had been working out something along this line when he left there a month previously and he knew how to do it. He said that when he had made the installation he would like to take the plane to New Guinea and hunt up a fight to try the scheme out. I gave him a go-ahead and told Connell to give him all the help he needed to expedite the job.

Brigadier General Howard Ramey arrived from Hawaii on January 18th. I spent most of the morning with him, going over the situation and the problems he would be up against. I went into the subject of skip-bombing with him at some length and told him to watch Ed Larner's work of training the 90th Squadron and let me know if they needed anything to speed up the day when they would be ready to go to work on a Jap convoy. Ramey liked the idea. I expected that his good steady hand would straighten out a lot of troubles in the Fifth Bomber Command. He and Whitehead were old friends and would work together. I told him to visit his units in the Townsville area briefly and be in Port Moresby by the 20th.

That evening I got a real shock. Major Bill Benn was reported missing since ten-forty in the morning when he cleared Port Moresby to look over the airdrome possibilities around a native village called Dona, which was about halfway between Buna and Salamaua. No enemy air activity had been reported all day at Lae or along the north coast of New Guinea. Whitehead reported that he had airplanes out all day searching but no trace of Benn's plane had been found. Whitey was afraid Benn had run into the mountains when the weather turned bad during the early afternoon. It hardly seemed a good explanation, as Benn was a superior pilot and had with him on this trip as co-pilot a Captain Young, who was another old experienced flyer. Both of them knew the Owen Stanley Mountains from flying over them dozens of times and knew that in bad weather they would have to go to 15,000 feet to play safe in clearing the range. Benn's loss hurt. He was the one who put across skip-bombing out here and if it hadn't been for that 63rd Squadron of his we might have been fighting the war in Australia instead of New Guinea. No one in the theater had made a greater contribution to victory than Bill Benn.

A couple of officers in from Washington told me that Colonel Merian Cooper was back home and probably available. I immediately wired Arnold to let me have him. I had known Cooper since 1917 when we learned to fly together at Mineola, Long Island. During World War 1 he was shot down and taken prisoner. After that war ended, he fought with the Poles against the Russians, who also shot him down and captured him. He eventually escaped from the Russian prison camp and got back to the United States, where he had been a big shot in the movie game and dabbled in a lot of other things as sidelines until this war broke out. After a short tour in the Intelligence Section of the Air Forces in Washington, he was sent to China, where he had been Chennault's Chief of Staff. Chennault had been loud in his praises of Cooper's energy and ability. "Coop" was sent back home to recuperate after a bad siege of dysentery, but if he had recovered enough to walk, I could use him. He was a real live wire, with more energy and imagination than most families.

The Japs started building up at Wewak. A lot of shipping had been picked up in the harbor there and the Nips had already built one airdrome and were feverishly working on three others in the vicinity. The Wewak field itself was occupied by fighters. The numbers had varied since they were first seen there about the time the Buna show ended, from twenty to forty, but there was no doubt that it was a permanent field from the supply dumps, buildings, and large antiaircraft gun installations that showed on the photographs.

On the morning of the 20th, a B-24 reconnaissance plane reported seeing twenty-five Zekes on the Wewak strip and a convoy of three small freighters outside the harbor. A few minutes later they reported being under attack by Jap fighters. Nothing more was heard from them. Four B-24s from the 90th Group managed to break through the bad weather off the coast and headed for Wewak to search for the convoy. They didn't find it, but twenty-five Jap fighters found them. Captain Faulkner, who was leading the B-24s, really did a job. Those four B-24s shot down twelve Jap fighters and damaged six more. All the B-24s were damaged, but they came back home. Two of the bombers had engines shot out during the combat, but Faulkner nursed his unit along, shifting the good airplanes around to protect the cripples. It was a nice bit of air leadership.

I got quite a kick that day out of a report of a MacArthur press conference. Some of the newspaper crowd told me about it. The General had finished his talk, when one of the correspondents said, "General, what is the Air Force doing today?" General MacArthur said, "Oh, I don't know. Go ask General Kenney." The newspaperman said, "General, do you mean to say you don't know where the bombs are falling?" MacArthur turned to him, grinned, and said, "Of course, I know where they are falling. They are falling in the right place. Go ask George Kenney where it is."

That was the best compliment I've ever received.

Tokio Radio said that night that if we didn't stop dropping those fiendish air-burst bombs on Rabaul the Japs would gas us out of New Guinea. We finally had those bombs working so that nearly half of them exploded about fifty feet above the ground. The rest went off on contact, with an instantaneous fuze. I sent word to Whitehead to drop more of them, as they were evidently effective. I didn't believe the Japs were going to drop any gas on us but I told Whitehead to play safe and see that his degassing measures were ready and that everyone had masks and protective clothing. He hadn't the clothing, so I ordered it sent up to New Guinea from the supplies in Australia. I also told Bostock to work up for me his plan of gas protection for the RAAF airdromes.

The Sanananda area was finally completely mopped up on January 23rd. A few Jap prisoners were taken but they were all service troops. The Nip infantryman, when the jig was up, committed suicide or yelled Banzai and charged into machine-gun fire and certain death rather than surrender. Even the service troops were sick and out of their heads from starvation or malaria or both when we captured them. The diaries were beginning to come in and the translators were busy extracting information. Every Jap kept a diary and wrote down everything he knew in it. A number of them had mentioned cannibalism, but the Australians said that they found plenty of real proof of it during the Sanananda mop-up. No wonder the troops refused to look upon the Japs as human beings. To the American and the Australian, the Nips were just vermin to be exterminated.

After adding up all the columns of figures showing what I had to fight a war with, I wrote a long letter to Hap Arnold. I knew he was doing what he could to get me airplanes, but the fact remained that I was going downhill all the time. During the past three months I had received from the United States a total of eighty-nine airplanes. In that same period I had lost, from combat, accidents, wear and tear, and all other causes combined, 146. My total strength was 537, of which around 200 were constantly in repair, getting bullet holes patched or shot-up engines replaced. Some day, someone had to look my way or I'd be out of business. I wired Arnold asking what had happened to six B-17s and two B-26s which his radio of January 2nd had said were en route. The next day I heard that thay got as far as Fiji and then were taken over by SOPAC on account of the emergency. I immediately sent another wire to Hap, telling him what I had heard. I knew I wouldn't get the airplanes, but I thought maybe Arnold would do something about that agreement that would permit my replacements to get to me.

For the past week, at the noon briefings in the War Room, the Australian intelligence people had been talking about 2000 or more Japs in the Sanananda area. I had insisted that there were not over half that number and at one time I said that I would eat all the Japs that were killed over 1000. The Australians suggested that they would rather have me buy beer if the count reached 2000. I agreed. The official count of buried Japs in that area finally passed the 2000 mark. There were only twelve Australians, but they certainly did drink a lot of beer.

Those Jap troops that landed at Lae on the 8th of January didn't wait long before they started to make trouble for us. They moved down to Salamaua, secretly cut trails west toward Wau, where the Australians had been maintaining about 1000 men in the old gold-fields area, and suddenly on the 29th drove in the Aussie outposts and moved forward to seize Wau and the little airdrome there which we used to keep the Australian garrison supplied. The Aussies called for help and we flew in 2000 Australian troops, landing under machine-gun fire from the Jap forward positions, which were just east of the airdrome. The troops actually landed under fire and were in action as soon as they got out of the airplanes.

The quick reinforcement saved the situation. The Aussies drove the Japs back out of machine-gun range of the strip and the next day we landed a regiment of 25-pounders, another 500 troops, and enough ammunition and food to keep them going for a while. The A-20s and the Beaufighters worked with the ground troops, strafing Japs and their supplies all the way from Wau to Salamaua. On the 31st we hauled in six Bofors guns, more personnel and supplies. During the three days, 194 planeloads of troops, artillery, and supplies had gone into Wau—a total payload of over a million pounds. Wau was now secure, but General Herring decided to drive on Salamaua.

Some of the SOS from Sydney came in to complain about my personnel on leave in Sydney. They said they thought it was time "these brats grew up and behaved themselves." I sounded off to the effect that I didn't want them to get old and fat and bald-headed and respectable like some of the people present because I was quite sure that, if they did, they would no longer shoot down Nip planes and sink Nip boats. Furthermore, that Sydney leave was only for such combat personnel and it was too bad to disturb the slumbers of the SOS, but I was for the kids. I looked up at this point to see General MacArthur standing in the doorway. I stopped talking, except to say, "Good morning, General." He grinned and said, "Leave Kenney's kids alone. I don't want to see them grow up either."

The Intelligence outfit brought me a translation of part of one of the diaries taken from a dead Jap at Buna that gave a good illustration of the curious, fanatical creature we were fighting against.

The writer related the story of the landing at Buna last July under air attack and the headlong drive over the Kokoda trail. They made good progress at first but then the constant air attacks drove away the native bearers and stopped food and ammunition from getting forward. The drive finally stopped and then the Australians began to push the Japs back. They were told to withdraw to Kokoda, where they would get food and make another stand. Many were so weak that they fell down on the trail and hadn't the strength to get up, so they lay there until they died. At Kokoda the survivors did not find the food supplies they had expected. They ate a few raw potatoes that they found in the abandoned and already picked-over native gardens. The writer got malaria and was out of his head most of the time but he stumbled on. He was finally told to report to the Soputa field hospital for treatment. He walked the trail from Kokoda for three days and arrived at the Kumusi River to find that our bombers had destroyed the Wairopi bridge. He then tried to use a log as a raft to cross the river. Almost across, he fainted and fell off the log, but the cold water revived him and he managed to get ashore. He then took off his clothes, dried them in the sun, and after eating some leaves and a piece of half-dried coconut, he staggered on for two more days and finally reached Soputa. The field hospital was full so he was simply given quinine and bedded down on the ground. After a brief period of treatment, the Jap doctors told him he was cured and off he went to Buna, where he was given another rifle and a chance to die fighting the Yanks and the Aussies.

Finally, after complaining daily for a month about the terrible and continuous bombing that almost drove him insane, he was so badly wounded that he knew he didn't have long to live. The last entry in the diary was a message to his wife urging her to remain virtuous and bring up their son so that some day he would be fit to take his father's place and fight for the glory of the Emperor and the Greater East Asia.

February 6th was a big day. We had an Australian with a portable radio stationed on a hill overlooking the Lae airdrome, who reported to us every time a Jap plane landed at or took off from the field. Early that morning he sent word that the Nips had between fifty and sixty bombers and fighters on the airdrome. Seven B-25s, one B-24, and five Beaufighters, escorted by twenty-two P-38s, sixteen P-39s, and eight P-40s, took off to see what they could do about it. Most of the Nip planes were gone by the time we reached Lae, but we set a lot of fires, burned up seven Jap planes on the ground, and left two beautiful craters from 2000-pound bombs right in the center of the runway. Just as they had finished and had turned south to return home they ran into a Jap force of forty to fifty bombers and fighters returning from a raid on Wau. They had bombed Wau and caused some slight damage and inflicted a few casualties but they certainly paid for it. Our fighters had a field day. The final score was: Jap losses—twenty-five planes definitely shot down, ten probably shot down, and five damaged; our losses—none. Three of our planes were slightly damaged, but all returned to base and were back in commission the next day. In the meantime, our bombers sank a Jap cargo vessel near Finschaven and got direct hits on three more at Ambon over in the Netherlands East Indies. The A-20s continued supporting the Australians at Wau by strafing and bombing the Nips in that area and our troop carriers carried over 150 tons of steel mat across the Hump to surface the runway at Dobodura so that we could station fighters there.

General MacArthur's communique told the story of the show over Wau and ended saying, "Our losses were light." I protested saying that we hadn't lost a single airplane. The General said he thought we'd better say our losses were light. He wanted the communique to tell the truth but he wanted people to believe it. To illustrate his point he said that at one time his father was aide to General Sherman when the General was sent out to a conference with the Sioux Indian chiefs who were rumored to be about to go on the warpath. Wild Bill Hickok was the interpreter. At the conference, the chiefs were all sitting on the ground in a half circle smoking as General Sherman opened the proceedings. Sherman said to Hickok, "Bill, tell these people that they shouldn't fight the white man. There's no sense in it. The white man is too smart for them. He invents things that accomplish results they never dreamed of. Tell them about how the white man invented the railroad train that will haul all the buffalo meat they could shoot in a month and haul it three times as far in one day as their fastest horses could go. Tell 'em about the railroad, Bill." Hickok told the chiefs about the railroad. When he had finished, they grunted a few times and were silent. "What do they say, Bill?" asked Sherman. "General," said Hickok, "they don't believe you."

"They don't?" exclaimed Sherman. "All right, tell them about the steamboat. Tell them how the white man made a big boat driven by steam that would carry the whole Sioux nation up and down all the rivers." Hickok told the Indians about the steamboat. The same thing happened as before. "General, they don't believe you," reported Hickok. Sherman got red in the face. His temper was rising. He played his last card.

"Bill," he said, "tell them about the telegraph. They'll understand that, as it's something like sending signals by smoke. Tell them how I have a little black box out here and the great White Father in Washington has a little black box. When I talk into my box the great White Father in Washington hears me, and when he talks into his little black box I hear him. Tell 'em about the telegraph, Bill."

Hickok stood there staring at the General. "Well, what's the matter now?" yelled Sherman. "General," said Hickok slowly, "I don't believe you, either." "So," concluded General MacArthur, "I think we'd better say—'Our losses were light.'"

On the 9th it was announced by both SOPAC and Tokio Radio that the Guadalcanal operation was over. The Japs said that they had just held this advanced post in order to cover the development of strategic bases in the rear. They said that was now accomplished so they had withdrawn their forces. It was a pretty story to cover up the fact that they had lost 50,000 men, hundreds of aircraft, and a lot of shipping in an unsuccessful attempt to drive the Yanks out of the Solomons.

My guess was that we would soon see some action to offset the effect at home of admitting a withdrawal. New Guinea was a good bet for trouble in the near future.

For some time, Whitehead had been working up a plan to really take Rabaul apart with a series of big attacks. First he would burn out the town, which housed a lot of Japs and which was loaded with supplies of all kinds. Then, after plastering the airdromes at Vunakanau and Lakunai, he would go after the shipping in the harbor, which was holding steadily to around fifty vessels constantly coming or going or being unloaded.

We were ready and just waiting for the weather to get a little better, when Admiral Halsey, who had succeeded Admiral Ghormley in command of SOPAC, sent a few staff officers over to confer with General MacArthur on their next operation, which was to occupy the Russell Islands, fifty-five miles northwest of Guadalcanal, on February 21st. Halsey wanted us to hit the shipping at Buin-Faisi. on the south end of Bougainville Island the night of the 18th-19th and the next four nights, in order to discourage any idea the Japs might have about sending troops to the Russells to dispute Halsey's taking over the islands. General MacArthur agreed to Halsey's proposal.

I sent word to Whitehead to carry out the attacks on the town of Rabaul as planned and to hit the shipping in the Buin-Faisi anchorage with a full squadron each night for four nights beginning the 18th–19th, but to wait until we saw what we had left after the 21st before we went on with the Rabaul operation.

The weather continued to prevent putting this plan into effect until the night of the 14th-15th, when we smashed Rabaul with the biggest attack so far. Thirty-two B-17s and four B-24s dumped 50 tons of demolition bombs and over 4000 incendiary bombs into the town, setting huge fires all over the place.

The next night we attempted to repeat the show but, after seventeen B-17s had struck Rabaul town with 35 tons of demolition bombs and 1400 incendiaries, the weather closed down and stopped any further attacks that night. From the reports, however, it looked like another good night's work. We lost no airplanes on the two operations, but the crews all reported that the Jap had about quadrupled the number of searchlights and antiaircraft guns around Rabaul during the past month or six weeks. Several vessels equipped with searchlights were moving around both nights, sweeping their beams at low altitude and evidently looking for skip-bombers. If we were to do any more skip-bombing in there at night, we would have to detail some airplanes to take out those lights before we started working on the vessels in the harbor.

The weather continued bad for the next couple of days but cleared in time for us to carry out our commitment to Halsey on the night of the 18th–19th, when twelve B-17s, led by Ken McCullar, skip-bombed and sank three vessels, all estimated to be 7000- to 8000-tonners, and damaged a fourth. McCullar got one of the ships himself.

The next night, eight B-17s and six Catalinas attacked the Buin area. The Japs had pulled out all their shipping, so the planes bombed the Buin airdrome, destroying a number of aircraft and burning up large stores of fuel. We radioed Halsey that we were switching the attacks to Rabaul, unless the Jap shipping returned to the Buin-Faisi anchorage.

Lee Van Atta, INS war correspondent in Australia, came into my office on February 19, bringing his boss Barry Faris, INS managing editor, who had just arrived that morning from the United States. Faris had just spent an hour with General MacArthur and had joined the rest of the General's admirers.

I spent the rest of the morning discussing with both Faris and Van Atta the situation as I saw it, what a terrific job we had on our hands, and the meager resources we had to work with. Faris said he was glad to get this background and that he would see that Van Atta's dispatches got a better play from then on. Lee was writing some good stuff. He was a likable youngster only nineteen years old, but he knew how to put the color into a radiogram. I was glad to hear that he was going to get a break.

I took a recommendation for Whitehead's promotion to the rank of major general up to General MacArthur and laid it on his desk without saying a word. He glanced at it, initialed it for transmission to Washington, smiled, and said, "All right. What else have you on your mind?"

I told him I was leaving for New Guinea for about three days to inspect and discuss with Whitehead our further plans for Rabaul. I didn't like that big concentration of shipping, reported as seventy-nine vessels by that morning's reconnaissance. Furthermore, the Nips had been repairing Gasmata airdrome again. They liked to have that place available for fighters when a convoy was making a run from Rabaul to Lae. I suspected that we would see another attempt soon and this time I would have Major Ed Larner and twelve of my "commerce destroyers" waiting for them.

The General said to come in and talk the situation over with him as soon as I got back. I landed at Port Moresby late the 20th.

I finally had a showdown on the food situation. After trying all the legal means that I knew of to get the rations in New Guinea improved, without any luck, I ordered fresh meat and vegetables flown from Australia to the airdromes in New Guinea so that the kids could have fresh food oftener than once every ten days, which was the best the SOS seemed able to do. Naturally, this food was for the air units. The base commander at Port Moresby, however, knew his regulations. He was the one charged with purchase, storage, and issue of all food supplies. Whitehead was ordered to turn all air shipments of food over to the base for "equitable distribution" to everyone. I told Whitey to distribute the food to the Air Force units and simply report the amounts and kinds of rations to the base so that they could keep their books straight. When Whitehead refused to deliver the stuff, the base commander suggested that he would send armed trucks to the airdromes and if necessary seize the supplies. I told Whitehead, if anything like that happened, to arm his trucks, too, and offer to shoot it out. At this point the base commander screamed to the boss SOS boys in Australia, who then complained to me. I told them that I would stop the air haul of fresh food just as soon as they would feed us a minimum of three fresh-meat issues a week. If they reached that level of supply and one of their ships for some reason broke down or happened to be late getting in to port, I offered to haul for all the troops in New Guinea, ground as well as air, during such emergency. However, in the meantime, I refused to argue about whether what I was doing was in accordance with their regulations or not. Those kids of mine had to be fed properly. They had not been fed properly and my doctors were beginning to get worried about scurvy. It was the job of the SOS to feed us properly, but if the SOS could not do it, I could and would.

I didn't hear anything more about the matter but I suspected that in some quarters my popularity had waned.

In the meantime, the kids in New Guinea were looking much better. The sick list was no longer alarming, the morale was up, and everyone was working harder and whistling while they worked.

On February 21st, SOPAC troops occupied Russell Island without opposition.

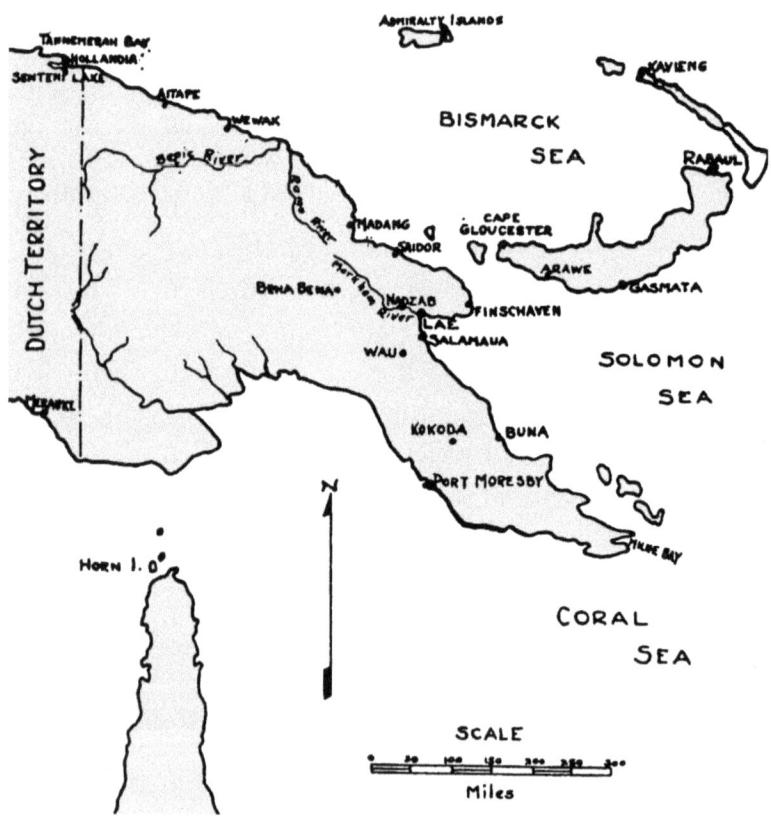

7. BATTLE OF THE BISMARCK SEA
March, 1943

ON FEBRUARY 25th we got some information indicating that a big Jap convoy was scheduled to arrive in Lae sometime early in March. Several cargo and transport vessels escorted by destroyers appeared to be coming from Rabaul and some others were coming from Palau. There might even be another increment from Truk. Both Madang and Lae were possibilities as unloading points for all or part of the convoy. The information was rather sketchy, but this was definitely to be on a much bigger scale than the convoy run into Lae on January 7th, which had consisted of five destroyers and five merchant ships. This looked as though it would be at least twice as large.

I sent a courier message to Whitehead, giving him what information I had, and told him to have his reconnaissance cover the Wewak-Admiralties-Kavieng-Rabaul area to insure an early pick-up of the convoy and then to hammer it day and night until it was sunk. I told him that I expected the Nips would throw at least a hundred fighters into Lae, Gasmata, and perhaps Alexhaven and Madang, as operating airdromes to cover the convoy. I advised him to get his P-38s and the B-25 commerce destroyers over to Dobodura, so that if the weather around Port Moresby was bad we would not be prevented from attacking the Nip vessels for that reason. I also told him to call off the big attack on Rabaul that we had planned and to go easy on all bomber missions from now until March 5th, so that the maintenance crews could get the squadrons up to maximum strength.

I talked with General MacArthur at some length about the information we had and discussed my tentative plans to really take the Nip this time. The General believed that the Jap intended to reinforce Lae with a full division preparatory to undertaking the offensive in New Guinea. He thought the convoy or convoys would be composed of both troop and supply vessels, heavily escorted by surface craft and airplanes, and very probably would be twice as large as the January convoy.

He told me to be sure to conserve my strength for this effort as the landing of a fresh Japanese division in New Guinea at that time would be a very serious matter, with our 32nd Division and the Australian 7th Division pretty well burned out from the Buna campaign, and part of the American 41st and the Australian 9th Divisions also needing a rest. He said that until this new threat was removed, no other mission would take precedence.

He told me that I was to go to Washington with Sutherland and Chamberlin the following week to present our plans for the future to the Joint Chiefs of Staff and get the means to carry out those plans if they were approved. We would need more airplanes, troops, and everything else if we were to go anywhere. SOPAC representatives and those of Central Pacific from Hawaii would be there, too, so that all the coming operations could be coordinated.

I told the General that the weather forecasts indicated bad weather for the first three or four days of March along the north coast of New Britain and I believed the Nips would use that weather for cover. Our plan was to locate the enemy convoy as soon as possible and start hitting it day and night with the heavy bombers until the Jap ships arrived at the point where they were in range of our commerce-destroying B-25s. Then we would have an all-out, coordinated attack, with the heavy bombers bombing from eight to ten thousand feet a split second before the B-25s started their skip-bombing attack. Following the B-25s would come the A-20s and the Beaufighters. The combined attack would be covered by all the P-38s we could put in the air. In the meantime, to cut down the Jap fighter cover as much as possible, we would attack Lae with all the shorter-ranged aircraft. I also told him that I expected to have at least one full-scale dress rehearsal of the combined show run off during the next day or two, to check our timing for the real thing, and that I planned to go to Port Moresby the next day to go over the final arrangements with Whitehead. I asked him not to send me to Washington until this party was over. The General agreed and said he would keep me posted if new information came in.

The next day I flew to Port Moresby. Whitehead and I went over all the information at hand and tried to guess how we would run the convoy if we were Japs. We plotted all the courses of all Jap ship movements reported for the past four months between Rabaul and Lae, checking both the route around the north coast of New Britain and the one around the south. The courses followed regular grooves. The weather forecaster still insisted that during the first three or four days of March the weather would be very bad on the north coast and quite good along the south coast of New Britain. We decided to gamble that the Nip would take the northern route.

With the range of our equipment, this meant that the heavy bombers would do all the attacking from the time the convoy was discovered until it was through the Vitiaz Straits and off Finschaven. Then we would throw the book at it. I told Whitey to plan, practice, and rehearse the show on February 28th, and then hold everything in readiness until the reconnaissance picked the convoy up. After Whitehead and I got through figuring what the Jap would do, we set the time as ten o'clock in the morning some day during the first week of March and the place of the big combined attack just off Finschaven. The P-40s, P-39s, and some of the A-20s which couldn't reach Finschaven would keep Lae airdrome beaten down.

A reconnaissance airplane flying in terrible weather reported seven vessels, three freighters, one transport, and three unidentified types, all small to medium size, proceeding at eight knots about 100 miles southwest of Rabaul, headed west at eight o'clock that morning. This might have been it, or part of it, but the weather would not permit an attack and we decided to try to keep it under surveillance a while longer before committing our bomber effort. Those vessels were not big enough to carry any force big enough to worry us. It might have been a supply convoy. If so, we would have been ready for it when it got a little closer.

Whitehead and I spent most of the next day inspecting the bomber and fighter units in the Port Moresby area. Everyone sensed that we were getting ready for something big. The crews were working like mad getting every airplane in shape so that we could strike with everything we owned when the time came. We told them that it looked as though a lot of Japs were headed for New Guinea to run us out and that this time we wanted to really take them. Whitehead's staff, headed by Colonel Freddy Smith, had worked out every detail of the operation down to the last item. I didn't make any mistake when I sent that boy Smith up last fall from Brisbane to help Whitey. Some day, not too far away, I intended to get him a star.

Major Ed Larner promised me that his squadron of B-25 commerce destroyers "wouldn't miss." I saw a couple of them practicing on the old wreck on the reef outside Port Moresby. They didn't miss. It was pretty shooting and pretty skip-bombing. I told Ed to warn that cocky gang of his that they were not taking on any high-school team this time. They were playing Notre Dame. Ed grinned and said he would give them my message. We really had something, I was sure. I had a hunch that the Japs were going to get the surprise of their lives.

Whitey ordered a rehearsal of the whole low-altitude, high-altitude, fighter-covered show for the next day, in order to check the split-second timing, and told the commanders that we would then let the planes sit on the ground until it was time to pull the trigger. Our warning system was good for a half hour's advance notice now, but to play safe we decided to keep some fighters up on patrol over the Port Moresby area until the party was over.

Late in the afternoon a reconnaissance plane picked up seven Jap vessels about a hundred miles west of Kavieng, headed west. It looked like the same convoy we sighted the day before, now on the regular course, which passed just south of the Admiralty Islands and then turned south toward Vitiaz Straits. This might have been the convoy our information referred to, but I still doubted it. The weather was too bad to make an attack, but if those Jap ships should have started going through the Vitiaz Straits they would have gotten into better weather and also within range of Ed Larner's B-25s.

On the 28th I returned to Brisbane and spent the afternoon with General MacArthur and the staff on the plan we were to present to the Joint Chiefs of Staff when we went to Washington. The General said he wanted me to help Sutherland and the rest of the staff sell the plan but he also hoped I'd get some airplanes headed out our way. I promised to do both jobs. I told him of our final scheme to knock off the Jap convoy and how the kids were all pepped up over the prospect. He said he thought the Jap was in for a lot of trouble and told me not to forget to keep him informed as fast as any news came in.

The weather was so bad north of Vitiaz Straits all that day that we had no sightings by the reconnaissance planes.

On March 1st, at four o'clock in the afternoon, one of our recco planes picked up a convoy of six Jap destroyers and eight cargo or transport vessels about 150 miles west of Rabaul, headed west. The weather then closed in and we were unable to find the Nip vessels again until eight-fifteen the morning of March 2nd, when a recco plane reported fourteen vessels about 50 miles north of Cape Gloucester, New Britain, headed south toward Vitiaz Straits. Two hours later, twenty-nine heavy bombers hit the convoy.

One large merchant vessel was split in two by direct bomb hits, rolled over, and sank. Another large merchant vessel and one medium-sized cargo ship were reported badly damaged, on fire, and the larger one in a sinking condition. A destroyer was also hit and set on fire. At eleven o'clock, when the bombers left the scene, the convoy had proceeded, leaving behind the two burning vessels which had stopped. Two destroyers were picking up survivors of the vessel that had sunk.

During the attack, our bombers were intercepted by thirty Jap fighters, three of which were shot down. Ten of our bombers were holed, but all returned to Port Moresby.

About three-thirty that afternoon the convoy had been augmented, as our recco plane reported that it now consisted of eight light cruisers or destroyers and nine merchant vessels. At five-thirty, two of the Jap destroyers left the convoy and headed south at high speed, but an hour later when nine B-17s made their attack, the convoy consisted of six cruisers or destroyers and there were now ten merchant vessels. At least five and probably seven vessels had been added to the original convoy. The nine B-17s got two direct hits amidships on a medium-sized vessel, which was left sinking. Another large merchant vessel was stopped dead in the water after two 1000-pound bombs had gone off close alongside. A medium-sized cargo vessel was badly damaged by two very close misses of 1000-pounders which lifted the stem clean out of water. During the attack, between fifteen and eighteen Jap fighters intercepted. One was shot down. Four of our bombers were holed but all nine returned to their base.

In the meantime, one of our reconnaissance planes bombed and sank a small transport in Wide Bay, about 50 miles south of Rabaul. We had not done too badly in the opening long-range skirmish, in spite of the weather and the Jap fighters.

Just before daybreak on March 3rd the recco boys reported eight cruisers or destroyers escorting seven merchant vessels, headed south through Vitiaz Straits. The two destroyers that had left the convoy the afternoon before had probably carried survivors of the sunken ship to Finschaven or Lae and had now rejoined the convoy. At eight-thirty, one destroyer low in the water was seen about forty miles north of the Jap convoy, headed back toward Rabaul.

At ten o'clock the big brawl began about 50 miles southeast of Finschaven, right where we had planned it.

Eighteen heavy bombers and twenty medium bombers attacked from 7000-feet altitude. As the last bombs were dropped, thirteen Australian Beaufighters swept in at deck height, strafing the whole length of the convoy, as Ed Larner with twelve of my new B-25 commerce destroyers skip-bombed, followed by twelve A-20s light bombers, also down "on the deck." Sixteen P-38s provided the top cover.

Ed Larner's squadron dropped 37 500-pound bombs, scoring 17 direct hits, and the A-20s, which also skip-bombed, scored 11 direct hits out of the 20 100-pounders they let go.

Twenty minutes from the time the attack started, the battle was just about over. Every Jap merchant vessel was sunk, sinking, or so badly damaged that it was certain they would never reach land. One of the destroyers had been sunk and three others were in bad shape from direct skip-bombing hits. All had been heavily strafed.

Our formation was intercepted by thirty Jap fighters. One of our B-17s, piloted by Lieutenant Moore, was hit and set on fire as he was coming in on his bombing run. With one wing enveloped in flames, Moore flew steadily on to the bomb-release point, but just as the bombs came out of the bomb-bay, the B-17 lost its wing and plunged into the ocean. Seven of the nine-man crew bailed out, but about ten Jap fighters dove down and shot all seven as they were hanging in their parachutes. The P-38s were heavily engaged up above with the Jap fighters, trying to keep them off the bombers. Captain Ferrault and two wingmen, Lieutenant Easton and Lieutenant Schifflett, pulled out of combat and went down after the Nips that were shooting up the B-17 crew. The three P-38s were shot down and we lost all three pilots, but they took five Japs along with them. The final result of the air combat was that we lost a B-17 and three P-38s, with a total of twelve men, and a gunner was killed on one of Ed Larner's B-25s which was so badly shot up that it collapsed on landing at the home airdrome, luckily with no one else hurt. Of the thirty Jap fighters that had intercepted, twenty-two were definitely destroyed, two probably destroyed, and four damaged.

Thirty-eight of our fighters in the meantime attacked the Jap airdrome at Lae, getting six more Japs definitely and three probables. A small cargo vessel in Lae Harbor was attacked, set on fire, and forced to beach where it burned to the water's edge. We had no losses.

At three in the afternoon, the second act of the destruction of the Jap convoy took place.

All squadron reports agree that, by this time, the convoy consisted of two destroyers and either four or five burning merchant vessels, all stationary. One of the destroyers was very low in the water but the other was circling around at high speed. Several miles to the north, a third destroyer was headed back toward Rabaul.

Sixteen heavy bombers, twelve mediums, ten of the B-25s, still led by Ed Larner, and five Australian light bombers, all covered by eleven P-38s, made up the attacking force.

When the attack was over, three Jap merchant vessels were still afloat but all were dead in the water, on fire from stem to stern and sinking fast. One large destroyer was stationary and on fire. One small destroyer was stationary and low in the water, with the stern awash and sinking.

The next morning all that remained of the convoy was one destroyer, stationary and burned out. One of Larner's commerce destroyers skipped a bomb into the Jap vessel and sank it. The Battle of the Bismarck Sea was over.

Ten Jap fighters were definitely shot down and nine others probably destroyed in air combat during the day and six more destroyed on the ground at Lae. All our aircraft returned home. One B-17, three Beaufighters, and one P-38 were damaged. There were no casualties to our crewmen.

A recapitulation of the results of the Battle of the Bismarck Sea showed:

Japanese losses:
- 6 destroyers, or light cruisers, sunk
- 2 destroyers, or light cruisers, damaged
- 11 to 14 merchant vessels sunk (whatever the number actually in the convoy, all were sunk)
- 1 merchant vessel destroyed in Wide Bay
- 1 merchant vessel destroyed in Lae Harbor 60 aircraft definitely destroyed 25 aircraft probably destroyed 10 aircraft damaged
  The Jap personnel losses were hard to estimate, but they might have been as high as 15,000.

Our losses:
- 13 men killed
- 12 men wounded 4 airplanes shot down
- 2 airplanes crash-landed at the home base.

On the afternoon of March 4th, after the horse had been stolen from the barn, one hundred Jap airplanes, mostly fighters, strafed the Buna area and went back to Rabaul. There was no interception, as everything we owned was out finishing off the convoy or attacking the Lae airdrome. The Japs fired a lot of ammunition but did practically no damage. It was a good thing that the Nip air commander was stupid. Those hundred airplanes would have made our job awfully hard if they had taken part in the big fight over the convoy on March 3rd.

At three o'clock on the morning of the 4th, I woke up General MacArthur to give him the final score which had just come in from Whitehead. I had never seen him so jubilant. I told him I was now ready to go to Washington and planned on leaving at six o'clock that morning. He said okay, made some nice remarks about me, and started outlining his communique. He told me to pass along a radio message which he addressed to me:

"Please extend to all ranks my gratitude and felicitations on the magnificent victory which has been achieved. It cannot fail to go down in history as one of the most complete and annihilating combats of all time. My pride and satisfaction in you all is boundless.

MacArthur."

I forwarded his message to Whitehead with one of my own:

"Congratulations on that stupendous success. Air Power has written some important history in the past three days. Tell the whole gang that I am so proud of them I am about to blow a fuze.

Kenney."

8. FIRST TRIP TO WASHINGTON
March, 1943

At six o'clock on the morning of March 4th I left Brisbane en route to Washington. With me on the plane were General Sutherland, General Chamberlin, his assistant for Operations (G-3), Colonel Larry Lehrbas, a former Associated Press man now in General MacArthur's public-relations section, Major Godman, Sutherland's aide, and my aide, Captain Kip Chase.

We stopped at Nandi, Fiji Islands, overnight. General MacArthur's communique was on the radio. No one believed it until I showed them some of the operations reports and told them the story.

We left Nandi on the morning of the 5th but as we crossed the International Date Line on the way, we landed at Canton Island on the afternoon of the 4th and at Hickam Field, Oahu, on the 5th. Admiral Nimitz, the boss Navy man in the Pacific,

General Delos Emmons, the Army Commander of the Hawaiian Department, General Richardson, commanding the ground forces in Hawaii, and General Willis Hale, the Air Force commander there, and a number of lesser lights greeted us as we landed. The first thing they wanted to know was the real story about the Bismarck Sea Battle. Everyone seemed skeptical at first, but when I showed them the operations reports it was easy to convince them that a truly spectacular victory had been won. Tokio Radio had helped by announcing that the Japs lost seven warships during the three-day engagement. They hadn't seen fit to discuss their losses in merchant shipping but they spoke of a tremendous air victory over me. According to Tokio, the Japs shot down more airplanes than I had in New Guinea at the time.

We stayed in Hawaii for a couple of days and arrived in San Francisco on the 8th. The whole of San Francisco Bay was shrouded in fog, but after flying around on top for two hours, the fog lifted at the Municipal Airport enough to let us in. We registered at the St. Francis as "Captain Chase and party," as our orders were marked secret and I asked Dan London, an old friend of mine and the manager of the hotel, to keep our names from getting to the press. Dan threw a small party for me and told me that his chef, a French boy named Georges Raymond, had been drafted and was on duty somewhere in Hawaii. London said Raymond was not cooking and he didn't think the Army even knew he was a cook. I made a note to see what I could do on the return trip to grab hold of him. I needed a good cook badly if I was to last through the Pacific war. I knew from eating many meals at the St. Francis that this lad was a superior chef.

The following day we landed at Wright Field, Dayton, Ohio, where Generals "Tony" Frank, Hugh Knerr, and K. B. Wolfe greeted me. We stayed overnight and the next day, with my wife Alice and daughter Julia on board, we hopped over to Washington.

General Arnold and all the air crowd gave me a big greeting. The papers were all full of the Bismarck Sea story, as it was the best news to come out of the Pacific for some time. After reading a few of the stories, I almost began to believe I was as big a shot as the papers said.

For the next two weeks we met practically every day with the Joint Chiefs of Staff or their representatives, presenting our plans for the next phase, listening to presentations of the South and Central Pacific Theater commanders, trying to coordinate all the various schemes and ideas and get aircraft, shipping, fuel, munitions, and troops assigned to us to carry on the war.

General Marshall (Army), General Arnold (Air), Admiral King (Navy), and Admiral Leahy, the Chief of Staff for the President, were the Joint Chiefs who listened to our stories and then turned the job over to the next level, the working members. Admiral Cook (Navy), General Al Wedemeyer (Army), and General Stratemeyer (Air), with batteries of assistants, consultants, and advisers, were the ones we took up the details with.

The Army had an expert to tell us how tight manpower was and that, with the necessity to defeat Germany first, it was going to be extremely difficult to give us any troops. The Navy expert pointed out that the German submarines were sinking shipping about as fast as it was being built and, accordingly, vessels to transport troops or supplies to the Far East were scarcer than hen's teeth. The Air experts were the most pessimistic of all. They warned us that they had monthly commitments in Europe already that exceeded the factory output and that it was simply impossible "to pay dividends out of a deficit."

In the meantime, the representatives of both the South and Central Pacific Areas were watching the situation very closely to see that, if anything was passed out, they would get their share of it before MacArthur's crowd from the Southwest Pacific grabbed it off.

The conferees listened to our plan to move west and north along the New Guinea coast, capture Lae, and then continue on, seizing both sides of the vital Vitiaz Straits. When the SOPAC forces had advanced north through the Solomon Islands and had gained control of Bougainville Island, we would then be in position to make a two-pronged attack to capture Rabaul.

The conferees were dubious about our ability to supply our forces in the Lae area, as it was doubtful if sufficient shipping could be made available for that purpose. Sutherland made a point of the possibilities of using the road across the Owen Stanley Mountains that the Aussies were then working on, which was to run from a point on the south coast of New Guinea about eighty miles northwest of Port Moresby to the Wau area. My personal opinion was that this road would never be finished in time for this war. I didn't tell the conferees that but assured them that if necessary I could handle the whole supply problem by air lift. The existing drain of air supply into the Buna area would soon be over, and after Lae and Salamaua were taken, we wouldn't need to maintain garrisons in the Wau area by air supply as we were then doing.

In spite of the demonstration of what could be done along this line with a handful of cargo airplanes, which we had just shown them during the Buna campaign, neither the Army nor the Navy seemed to appreciate the real capabilities of air lift or how flexible a means of supply it was. They listened, but I don't believe they were entirely sold. This business had to be seen to be believed. Of course, I would have to get some more cargo airplanes out of Arnold before I could do much demonstrating.

Mr. Stimson, the Secretary of War, Judge Patterson, the Assistant Secretary of War, and Bob Lovett, the Assistant Secretary for Air, wanted to know all about the show in New Guinea and were most sympathetic about helping me out, but in the final analysis, Hap Arnold would have to give me the airplanes and the crews to fly and maintain them. He in turn had to deal with the Joint Chiefs of Staff of the United States, whose policies were decidedly influenced by the Combined Chiefs of Staff of the United States and Great Britain, and they in turn heard from and listened to President Roosevelt and Winston Churchill. Every once in a while the Russians put their demands in the pot, and the rest of the Allies were afraid the Russians would be knocked out of the war if they didn't get what they had asked for. No wonder Napoleon once said he would rather fight against allies than against any single opponent.

Sometimes our discussions became quite heated. The location of the boundary line between SWPA and SOPAC was a source of almost endless discussion in and outside the conference room. SOPAC wanted to move it west and SWPA representatives looked upon that as not only an infringement but an attempt to keep SOPAC in business. Along with increased areas of responsibility, of course, extra means to carry out the job would have to be provided. The bigger your area, the more excuse you had for getting a bigger cut of the manpower, shipping, aircraft, and supplies that were available for assignment to the Pacific.

The whole business of rigid and fixed boundary lines, particularly between SOPAC and SWPA, seemed silly to me. There should have been one Supreme Commander out there with an Army, a Navy, and an Air man working for him. Those conferences would certainly have gotten along faster if we had had such an organization. Decisions by the Joint Chiefs of Staff were difficult, if not impossible; they tended to back up their assistants, who themselves often could not agree and who in such cases passed the decision back to the Joint Chiefs themselves. I hoped that some day we would have sense enough to get ourselves a single Department of National Defense, with sub-branches for Army, Navy, and Air.

Everyone was really stubborn about giving me airplanes, or even replacements for my losses. I warned them that if they didn't keep me going, we would be run out of New Guinea, as supply of our troops there was impossible unless we maintained at least local control of the air. I suggested that maybe they might let me have ten per cent of the aircraft factory output and let the rest of the war have the remainder. The answer was still No. The European show did not like the B-24, or the P-38, or the P-47 Republic Aircraft fighter. They preferred the B-17 as a bomber and the P-51 Mustang as a fighter. I told Arnold I was not that particular. All I wanted was something that would fly. I thought that maybe I could get somewhere with this argument.

First Trip to Washington

Before I left Australia, I had mailed Arnold a set of drawings showing the changes I had made in the B-25 to make it a skip-bombing, strafing commerce-destroyer. I asked him to have Dutch Kindelberger, who ran the North American factory at Los Angeles and produced the B-25, make these changes in his production and save me the burden that I was forcing on Bertrandais' Townsville depot.

One day, during a lull in the conferences, Arnold told me to come to his office. On arrival there I found a battery of engineering experts from Wright Field who explained to me that the idea was impracticable. They tried to prove to me that the balance would be all messed up, the airplane would be too heavy, would not fly properly, and so on.

I listened for a while and then mentioned that twelve B-25s fixed up in this manner had played a rather important part in the Battle of the Bismarck Sea and that I was remodeling sixty more B-25s right now at Townsville. Arnold glared at his engineering experts and practically ran them out of the office. He told me to send a wire to Pappy Gunn, my uninhibited experimenter, to come back to Los Angeles, report to Dutch Kindelberger, and show him how to do the job.

At dinner one evening with Charlie Wilson and some of his war-production people, I met Arde Bulova, the watch man. I told them a lot of stories of the show in New Guinea and, after dinner, Bulova said he would like to give me a hundred watches to give to the pilots for shooting down Japs. I told him I would take the watches but that I awarded decorations for air victories. I suggested that I award the watches to the hundred best ground-crew chiefs in the Fifth Air Force, the men who kept the airplanes in shape so that the combat crews could earn medals for shooting down Japs and sinking ships. He thought it was a fine idea. The next morning the hundred Bulova watches were delivered to me at the hotel.

On March 17, I was invited to pay a visit to the White House. I talked for some time with President Roosevelt, who wanted to hear the whole story of the war in our theater in detail as well as a blow-by-blow description of the Bismarck Sea Battle. I found the President surprisingly familiar with the geography of the Pacific, which made it quite easy to talk with him about the war out there. He wanted to know how I was making out on getting airplanes. I told him that so far my chances didn't look very good. When he asked why, I said that among other reasons given me was that he had made so many commitments elsewhere that there were no planes left to give me. The President laughed and said he guessed he'd have to look into the matter and see if a few couldn't be found somewhere that might be sent me. He said that if anybody was a winner, he should be given a chance to keep on winning.

He asked about General MacArthur's health and asked me to be sure to remember him to the General. Although there was a lot of speculation in the United States about MacArthur as a political opponent next year, the President did not mention the subject once during our conversation. MacArthur had told me several times that he was only interested in winning the war and was not in the market for any political office. He had already made public statements to that effect but people did not seem to believe them. I kept telling everyone that if they would leave him alone, he would win that war for them. I believe that the President knew that General MacArthur was not going to oppose him if he did run again. He certainly seemed to admire what the General was doing and said so emphatically several times during the conversation. By and large I enjoyed the meeting very much. Furthermore, I got a hunch that I was going to get some airplanes.

The Joint Chiefs conferees finally began to show signs of loosening up a little. It looked as though we were going to get a little shipping and some ground forces, both combat troops and service troops, especially more engineers that we needed desperately to build airdromes and roads.

Arnold called me to his office on the 22nd and told me that he had gone over the whole picture and had squeezed everything dry to give me some help. He said I was to get a new heavy-bombardment group, two and a half medium groups, and three more fighter groups. One of these fighter groups would be equipped with P-47s, which no one else wanted. I said I'd take them. The other fighter group would have to be manned by me. Hap said he would give me P-38s to equip it if I could furnish the pilots and the mechanics. I said to give me the planes and I'd find the men if I had to dissolve my own headquarters staff to get the people to fly them. He also promised me another troop-carrier group of fifty-two C-47s, some depot and service outfits, and other odds and ends to balance the organization. Altogether it looked as though, by the end of 1943, I'd have about five hundred more aircraft than I had when I left Australia.

On the 25th I went to the White House again, this time to be present when the President gave Kenneth Walker, Jr., the Congressional Medal of Honor awarded posthumously to the youngster's father, Brigadier General Kenneth Walker.

The kindly, fatherly way the President handled the situation, putting young Walker, who was about seventeen, at ease and sending him away with his eyes shining with pride over the nice things F.D.R. had said about his father, was something to watch and listen to. The Roosevelt

First Trip to Washington

charm was no myth. He did a swell job. I could understand a little better why they kept on electing him President.

After the ceremony, the President asked me to stay for a while to talk about the airplane situation. I told him about the latest promises I had been given. He asked me if I was satisfied. I told him that if I got a million more planes, I'd probably want more but that I was extremely grateful and glad to get what I had and was quite sure we would get real dividends out of them. He laughed and said, "I'll be watching the reports as soon as you get them out in the Pacific."

When I returned to Arnold's office, I learned that Brigadier General Howard Ramey, the head of my 5th Bomber Command and Walker's successor, was missing. I asked Hap to give me Colonel "Big Jim" Davies to help me out. Davies was one of the old Philippine and Java veterans that I had sent home the previous fall to rest. I had seen him in the hall that morning and Jim had asked me if there was any chance of getting back with the Fifth Air Force. He looked in the pink of condition and anxious to go to war again. Arnold said I could have him.

Jim would be a big help. Men liked him, and his outfits had always had a good record. That bomber command needed some good leaders. Their losses were higher than in any other combat branch of the Air Force and when a bomber went down, it took a big crew along with it.

Colonel Koon, then commanding the 90th Group, was good but he was tiring fast. Colonel Roger Ramey of the 43rd, my other heavy group, looked good, meant business, and was a likable leader. I decided that after Davies had been back in New Guinea for a while, I'd size him up in comparison with Ramey for the job of bomber commander. Then I'd make one of the two the commander and the other his chief of staff. In the meantime, I had wired Whitehead to hold the job himself until I got back to discuss the matter with him.

On the 28th, the conference ended. Our plan (Elkton) was approved with a few minor changes. It looked as though it would be late summer or early fall before we got the troops and equipment to start rolling into Lae. SOPAC was to keep pushing ahead in the Solomons and would be landing on Bougainville Island, probably around November 1943. MacArthur and Halsey were to coordinate their operations with each other, but MacArthur was to have "strategic direction" of the efforts of both SWPA and SOPAC. That was by no means the same as having command nor did it wash out the silly boundary line between the two theaters, but it put the timing of the operations in MacArthur's hands, which was quite important.

I was glad it was over. Even though I was going back with a lot of airplanes, I had been away so long that I felt as though I'd been playing hookey from school.

From the reports that had come in from the Pacific, Whitehead had been doing a superb job while I had been gone. The Japs had been throwing a lot of air attacks at him but Whitey had been on the job and had made them pay heavily for the really trifling damage they had inflicted. Wewak was building up rapidly but our continuous attacks had cost the Nip some precious shipping. In addition to his other activities, Whitey had been supporting an Australian drive toward Salamaua from Wau and pounding the Jap airdromes around Rabaul. With a small staff and dwindling numbers of airplanes, I expected he would need a rest by the time I got back. I'd have to take care of that lad. He was my strong right arm. I considered him the greatest operator in the business. When I gave Whitey a mission and he once said, "Okay," I could relax and quit worrying. The job would not only be done but it would be done better than anyone else could do it.

On March 29th I left Washington and, taking Alice and Julia with me, flew to Wright Field, Dayton, Ohio, to confer with the supply people there who were responsible for keeping me going with the engines, propellers, and other bits and pieces that airplanes had to have to operate. General Tony Frank, a grand guy I had been associated with for years, was running the show at Dayton and he had the right idea. His slogan was, "If any of those combat people ask for anything, any one of you can say 'Yes,' but remember that I am the only one that can say 'No.'"

On March 31st we flew to San Francisco, where we spent a day making last-minute purchases, and we pushed off on April Fool's Day for Hawaii, arriving there early on the 2nd.

As soon as we landed, I told Chase, my aide, to locate that chef of Dan London's right away. Chase found that he was over on the island of Maui with a coast-artillery unit and, as far as he could find out, they didn't know he was a cook. With this information as a background, the plot unfolded to the effect that I had known Private Georges Raymond's family for years and Georges himself ever since he was a child. For purely sentimental reasons I wanted to be able to tell his parents from time to time that I was taking good care of their son and that he was well and happy. In other words, I would certainly appreciate it if he could be flown over from Maui in time for me to take him along to Australia with me or ship him out to Brisbane on the next plane, tagged for delivery to me. As it would leave the Hawaiian Command short one private, I was willing to have them sidetrack the next one who came through on the way to Australia, although I didn't expect that they would take such an advantage of me. After all, he was just a plain private and surely his loss would not detract very much from their war effort. The story worked. As we were leaving the next morning, Raymond could not be flown over in time to get aboard my plane, but Emmons promised that he would have the lad follow me to Australia as soon as there was a vacant seat on a plane headed that way.

From Hawaii we flew to Palmyra Island, then to Nandi in the Fiji Islands, where we gassed up, and then on to Brisbane landing about noon on April 6th, ready to go back to work.

9. SLUGGING MATCH: DOBODURA
April—May, 1943

ON ARRIVAL in Brisbane, I learned that Whitehead really had been going to town during the past few days attacking Jap shipping in Kavieng. The B-17s had been skip-bombing the Nip vessels right in the harbor itself in broad daylight and, without losses to themselves, had sunk or badly damaged a heavy cruiser, two light cruisers, three destroyers and two transports. General MacArthur had sent Whitey a radio:

"The heavies did a swell job. Congratulations. MacArthur." I reported the results of our Washington trip to the General and told him that while we would be pinched for aircraft for a few months yet, it looked as though by August we would be ready to go places. MacArthur looked fine. The good news from our mission and a couple of months' rest from the anxiety of the Buna campaign had combined to make him look younger and more alert than ever.

During the afternoon a radio came in from General Delos Emmons, saying that "the transfer of Private Georges Raymond now on Maui" to me had "high priority air transportation." It must have been "High," for just as I had finished reading the message, Private Raymond came in and reported. I said, "Good afternoon, Corporal." Raymond looked puzzled and explained that he was only a private. I said, "Here is the key to Flat 13 at Lennon's Hotel. There is a kitchen there. Go to the manager and get dishes and stuff and linens. Then see the Quartermaster people and go to the markets, buy me some food, and be ready to feed me, beginning this evening. Don't worry about what to cook. I eat anything. I like my steaks rare and everything else well done. Don't forget to stop on the way and buy some corporal's stripes. If at the end of the month you are still a good chef, I'll make you a sergeant. If you are not, I'll make you a private again and send you back to Hawaii."

Raymond grinned, saluted, and walked out. That evening the meal was so good that I had to restrain myself from making him a sergeant right away instead of waiting for the end of the month.

On April 10th I flew to Port Moresby and told Whitehead go back to Australia for the rest of the month and relax. He was really tired and needed the rest. Colonel Freddy Smith, who was now commanding the air units on the other side of the Owen Stanleys around Dobodura, came over for a conference during the afternoon. He had the 49th Fighter Group and most of the 3rd Attack Group over there. One field was completed and several others under construction. Freddy was also tired. I hated to let him go but I didn't want him to break down, so I decided to send him home for a couple of months to rest up before he came back to New Guinea. Whitehead left that night for Australia.

During the past week the Japs had brought in a lot of aircraft and crowded every field around Rabaul. The weather had prevented getting complete photographic coverage, but it was estimated that there were at least 150 fighters and 100 bombers in that area. In addition the normal Nip strength around Kavieng and in Bougainville remained about the same. I hoped the Nip intended to try some more convoy runs down our way within reach of my B-25s, but my guess was that he would start a series of big air attacks to see if he could regain air control. The newspapers back home had come out with the story that I had been back for a series of conferences, and all of them hinted rather strongly that I came back to get more airplanes. Maybe the Jap had figured that I got some promises and had decided to try to take me out before my air reinforcements arrived.

The next noon between forty and forty-five Jap fighters and bombers attacked our shipping at Oro Bay, a little port about fifteen miles southeast of Buna which had become the main supply point for the area. One small coastal vessel was sunk and some other minor damage inflicted. We intercepted with fifty fighters and shot down eleven Jap fighters and five bombers. We had no casualties but one of our P-38s crash-landed at Dobodura.

That raid on Oro Bay puzzled me. The Nip could have used at least three times as many airplanes in it and still had a lot left over. He made no attempt to hit our Dobodura airdromes, although his reconnaissance must have told him that we had over a hundred aircraft there. The shipping target at Oro Bay was not big enough to warrant going after. Milne Bay, on the other hand, was full of shipping. On the guess that his next target would be Milne Bay, I ordered practically all my fighter strength concentrated at Dobodura and Milne Bay, leaving only eight P-38s and twelve P-39s at Port Moresby for local defense of the port and the airdromes. Our radar-warning service was pretty good but there were a few blind spots that worried me, for if I ever got caught on the ground, I'd be through.

Our information indicated that the Japs were going to run a convoy of five destroyers and four merchant vessels into Hansa Bay, sometime during the next two or three days. Our reconnaissance planes looked for them all day, but the weather made searching difficult. I decided that if by midnight that night they didn't find that convoy, I'd send some bombers against the Rabaul airdromes to cut down on the size of any other raids the Nips might try on me.

Shortly after midnight I sent nine B-17s off to hit Vunakanau, Lakunai, and a new field at Rapopo, in that order. On account of the continued bad weather, only four got through. They all reported starting a lot of fires but could give no assessment of damage.

About three o'clock on the morning of the 12th, a recco plane picked up two large and two medium-sized merchant vessels escorted by four destroyers, coming into Hansa Bay. We had eighteen bombers loaded and ready to go so I ordered them off. Eleven cleared but the next bomber, flown by Major Kenneth McCullar, barely got off the ground and then nosed over, hooked a wing, and burst into flames as it crashed. We lost McCullar and the whole crew. The bombs went off from the heat and the runway was blocked for the rest of the bomber take-offs. The story was that just as Ken's B-17 was ready to leave the ground it collided with a kangaroo or a wallaby, which hit the turbo-supercharger exhaust, setting fire to the main gasoline tanks in the wing. Whether or not this was the real explanation, McCullar's loss was a tough blow for his squadron, the 43rd Group, and the whole Fifth Air Force. His courage, his ability, and his accomplishments had become an inspiration to the rest of us. His exploits were already legends that would be told and retold long after this war was over.

The bombers found the Jap convoy at Hansa Bay. The two large merchant vessels and one of the smaller ships were left in flames and one of the destroyers was hit and left listing badly. The mission was heavily intercepted by Jap fighters, six of which were shot down. All our planes returned home.

About nine in the morning we got a radar pickup on a big Jap formation coming from the direction of Rabaul, about a hundred miles from Milne Bay and headed that way. Every fighter at Dobodura and Milne Bay was ordered into the air for an interception, which I expected to take place over Goodenough Island. Then the radar lost the Nips due to a skip or some other interference and, the next thing I knew, in came a report that over a hundred Jap planes were crossing the Owen Stanleys at Kokoda. We immediately ordered the eight P-38s and the twelve P-39s at Port Moresby to intercept and radioed the rest of the fighters, now on a wild-goose chase toward Goodenough Island, to head toward Lae, where we hoped the Nips would land to refuel after their attack on Port Moresby.

The Jap planes came in sight of my headquarters at ten-twenty. Forty-five of them were bombers, flying in one excellent mass formation of twenty-seven, followed at a distance of about half a mile by another group of eighteen, while above them were between sixty and seventy fighters for protection. Their altitude was between 18,000 and 20,000 feet. The P-38s, flying practically abreast, took on the leading bomber formation in a head-on pass and shot down three of them right away. The bombers immediately broke formation, shed their bombs, and started a big left turn, heading back toward Lae. The P-38s then knocked down two more of the second Jap bomber formation, which held together until it passed Laloki airdrome, where it started bombing. They began to break, unloading bombs indiscriminately, and followed the first outfit. In the meantime, the twelve P-39s were tangled with so many Nip fighters that it was impossible to follow the combat. Every once in a while an airplane would burn and start spinning down and I'd hope it wasn't one of mine. In a few minutes, however, the show at Port Moresby was over and we could guess whether or not the gang on the other side of the range could catch the Nips on the way back. A lot of the fighters from Dobodura were too low on gas but ten P-38s and twenty-two P-39s caught up with the retiring Japs. Although they didn't have enough fuel to stay with them very long, they did pretty well. The Japs lost twelve bombers and ten fighters definitely destroyed, and four more bombers and three fighters were listed as probables. The antiaircraft gunners at Port Moresby claimed three Jap bombers, but I think one of them was a B-17 recco plane which blundered home in the middle of the fight, got shot at by our antiaircraft guns, ducked behind a mountain until the shooting was over, and then came in and landed. The pilot told everyone who would listen that he was going to invite some of those "ack-ack" boys to go with him on his next trip to Rabaul, where they would find out how good ack-ack can be. General Van Valkenberg, the antiaircraft commander, didn't think it was a joke and told me so. I told him I didn't think it was a joke for his gunners to shoot at a B-17, either. There were no further comments.

We didn't get off scot-free, by any means. In the air combat, we still had a P-39 missing. Another was shot down in the fight over Port Moresby, but the pilot bailed out and landed alongside an Aussie jeep about ten miles from his own airdrome. He was back home in half an hour. At Laloki strip we lost two B-25s, with two more out of commission for two weeks and two others slightly damaged. One Aussie Beaufighter was destroyed and another badly damaged. Two A-20s and two C-47 cargo planes would also be out of action for a couple of weeks. A lucky hit for the Nip was a random bomb that hit an Australian Army motor-fuel dump and the Aussies were now short about 5000 drums of gas.

I got badly fooled and was lucky to get out of it as well as I did. I still couldn't figure why the Nips didn't hit Milne Bay and the shipping there. They ought to have known that at ten o'clock in the morning we wouldn't have many planes on the ground and that our warning service was good enough to get our fighters up to intercept them. I gambled that the following day they would hit Milne Bay with another big show if the weather was right. However, I dared not skin the Port Moresby defense so thin this time, as the fool Japs might try to repeat the attack. What really burned me up was that, if I had guessed right, I would have had nearly a hundred fighters take on that Jap show and we would have made a killing worth writing home about.

After the fight, the 49th Fighter Group added up their private score toward that magnum of brandy. Lieutenant Grover Fanning was credited with shooting down Number 200. The kids were saying that it wouldn't be long now, if the Japs would just keep sending over shows like the one they had just tangled with. Only three hundred to go.

In the evening I attended the premiere of "Hells a Papuan," the soldier show that the kids, assisted by Red Cross girls coaching, had been rehearsing for the past six weeks. The show was put on just outside my headquarters on an outdoor stage. It was a knockout. It was scheduled to run for five nights, and I believe would have run indefinitely if it could have been put on in New York. The orchestra was all name-band players, including the leader, who complained that between the rehearsals and shows he wasn't giving enough time to the P-38 he was a mechanic on. The singing was excellent and while some of the lyrics might have had to be cleaned up in some towns back home, they were really clever. The scenery was a tribute to improvisation and it would not have had to be changed for any stage anywhere. The show was so good that I saw it again the next night and enjoyed it even more than the first time.

We waited around all the next day for the Nip to repeat his attack but the weather was too bad for either of us to do any flying. I wanted to have the bombers take another shot at the convoy which was still in Hansa Bay, but every reconnaissance flight that we sent out reported such lousy flying conditions north of the mountains that I called it off.

On the 14th the Nip pulled another big show. This time I guessed right, but the Japs had a lucky break on the weather, as Dobodura was fog-bound and sixty of my fighters there were held on the ground while the raid was on. The attack was made by thirty-seven bombers, twenty-five dive-bombers, and thirty fighters, and was aimed entirely at the shipping in Milne Bay. They hit three vessels but only one small one was seriously damaged.

We intercepted with eight P-38s and thirty-six P-40s and got fourteen Jap bombers and three fighters confirmed as destroyed, with five more bombers and one fighter listed as probables.

We lost one P-40 missing, and a P-38 crashed on landing. Four other P-40s were pretty badly shot up. The pilot of the missing P-40 was our only casualty.

Four of our B-17s managed to get through to Hansa Bay, where they made three skip-bombing hits on a Jap cargo vessel and sent it to the bottom. The cargo had probably been unloaded but that was one more vessel the Nips couldn't use again.

The next day I hopped over to Dobodura to inspect the units there and make sure that Freddy Smith was packing up to go home. Even after you told some of these rascals that they had been ordered back for a rest, you had to practically load them on board an outgoing airplane to make sure. Freddy knew he was tired and was reconciled to going home but he seemed a bit leisurely about getting started.

The flat plain of the Dobodura area was bustling with activity. The engineers were building a road from Oro Bay to our main airdrome at "Dobo" itself, with branch roads out to five other fields that were under construction. An unbelievable amount of materials had been flown over the Hump, including 3000 tons of steel mat and building materials for both groups to construct a medium-sized village apiece. The personnel were in tents and quite well taken care of by a host of native "orderlies," who wandered around keeping the place picked up and doing odd jobs. These natives, the Orikivi, who joined the Japs when they first landed at Buna, were supposed to be a tough outfit. If the Nips had treated them properly we might have had trouble with them, but after the Japs had put them on forced labor, looted their villages, and assaulted their women, the Orikivi decided that they had made a mistake and had been on our side ever since.

In order to pay them we resorted to all kinds of schemes, as we had to do everywhere in New Guinea, where the currency problem was complicated by the fact that almost anything was used for that purpose except money.

The natives worked around the camps, did our laundry, sprayed oil to keep down mosquitoes, worked on the landing fields, and helped out a lot in guiding our crews home when they were forced to parachute over the jungle. Sometimes this latter service involved feeding and even carrying them for weeks if they were disabled. Some way had to be figured out to reward them. Around Port Moresby or along the coast where they had been in contact with the white man for years, they were generally paid in trade goods such as tobacco, hatchets, knives, cloth, canned food or candy, and trinkets. Then we found that used safety-razor blades were highly prized, so everyone started carrying a few packages in the airman's emergency kit. Cheap costume jewelry, like the brightly colored pins the girls back home wear and discard after a few times, got a big play as soon as we tried this method. It was still good, although the market had become a little oversold. Up in the Bena Bena plateau country, however, while trade goods were nice things to have along, the best currency of all was what was known as a gold-lipped shell, which could be procured for almost nothing down along the south coast. It got its name from the bright gold coloring that extended all around the inner edge of the lip of the shell. Around Bena-Bena one of these shells would buy a woman, and two of them represented the standard market value of a full-grown pig.

Some time ago one of our lads, who said he knew something about farming, came back from there proposing that we start a big vegetable garden at Bena Bena, as the soil was exceptionally rich and the climate much better for vegetables than around Port Moresby. We had tried to start gardens around our camps, but the soil was too wet when it rained and too dry when it didn't rain and if anything did come up the bugs snipped the little shoots off as soon as they appeared.

The Bena Bena idea looked promising enough to try it, so we got Golden Bantam and Country Gentleman corn seed, young tomato plants, and other garden seeds and put about two hundred natives to work. The project was a huge success from the start and we were now flying around 5000 pounds of fresh vegetables in to our camps every day. As a native would work a week around the garden for one of those gold-lipped shells, we flew in nearly a ton of the stuff so that we would have enough to meet the weekly payrolls and handle emergency obligations for some time to come. The Australian commissioner, however, was so worried about the possibility of inflation if this hoard were dispersed too fast or stolen that he protested violently when he found that we had flown it in. We mollified him, however, by leaving it in his care, where it would be under guard, and just drawing on our "account" as we needed it.

The Orikivi boys all seemed to want GI undershirts but, as the supply was limited, they were being equipped rather slowly. As soon as they got one, however, a GI haircut for some reason replaced the standard fuzzy-wuzzy style, and when occasionally the boy acquired a pair of khaki trousers, he was hard to tell from some of the shorter members of our Negro engineer battalions, until you got close enough to see the tribal markings tattooed on his face. One of the aftereffects of the war will probably be the establishment of a good market in New Guinea for some trader enterprising enough to bring in a load of white undershirts and GI khaki trousers to exchange for copra or labor on the coconut and rubber plantations.

Wewak was being built up fast. Airdrome construction was being pushed, trails widened into roads, and bridges built over the streams. Antiaircraft gunfire was becoming intense and the Nip fighters met every daylight raid that we attempted.

On the 15th, three destroyers and six cargo vessels were sighted in the Wewak Harbor. I ordered the old reliable 63rd Squadron out just after dark, when the weather began to clear, and the lads did another sweet night skip-bombing job. One large vessel was sunk, one left sinking, and a third on fire. In addition, a small destroyer or gunboat was hit amidships and sent to the bottom.

The next day I lined up the 43rd Group and decorated everyone who had a medal coming to him. I told them that I was abdicating the job as temporary Bomber Commander in favor of Colonel Roger Ramey and that Colonel Jim Davies would be his Chief of Staff. Both were extremely popular, but as far as I could tell the Bomber Command was quite satisfied with my choice. Everyone had known for some time that I was weighing these two officers but until I made the announcement no one had known which one I would pick. Davies, of course, was disappointed, but I was sure he would work loyally just as Ramey would, if the choice had been the other way. They were that kind.

It was apparent that the Jap had given up any idea of again undertaking the offensive toward the Solomons or against our holdings in Papua. His shipping in Rabaul Harbor was down to thirty vessels. His air strength in the Rabaul area was over 250 planes on the 16th. After four days of bad weather, on the 20th we got some pictures of Lakunai showing thirty-eight fighters where there had been over 100 on the 16th. On the 22nd we got full photo coverage of Vunakanau, Lakunai, and Rapopo airdromes, showing a total of only thirty fighters and fifty bombers. No increase was noted in the Wewak area, so the Rabaul-based planes had probably been flown back to the aircraft carriers. I didn't know what the purpose of the recent raids was, but the small amount of damage inflicted was hardly worth the seventy-five or more aircraft and crews that it cost. The Nip just did not know how to handle air power. Just because he knocked us off on the ground at the beginning of the war, when we were asleep at Pearl Harbor and in the Philippines, he got a reputation for being smart, but the way he had failed to take advantage of his superiority in numbers and position since the first couple of months of the war, was a disgrace to the airman's profession.

As the decoration of the 43rd Group had used up all the decorations I had in Port Moresby, I sent my aide Captain Chase to Brisbane with the courier mail and told him to bring back some more medals. He came back with a report that, in order to save metal for the war effort, some bright boy in Washington had decided not to have any more medals made until the war was over. In the meantime, we could just pin pieces of ribbon on the lads as evidence that a medal would be given them later on.

When I was in Washington the month before, Donald Nelson had asked me to give a talk to his War Production Board staff about the Pacific war. I had done the job for him and figured this was the time to ask him to do something for me. I wrote him as follows:

Dear Donald;
Prior to leaving Washington I hoped to get around to taking advantage of your many offers to help me out, but I had so many things to do at the last minute that I did not make the grade. I really wanted to have a final talk with you to let

you know that I did not do too badly on getting some airplanes allocated out here.

I was glad to get back on the job and see my gang again. It is the same crowd I bragged about to you, except that I think they improved some while I was gone. Nothing for the big headlines has been done this month, but they have been steadily pegging away and have kept me pinning on decorations for the same type of exploits I told you and your staff about.

By the way, I did something for you that evening; how about returning the favor by giving me some information on what may seem unimportant but which really does amount to a lot in the long run? I have just learned that a decision has been made by someone to manufacture no more medals for the duration of the war but to issue ribbons instead. I suppose that the idea is to save metal for other war purposes deemed more essential. During the past eight months I have pinned hundreds of medals on members of the 5th Air Force who have distinguished themselves in operations against the Japs in this theater. If you could see the pride in the faces of these youngsters as they steal a look down at the medals after they get back in ranks, I believe you would feel as I do about this business of substituting ribbons for the real thing. I believe no better use could be made of the small amount of metal involved than to boost the morale of the recipients. That was what medals were intended for in the first place. No Air Medal has been produced yet, so I have to give them ribbons for that decoration. While, of course, they appreciate the ribbons, there is nowhere near the enthusiasm that you see in the case of the D.F.C., the Silver Star, the D.S.C., or the Purple Heart, for which a limited stock of the actual medals still exists. If I knew who was responsible for the decision I would take it up with him direct. Perhaps you know and can let me know whom to deal with. I hope you are not the one, but if you are at least I have told you how I feel about it.

Of course, if the War Department is responsible and made such a decision of their own free will, I won't buck it as—after all—I am really a soldier. However, I cannot imagine their coming to a conclusion of this kind. I think it must have been some conservation agency which did not consider the morale angle.

I feel quite strongly about the matter, for these youngsters of mine don't live too long at best in this game. For approximately three hundred hours of combat flying time, on the average, a man is dead or missing. I would like to have him look at a medal before he dies, and God knows I would like his wife or mother to receive a medal instead of a ribbon when I send his effects back home. The same thing, of course, applies to the other arms of the service. I hate to deprive any of our heroes of this small but tangible evidence that there is some appreciation of their efforts.

Sincerely,
George C. Kenney

Whitehead was on the plane that came up from Australia with Chase. Whitey looked fine and was "raring to go." I spent most of the night going over the situation with him and left for Brisbane the next morning.

During the evening, in General Mac Arthur's flat in Lennon's Hotel, we discussed the outlook for the next two or three months. We were at a low ebb right then as far as any decisive action was concerned. The ground forces were pretty well worn out in the Buna campaign. The two regiments of the American 41st Division that did not see action at Buna were now slowly working west along the north coast of New Guinea, but there wasn't enough strength to go very far. It would be another three or four months before the 7th and 9th Australian Divisions and the American 32nd were recovered from their malaria and recruited and trained for another major effort.

The Allied Air Forces were going downhill fast, and if the replacements and reinforcements I was promised in Washington didn't start coming soon I was going to be in a bad way. I had three groups of fighters, but instead of seventy-five planes per group, I'd have been lucky to get seventy-five planes in the air from all three combined. The old P-39s and P-40s were just about worn out and only fifteen new P-39s were on the schedule to come to me that month. The only other fighters referred to on that schedule were "100 P-47s being prepared for shipment." If nothing happened to delay them, that meant another three months before that new group got into action. My bomber units averaged about two-thirds strength. The Australian fighter and bomber squadrons were slightly worse off than those of the Fifth Air Force. Two squadrons of Spitfire fighters, flown by Australian pilots trained in England, had arrived and were going into action at Darwin. They were a grand lot of lads but I was afraid they were in for a rude awakening. They had already hinted that with the Spitfire they would show us how to shoot down Jap airplanes the way it should be done. They were obsessed with the maneuverability of the Spitfire and had been trained in the maneuver type of combat, or "dogfighting." That was all right in Europe as the Spitfire would outmaneuver anything the Germans had, but it was not good in the Pacific war, where the light Jap fighter outmaneuvered everything we had. I told Bladen, the Darwin commander, to warn them, but I was afraid those cocky rascals would do as they pleased until they learned the same lessons the rest of the fighter crowd have had to learn— the hard way.

I told General MacArthur that according to the present schedule of promises I would get another heavy-bombardment group in late June, a group of B-25s about the same time, and a group of P-38s in July. With that increase and with promised replacements for the worn-out equipment the rest of the units now had, we would be able to support the contemplated ground operations that summer and the capture of Lae and Salamaua in late August or early September. MacArthur was getting some more ground troops and the Navy was strengthening the 7th Fleet with both combat ships and amphibious craft, so that by early fall this year we should be ready to start the drive west along the north coast of New Guinea and get into position for the jump-off for the Philippines in 1944.

Just before I returned from New Guinea, Admiral Halsey came over to Brisbane to confer with General MacArthur. The next SOPAC operation would be to land on Rendova Island, Vanganu, and New Georgia, build new airdromes, and after the seizure of Munda, the main Jap air base on New Georgia, to develop that field into a real bomber and fighter base. This operation was set for June 30 and synchronized with SWPA landings to be made on the same date at Kiriwina and Woodlark Islands, and with an amphibious landing at Nassau Bay, on the New Guinea coast about twenty miles southeast of Salamaua, by part of the 41st Division, which was then to move in on Salamaua in conjunction with the Aussies coming from the Wau area. Development of airdromes at Kiriwina and Woodlark would give me fighter bases from which we could send escorts for the bombers and strafers on big-scale daylight raids on Rabaul. When that time came I could smash Rabaul, make it untenable for both aircraft and shipping, and isolate the place with an air blockade. Then if we wanted to, we could bypass Rabaul and start moving west. At that time we could turn the job of maintaining the blockade over to SOPAC.

I told the General that in order to conserve my air strength for the present, I wanted to carry out the normal reconnaissance missions and only strike in force with my bombers when a good target was presented. Even then, if the range was too great for the fighters to protect the bombers, I would not make any daylight attacks. General MacArthur agreed.

As an indication of how thin my "shoestring" was getting, that afternoon I issued orders moving three fighters from Australia to New Guinea. One of these, a P-40 without guns, rebuilt from parts from three different wrecks, was being used as a flying test-block to check engines being over hauled in Melbourne by the Aussies. Another P-40, also unarmed and salvaged from the scrap pile, was being used by one of my representatives in Sydney coordinating contracts for me with Australian industry. The third plane was a P-39 in Brisbane with a broken landing gear. I gave orders to fix up all three, put the guns on them, and send them up to General Squeeze Wurtsmith as reinforcements. Those three fighters were the last three left in Australia. That scraped the bottom of the barrel.

I got another shock on May 1st, when a radio from New Guinea announced that Major Ed Larner had been killed the afternoon before, landing at Dobodura. It was another unexplained accident. He was turning into the strip to land when suddenly his B-25 fell off on a wing, crashed, and burned, killing Larner and his whole crew, the same youngsters who had gone through the Bismarck Sea Battle and a dozen other combats without a scratch. It didn't seem right somehow for real leaders like Ed Larner, Ken McCullar, and Bill Benn to go out in accidents, but all three had gone that way after having looked the Grim Reaper in the eye and laughed at him a hundred times. A gallant trio of low-altitude tacticians, they will be remembered as long as I can find someone to listen to the stories of their exploits.

On May 2nd, twenty-one Jap bombers escorted by thirty fighters made a raid on Darwin. The Aussies intercepted with thirty-two Spitfires and knocked down five Jap fighters and one bomber. We lost thirteen Spitfires, seven of which were shot down; three landed at sea from engine failure; and three chased the Japs too far to sea and had to land in the water when they ran out of gas. Just what I had feared had happened. The Aussies had tried to outmaneuver the Nip fighters in the air. It couldn't be done, even with the highly maneuverable Spitfire.

I sent Bostock to Darwin to talk with the kids and tell them that, if they didn't stop that dogfighting business, I'd send them to New Guinea to serve with the Americans and learn how to fight a Jap properly. The Aussies who had been fighting the Nips for the past year or so were all sore at the Spit crowd for not taking advice. I guess they had to learn the hard way. We Americans haven't much to say along this line, either. We hate to take advice from anyone else and seldom do.

Colonel Merian C. Cooper arrived that day, fresh from Washington and anxious to work. He had completely recovered from the dysentery he had picked up while in China, and, in fact, looked as fit as the day I first met him back on Long Island, New York, in August 1917, when we were both first learning how to fly. A little over medium height, spare, active, with a pair of keen eyes in his lean face, a big nose, and a healthy chin, Cooper had changed little in the past thirty years, in spite of a career of roving over almost every part of the globe. People still tried to remember whether or not they had ever seen him without a pipe in his mouth, and you could still locate him at any time by following the trail of spilled tobacco as he constantly filled his pipe, took a few puffs, knocked out the ashes, and reloaded it. He would never look like a spit-and-polish soldier, but he could visualize and plan a military operation with the best of them. I was sure glad to get him. He always had more ideas in an hour than most people get in a month, and some of those ideas were good. He worked like a dog, planned well and thoroughly, and was loyal. He had been Chennault's Chief of Staff long enough to appreciate shoestring operations. His only trouble was that he never relaxed and, at the pace he normally set, would burn himself out in a year. I intended to keep an eye on him and send him home if he showed signs of cracking. However, as long as Coop remained conscious, I could use his head. I told him to stay around Brisbane for three or four days to find out what the score was and then report to Whitehead.

I wired Arnold that, according to the promises given me in Washington, deliveries for April were short 224 airplanes. During May I was supposed to get 276 more. If these schedules were not met, I would not only be in a bad way but our plans agreed on in Washington would have to be revised. It seemed almost impossible to get people back home to realize that while by superhuman efforts we were keeping a high percentage of our aircraft flying, over half my planes had been on continuous combat duty for from seven months to a year. They simply could not last much longer.

After delving into the general picture in the SWPA with our Allied Intelligence crowd, Cooper was quite disturbed about our situation. He said it looked to him as though the Jap could put me out of business any time and then the whole show would go to pot. If I didn't get help from home on airplanes soon, he might be right. I told him, however, that things were not half as bad as they had been. I admitted that if the Japs had a Whitehead to run their Air Force it would really be serious, but the Nip was dumb to begin with and did not understand air warfare. He did not know how to handle large masses of aircraft. He made piecemeal attacks and didn't follow them up. He had no imagination. As long as his set plan went, he was a great boy, but if anything unforeseen interfered and upset the plan, he got confused. If ever anyone was a sucker for a left, it was the Jap. Our kids were better flyers and better shots than his were, and now that we had learned to use them properly, our airplanes were better. One big thing in our favor was that the Jap had no good heavy bomber that could fight its way to the target and back. We had both the B-17 and the B-24. When surface means of supply were lacking or insufficient, we had the air cargo planes and knew how to use them. The Nip had no air cargo service and would not have known what to do with it if he had.

Whitehead finally became food-conscious. He wrote me from New Guinea that he and his whole staff were practically starving to death and that the only thing that would save the situation was for me to let him have Raymond, the chef I had "liberated" from Hawaii the month before and had just promoted to the grade of sergeant for distinguished services in my kitchen. I shouldn't have let Whitey eat one of Raymond's meals. Now I had spoiled Whitehead, but if I didn't keep him healthy I'd lose someone more important than a cook, so I sent Raymond to New Guinea with a note to Whitey to use him to train a couple of the Chinese I had sent up a week or so previously.

In order to get the ground crews for the P-38 group that Hap had promised to give me the airplanes and pilots for, I had been picking up stray individuals wherever I could find them ever since returning from Washington. One day I found out that there were over a hundred Chinese cooks, stewards, and messboys in Australia, refugees from British ships that had been sunk in the Far East during the war. Australian laws did not permit entry to the Chinese, so the Aussies had been quartering them in jail. I sent one of my medical officers to Sydney to see if he could squeeze about twenty of them through our physical requirements and, after making the necessary arrangements with the Aussies and the Chinese consul, we enlisted that number and sent them to New Guinea to work around Whitehead's headquarters and in his officers' mess. It not only released twenty men for my new fighter group but provided me with some excellently trained mess personnel. One of them, a good-looking, round-faced, five-foot lad with a perpetual grin, named Foo, had already attached himself to me as a personal orderly. In spite of anything I could do about it, he took such good care of me that I gave in and Foo became an important member of my staff for the duration of the war.

A few days later Whitehead wrote that Cooper and Raymond had arrived. It was hard to tell which one he appreciated more. He said Cooper was just the man he had been looking for and then devoted a half page to a description of how much his health and the health of his staff had improved during the day and a half Raymond had been cooking for them. It began to look as though I would lose my cook if I wasn't careful.

My frantic letters and radios to General Arnold finally bore fruit. Hap sent word to me that he would send me the P-38s for the new fighter group, the 475th, that I was to organize myself, in June instead of September as originally scheduled. He said the P-47 fighter was having engine trouble so that while the personnel had already left the States, he could not send me the airplanes until the middle of June.

The P-40 and P-39 situation would also remain static until June when I would get replacements at the rate of thirty P-40s and fifteen P-39s a month to the end of 1943.

The new medium group, the 345th, equipped with fifty-seven B-25s, was due in the theater during May, and another eighty-five B-25s were supposed to arrive during that same month as replacements for my 3rd, 22nd, and 38th groups. It would still leave me way short of requirements but I could at least remain in business. I was still short fifty-five bombers for my three heavy groups but Arnold promised to gradually increase my quota and bring me up to strength by late fall.

One bit of good news was relayed over the grapevine to me. What had caused the change was not mentioned but the edict had come from somewhere to SOPAC—no more hijacking of my airplanes.

Just as this leak had been dammed, however, Arnold's office tried to open another. A wire came in asking for Colonel Freddy Smith, whom I had sent back home for a couple of months' rest, and for Pappy Gunn. I immediately wired Arnold that I had no intention of giving up Smith and that I considered his prospects for advancement under me were better than anywhere else. As for Pappy, I said his services were urgently needed in Australia and asked that he be returned as soon as possible. I really needed Pappy, but anyhow I had to get him away from that bunch of aeronautical engineers. In a short time they would either have made Pappy scientific and ruined him or Pappy would have driven them insane and ruined them and maybe the whole aircraft-production program. Arnold replied that I could keep Smith and that Pappy Gunn was returning to Australia immediately.

On May 12th General Miff Harmon, General Nate Twining, Admiral Fitch, and some of their staff officers arrived in Brisbane from SOPAC for a conference on the coordination of our two air forces and the utilization of engineers and shipping during the operations to occupy and build airdromes on Kiriwina and Woodlark Islands. SOPAC agreed to furnish the engineers to build up Woodlark and to station a fighter squadron there which would operate on loan to me and under my control. The 6th Army from SWPA which now included all MacArthur's ground forces, commanded by Lieutenant General Walter Krueger, was to supply the occupational force for Woodlark and all the troops for

Slugging Match: Dobodura

Kiriwina.

That same day I acquired two lads that I spotted at once as aggressive leaders I would bet on any time. Colonel Jared Crabb, the commander of the new 345th Medium Bombardment Group, and Colonel David Hutchison, who was scheduled to take over the new photo-reconnaissance group that Arnold promised me during March, reported in. I made a mental note to watch both of them and told them to go on up to New Guinea to Whitehead.

Photo reconnaissance of the Rabaul area on the 13th showed 116 fighters and eighty-four bombers. A lot of shipping was due to start unloading at Oro Bay the next day, so we pushed the fighters across the mountains to Dobodura and waited for the Nips to show up. During the night six B-17s and four B-24s unloaded incendiaries and fragmentation bombs on Vunakanau, where the Jap bombers and some of his fighters were. The kids thought from the number of fires that blazed up that they had cut the Nip strength down considerably. The next morning it looked as though they had been right.

The Nip made his attack on May 14th, as we had expected, and we took him, as we had expected. He sent over twenty bombers, two dive-bombers, and twenty fighters. We intercepted with forty-three fighters and shot down eight bombers and eight fighters. Another five Nip bombers and one fighter were listed as probably destroyed. We lost one P-38 from which the pilot parachuted safely, and one P-40 cracked up on landing. The Jap attack destroyed a few drums of fuel and cut the telephone line between Dobodura and Oro Bay.

During the morning a little sawed-off tough-looking sergeant bombardier, with a bristly little mustache and a pair of soft blue eyes, came into the office to have me pin his decorations on him before he went home. He had put in 450 combat hours, had a Silver Star, a Distinguished Flying Cross, a Purple Heart, and an Air Medal with a couple of Oak Leaf Clusters coming to him, as well as a seat on a plane leaving for home at midnight. I pinned on his medals, shook hands, and wished him luck; we exchanged salutes and said goodbye.

About two o'clock that afternoon my secretary came in and told me that the sergeant was back, a little tight, quite excited about something, and very anxious to see me. I said to send him in.

The sergeant burst into the room waving a newspaper on the front page of which were the headlines announcing the sinking of the Allied hospital ship Centaur by a Jap submarine and the loss of several hundred nurses, doctors, patients, and crew.

The mustache was standing straight out, more bristly than ever. The soft blue eyes were almost black.

"General," he said hoarsely, "did you see this? Did you read what those so-and-so so-and-so's just did? General, I don't want to go home yet. I want to go back to my squadron in New Guinea. I'm the best damn bombardier in the whole Fifth Air Force and"—he lowered his voice to a confidential whisper—"I know where there is a Jap hospital ship."

I told him the answer was no, that he was tired, he needed a rest, he had a seat on a plane going home that night, and he was going to take it if I had to have him carried on board. I advised him to go have fun and forget it.

"Aw, General"—the eyes blinked and a couple of big tears rolled down his cheeks—"you're treating me just like a little kid." Then suddenly he drew himself up, saluted, and said, "Goodbye, General, and thank you," and left.

I felt like going along to console him but I had too many papers to shuffle.

That morning, flying at high altitude in spotty weather, one of our B-24 planes out on reconnaissance reported a large convoy of around thirty vessels hugging the coast of New Guinea and coming through Vitiaz Strait. There are a lot of rocks in that locality that do look like ships from high altitude, especially if the weather is hazy. The current through the straits adds to the illusion by making what looks for all the world like a wake streaming behind each of the rocks. We had had reports like this one before and this particular crew was new to the theater. Whitehead sent out a B-17 with an experienced crew from the old reliable 63rd Squadron to investigate at low altitude, but played safe by putting the whole bomber force on alert. A couple of hours later the B-17 back a radio reading: "The presence of the convoy of rocks in the Vitiaz Straits reported by a recco plane of the 90th Bombardment Group has now been officially confirmed."

The 90th Group didn't see the joke at all and vowed vengeance at the first opportunity. The 43rd Group, and particularly the 63rd Squadron, considered themselves the only really dependable bombardment outfit in the Fifth Air Force and scornfully disputed any claims on the part of the 90th that they were anything more than a bunch of newcomers and nice lads who were trying hard and who someday might amount to something if they stayed around long enough to learn from their betters.

What infuriated the 90th crowd was that the 43rd left no doubt as to who they meant by "betters."

The Jap supply situation at Lae and Gasmata evidently had become critical about this time, as we picked up some messages indicating that the Nip had decided to send no more shipping to these points on account of the danger from our air attacks. In lieu of sending vessels, barge routes were to be set up which would transship supplies and troops from Rabaul to Gasmata and from both Rabaul and Wewak to Lae. I sent a message to Whitehead to go into the barge-hunting business as soon as the Nips started it. If we could cut off his supplies, he would not try to reinforce Lae, and, furthermore, the airdrome there would not be of much use to him if he couldn't get gasoline to it.

By the 24th Rabaul's air strength had jumped to 120 fighters and 130 bombers, and on the four airdromes in the Wewak area we counted 107 more aircraft. Continuous barge traffic was observed along the north coast of New Guinea, from Madang to Finschaven, heavily covered by fighters, who interfered considerably with our reconnaissance. Whitehead immediately went to work on the Rabaul airdromes and started combing the coastlines for barges. By the 29th the photos showed only fifty-nine fighters and eighty-five medium and light bombers in the whole Rabaul area, although the Nip had flown in around a hundred aircraft during the previous week. Daylight barge movements stopped along the north coast of New Guinea. The constant air attacks were evidently worrying the Nips, besides destroying too many of their barges. They were confining all movements to darkness. During the daytime they pulled the barges into coves and river entrances, where they tied them together in groups of five or six and camouflaged them by covering the decks with palm leaves to make them look like islands. Sometimes the palm leaves would form a cover over poles laid from the decks of the barges to the shoreline and be indistinguishable from the land, except to an airplane flying a hundred or so feet above the water.

A further indication that the Nip was worried about his supply line to Lae was the widening of trails into motor roads along the coast from Wewak to Madang, and what looked like the same kind of activity from Madang south, along the old native trail across the Bismarck range, to connect with the excellent trail which ran along the Ramu and Markham River valleys into Nadzab and on to Lae itself. Construction of a road along that route was a tremendous job and it would be difficult to run traffic on it in the face of air attacks, but our reports all began to indicate that the Jap had mobilized every native he could get his hands on and had started them working on the road bossed by Nip construction engineers.

On May 31st I flew to Port Moresby to take over from Whitehead, who had to go back to Brisbane to get an infected toe treated. It had swollen up so that he couldn't even put a sock on that foot. I told him to stay in Australia long enough to get a good rest.

While wandering around the area looking over the combat units one noon, I ran into Edwin McArthur, the Metropolitan orchestra leader, playing at one of the messes. He had arrived at Brisbane about the first of May and had come in to see me and asked what to do. He had been sent over by the USO crowd, but the SOS in Australia didn't seem to know that he had arrived. I made arrangements for him to fly to New Guinea when I found out that he had an accordion and could really do things with it as well as play a piano. I had flown several pianos to Port Moresby, so I asked McArthur to play the new hits, as well as easy-to-sing songs like "White Christmas" and the old-timers such as "Kentucky Home" and "Swanee River." I suggested that, with the accordion, he might get the kids to sing while they were waiting for the evening movies to start. I wanted him to entertain them and not try to uplift them. He promised to do his stuff.

Now I found him up to his ears in work and doing a grand job. With that little accordion of his he was wandering around at mess time, just before the evening movies, or anytime when he was liable to find a gathering of kids to play for. He played quite well and injected so much of his own likable personality and enthusiasm into it that in spite of themselves the audience would start in singing. From the singers McArthur would form quartets, quintets, or sextets and work up to some real barber-shop stuff. The men were crazy about him. He lived with them, ate in their chow lines, and listened to their stories and their gripes. The SOS crowd still hadn't found out that he was in the theater or I would have been accused of stealing his services for the Air Force. He told me he would like to organize a traveling orchestra and song team out of the talent in the Fifth Air Force. I told him to go ahead if he could get the personnel without stealing any crew chiefs or air gunners or radio operators. I promised to help him get the instruments and to have the show flown all over the theater as soon as it was ready to perform.

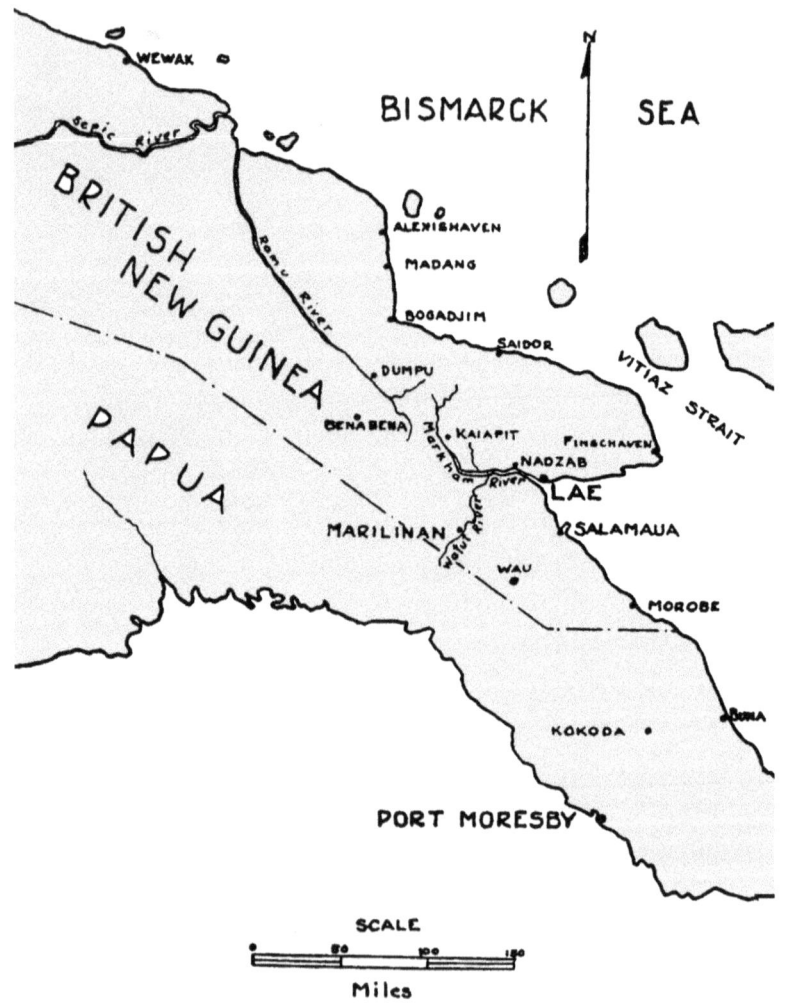

10. AIR SUPREMACY: MARILINAN AND WEWAK

June—August, 1943

Ever since we first began to talk about capturing Lae and Salamaua, back during the Buna campaign, we had been looking for a place to build an airdrome close enough to Lae so that our fighters could stay around to cover either an airborne or seaborne expedition to capture those important Jap holdings. The Dobodura airdromes were so far away from Lae that our fighters didn't have enough fuel reserve to stay over Lae for more than half an hour before they would have to go home. We had surveyed Kokoda, Wau, Dona, and a dozen other places but none were suitable. Whitehead finally located a flat spot along the upper Watut River, about 60 miles west of Lae and near a native village marked on the maps as Marilinan. It was a pretty name, the site looked good, and it was ideally located to support operations to capture Lae, Salamaua, and Nadzab. The only drawback was that, while it was forty miles away from the fighting zone around Wau, that forty-mile distance was into territory supposed to be controlled by the Japs.

We decided to have a small ground party walk the trail from Wau to Marilinan, look the place over, and then, if it still looked good and the natives could be induced to cut the grass, we would fly in some troops and grab the place as we had done at Wanigela Mission. Then the next step would be to fly in the engineers and their equipment and get going on the construction of an advanced air base. The Aussies told me that there used to be a small grass field there about 2000 feet long, which had been used by the gold miners, but that it had been abandoned about a year before the war and they didn't know what shape it was in.

On June 7, our reconnaissance party returned from Marilinan. They had found the old grass strip which was in good shape but overgrown with kunai grass about six feet high. They had gotten the natives to cut a strip a hundred feet wide the whole length of the field. The natives said that Jap patrols came in there about once every couple of weeks and that the last time was about five days before.

Squeeze Wurtsmith said he would like to fly up there the next morning in a P-40 and take a look at the place. I told him to go ahead but to have someone else follow him with an A-24 two-seater, which could land and take off in a much shorter space than a P-40. That would give him a chance to get home, in case the P-40 cracked up or the field was not long enough or good enough to attempt a take-off. I told Squeeze to load up the empty back seat of the A-24 with trade goods to give the natives a good start on preferring our business to dealing with the Nips. Lieutenant A. J. Beck took the A-24.

The next day General Squeeze Wurtsmith and Lieutenant Beck landed at Marilinan and made a complete reconnaissance of the whole area. The old strip could be lengthened a little more but would only be satisfactory for cargo airplanes like the DC-3. Wurtsmith had to have some more grass cut to get off as the natives had cut only about 1200 feet. Squeeze claimed he made the shortest landing ever made with a P-40. However, four miles north of Marilinan, at a little village called Tsili-Tsili, they found an excellent site, which looked as though it could be made into an airdrome with double runways 7000 feet long and with plenty of room for dispersed parking areas.

Whitehead had been wanting for some time to build an airdrome up near Bena Bena at a place called Garoka. We didn't have the cargo planes to support a real show up there but I told Cooper to send someone to Garoka and get the natives to start clearing a strip right away. I didn't care about making anything more than a good emergency field, but I wanted a lot of dust raised and construction started on a lot of grass huts so that the Japs' recco planes would notice. Also I ordered some indications of activity at Bena Bena itself and told Cooper to be sure to explain to the natives that we were playing a good joke on the Japs, that we were trying to get them to send some bombers over. If the Nip really got interested in the Bena Bena area, he might not see what we were doing at Marilinan. In about a week I figured the Nip should get worried and start bombing.

In the meantime, we would fly into Marilinan enough Aussie ground troops to cover all the trails coming into the area and then move in the 871st Airborne engineers to build the main base at Tsili-Tsili. Squeeze liked that name but both Cooper and I liked the sound of Marilinan better. Besides, Tsili-Tsili might have suggested to some people that it was descriptive of our scheme of getting a forward airdrome.

General Herring, the Australian ground commander in New Guinea, gave me a battalion of 1000 Aussies and some machine guns. They liked the idea, too, for while at Marilinan and attached to the Fifth Air Force they would get American rations, which were better than their own.

I didn't even try asking the Americans for ground troops to work under my orders. Our Army thought that the Air Force was an auxiliary arm, and to put infantry under an air commander would have been unthinkable. The Aussies had had a separate Air Department for some time, so to them there was nothing wrong with the idea.

That evening I heard the latest story on Pappy Gunn.

One day in May, just after his return from showing the North American Aviation Company how to install a lot of fifty-caliber guns on a B-25, Pappy came into my office in Brisbane and asked how he could get a pair of pilot's wings. I told him to go down to the Post Exchange store and buy a pair. Much to my astonishment Pappy told me he was not rated as a pilot in the Air Corps. It seemed that when the war broke out he was a civilian, flying a plane of his own in the Philippines. The Army had mobilized Gunn and his airplane and commissioned him as a captain. Something was said at the time about having Washington rate him as an Air Corps pilot, but nothing had happened.

Pappy flew his airplane, delivering courier messages, carrying supplies, and doing other miscellaneous jobs until the plane was wrecked. He was then evacuated to Australia, where his talents as an aircraft-maintenance man and his enthusiastic willingness as well as ability to install things in airplanes soon had him installed as a combination gadgeteer and test pilot.

To my knowledge Pappy had flown every type of airplane we had and had flown around 200 hours actually testing his ideas under combat conditions against the Japs.

I sent a wire to the Chief of Air Forces asking that Major Paul I. Gunn be given the rating of pilot. Some staff officer answered it for Arnold, saying that if I would send Pappy back to the United States he would be sent to the Air Corps Training Center and that, if he successfully completed the nine months' course prescribed by regulations, he would be given a rating. I followed this with a personal wire to Arnold marked "Eyes Only," which told Hap the story of Pappy's life, suggested a suitable disposition of the staff officer who had sent me the answer to my original wire, and again asked that Pappy be rated an Air Corps pilot.

Hap wired back rating Gunn as a pilot effective December 7, 1941. Pappy immediately went to our finance officer and drew eighteen months' back flying pay, which he put into large-denomination bills. He then secured them in his shirt pocket with a safety pin and got aboard as co-pilot on a B-25 on its way back to New Guinea.

Halfway across the Coral Sea, between Townsville and Port Moresby, the cockpit became a bit hot. Pappy took off his shirt and draped it across his lap. It was still too hot, and rather stuffy from the big cigar he was smoking, so to get better ventilation Pappy opened the side window and, before he could grab it, the money-laden shirt was sucked off his lap and out into the blue. He looked back to see if it had hung up on the tail surfaces. It hadn't.

Pappy took a few more puffs on his cigar, looked at it reflectively for a while, and pitched it overboard. When the air had cleared he carefully shut the window, turned to the pilot, and in a resigned voice remarked, "Well, come easy—go easy."

As far as Pappy was concerned the incident was closed.

Whitehead returned from Brisbane on the 10th. We talked over the situation and made our plans for utilizing the new bases to be built and the units we were to station there. Whitehead was to push the construction at Marilinan day and night, build up a big stockage of gasoline, and move the 35th Fighter Group in there with their P-38s just as soon as the place was ready to take them. The Jap build-up at Wewak worried both of us, but until Marilinan was ready as an advanced fighter base we could not make an all-out daylight attack with bombers, strafers, and fighters to really liquidate the place. The recco photos showed too many fighters on the four airdromes, at But, Borum, Dagua, and Wewak itself, for us to expose our meager bomber strength in an unescorted raid, and our fighters based at Dobodura and Port Moresby didn't have the range to go to Wewak and return. In the meantime, to make sure that the Jap air force at Rabaul wouldn't interfere with our landing operations in the Solomons and on Kiriwina and Woodlark Islands on June 30, we had to keep working on the Vunakanai, Lakunai, and Rapopo airdromes. I told Whitey to concentrate on those targets as primary night-bombing objectives and 15 hit Wewak also at night, but only when the weather interfered with attacking Rabaul.

Whitehead wanted me to argue the Aussies into making a drive to capture Salamaua right away and see if General MacArthur would not order Sir Thomas Blamey to do it. I told him that I had already discussed the proposition with Blamey and promised to support him to the limit, but he felt that it would cost too many troops and leave him with insufficient strength to take on the capture of Lae later on. I think that Blamey as Allied Land Force Commander wanted the Aussies to be in charge of all land operations along the north coast of British New Guinea and, if New Britain was to be occupied, to be charged with that job, too. He expected Krueger to be the big land-force commander when we got ready to go into the Philippines but hoped he would be in charge up to that time. If Blamey had to hurry up the Salamaua capture, he was afraid he'd have to use the 7th or 9th Division. Then they would not be available for the Lae show and he might have to turn that operation over to Krueger. In addition, he didn't have too much faith in the fighting ability of our infantry, and from the operations so far I couldn't really blame him. The trouble was that Blamey had only the 7th and 9th to depend on. The 3rd and 5th were pretty good but there would be trouble recruiting them up to strength, as Aussie manpower was already strained. The 7th and 9th, as regular divisions under two brilliant leaders like Vasey and Wooten, would get priority on recruits. As for the Americans, while our 32nd and 41st Divisions had not yet distinguished themselves, they would be much better the next time and more American divisions were on the way.

A lot of the GHQ staff and some of the American commanders wanted to sidetrack Blamey and let Krueger run the ground forces as soon as there were more American ground troops on the scene than Aussies.

General MacArthur sensed all the ambitions and jealousies in the picture and I don't think he shut his eyes for a minute as to the relative stake of training and combat potential of the Aussies and the Yanks. Of course, he would have liked to grab Salamaua right away, but he knew he would need the Aussies not only for the capture of Lae but also to roll on to the Vitiaz Straits. Krueger's troops would get valuable experience in the occupation of Kiriwina and Woodlark, but the Aussies, with part of the American 41st Division attached, would have to do most of the real fighting until the end of 1943.

The importance of capturing Salamaua as soon as possible was mentioned to Blamey, but it was left up to him to decide how hard he would press the campaign. Blamey suggested that if I could maintain the air blockade of the Huon Gulf area, in order to hold Salamaua the Japs would have to keep draining troops out of Lae until that place would be easy when we moved against it later on. The idea had some merit, but I decided then that the capture of Salamaua was not going to precede the Lae show by very much.

Following the recco report a few days previously by a 43rd Bombardment Group plane, which confirmed the fact that a 90th Group crew had mistaken a lot of rocks for a Jap convoy, the 90th Bombardment Group board of strategy went into a huddle to see what could be done to restore their own prestige and, if possible, to humiliate the overproud 43rd. The next day the 43rd Group received a very cordially worded invitation to have dinner at the 90th Group officers' mess. The invitation was accepted with especial alacrity as rumor had it that the 90th had just flown in several cases of Australian beer.

That evening as the procession of 43rd Group jeeps made the last turn in the winding road leading up the hill on which the 90th Group mess was perched, they were horrified to see just off the road an unmistakable Chic Sale structure with a huge sign on top of it, reading "Headquarters 43rd Bombardment Group." The Kensmen didn't say a word about it all evening. They helped their hosts eat an excellent meal and consume all the Australian beer and wound up the festivities by thanking them profusely for a fine party. The next morning just as the 90th was tumbling out of bed, a lone 43rd Group B-17, slipping in over the tops of the trees, suddenly opened fire on the Chic Sale with a pair of fifty-caliber guns shooting nothing but incendiary ammunition. The little building blazed up as the B-17 kept on going until it disappeared behind the hills. In a few minutes there was only one building in Port Moresby labeled Headquarters 43rd Bombardment Group and that was over at Seven Mile Airdrome where the group lived.

I passed the word to both groups that, while I had seen nothing, heard nothing, and knew nothing, I did not expect anything of that kind would take place again. As a matter of fact, there was no danger involved as the little building was off by itself so that the rest of the 90th Group camp was completely out of the line of fire, but if some other shooting affair started, the conditions might not be so favorable. Little silly things like that, which now sound like a species of insanity, were wonderful incentives to morale and set up a spirit of competition and a desire to outdo the rival organization that meant more hits on the targets, a quicker end to the war, and thereby a saving of American lives.

The night of the 10th–11th, thirteen B-24s raided Vunakanau and

Air Supremacy: Marilinan and Wewak 193

Rapopo. The next morning the photos showed plenty of wrecks and burnt-out aircraft in the revetted dispersal bays. On the 12th the Nips flew 100 fighters into the Rabaul area, bringing their total aircraft strength up to 271. That afternoon SOPAC destroyed twenty-five Jap fighters in a brawl over Russell Island. That night Whitehead sent twenty-three bombers to Rabaul. Again the photos looked good. Again the Nips flew in between eighty and a hundred more bombers and fighters and, on the night of the 15th just before we hit them with twenty-two more bombers, the pictures showed the Jap aircraft count in the Rabaul area as 281. About 120 more were spotted at the same time on the Jap airdromes on Bougainville and the Solomons, making a total of around 400 aircraft available to the Nips to strike at our forces in the Solomons or to combine with the gang at Wewak to take me out in New Guinea.

Our Rabaul raids evidently were pretty effective, as on the 16th the Nips dispatched a series of piecemeal attacks on Guadalcanal, involving throughout the day a total of only 120 aircraft. The SOPAC Army, Navy, Marine, and New Zealand flyers had a field day, putting 104 fighters in the air and shooting down seventy-seven while the antiaircraft guns got seventeen more. Six SOPAC aircraft were reported missing.

During the morning one of my B-17 reccos, flown by Captain Jay Zeamer and his crew, performed a mission that still stands out in my mind as an epic of courage unequaled in the annals of air warfare. This crew had already become known throughout the Fifth Air Force as the Eager Beavers, by asking for and volunteering for every mission that looked like a good fight. This day we needed some pictures of Buka strip. Zeamer and his crew got the mission and remarked before they took off that things looked promising for a little action. They got it. About ten miles from their objective, in blue-sky visibility, they saw the Jap fighters from the Buka strip taking off and climbing to intercept them. Instead of abandoning the mission, at least temporarily, as the crew of an unescorted recco plane would be justified in doing, this gang kept right on course and started their photographic run. About this time the Nip fighters, estimated at fifteen to twenty in number, opened the attack. The first one, making a head-on attack, was knocked off by Lieutenant Joseph R. Sarnoski, the bombardier, manning the front gun. The top-turret gunner, Sergeant Johnny Able, picked off a second. The B-17, however, was stopping some bullets and so were the crew. They were flying at 28,000 feet and sucking pure oxygen. A Jap bullet cut the main oxygen line. Zeamer pushed the ship over and dove to 18,000 feet while the gunners gasped and kept shooting. Just as he was leveling out, Zeamer saw a Nip fighter just off to his left. He had a fixed fifty-caliber gun on his side of the cockpit, which he had had mounted with the hope that some day he could get in on one of the shooting parties. Quickly kicking the rudder and making a diving turn to the left, he got the Jap in his sights and pressed the trigger. The Jap went down in flames.

As he straightened out another enemy plane made a head-on pass at the B-17. A. 20-mm. shell burst in the bombardier's cockpit, knocking Sarnoski back into the passageway under the pilot's cockpit. One fragment hit him in the stomach, wounding him fatally. Zeamer had both legs hit, so that the co-pilot, Lieutenant Johnny Britton, had to handle the rudder controls. Sarnoski called out, "I'm okay. Don't worry about me," and crawled back to his guns just in time to open fire on another frontal attack. He fired one long burst. The Jap plane blew up and disintegrated. Sarnoski collapsed over his guns—dead. The radio operator, Sergeant William Vaughan, had his radio shot away in front of his face. He was wounded. Zeamer had collected some more bullets and was now wounded in both legs and arms. The tail gunner, Sergeant "Pudge" Pugh, would fire a burst and then dash forward and help bandage up gunners who continued to man their guns. After forty minutes of this crazy combat, the Japs finally pulled away. They had lost five planes definitely. Most of the others were damaged. The B-17 had completed taking its pictures and headed back toward home. The photographer, Sergeant William Kendrick, also doubling in brass on both side gun stations, had been alternately taking pictures and shooting at Japs. A lot of lead had poured through his section of the ship but he had collected none of it. Zeamer had passed out from loss of blood. The copilot was being bandaged. The navigator, Lieutenant Ruby Johnston, was badly wounded and out of commission. The top-turret gunner, just slightly wounded and by now nicely bandaged, who doubled in brass as flight engineer, was in the co-pilot's seat, flying a B-17 for the first time in his life—keeping the sun at his back—which was roughly the direction to the home airdrome, 580 miles away. No radio, no compass (Jap bullets had knocked that out, too), no brakes, no flap control, just a shot-up B-17, one man dead, 5 men wounded, the pilot unconscious, and the sergeant engineer at the controls—but they had the pictures they had started out to get and could now paint five more little Jap flags on the side of the fuselage.

Zeamer woke up occasionally, gave directions about the course as he recognized islands and coral reefs along the route, and kept lapsing back into unconsciousness again.

Air Supremacy: Marilinan and Wewak   195

The coast of New Guinea came in sight. Then Cape Endiaidere. Twenty-five miles more to the field at Dobodura. Zeamer came to. His legs were gone, but he could lift one of his arms to the wheel. The co-pilot had revived and was now back on the job. He could handle the rudder. There was no time to circle the field and worry about heading into the wind. Zeamer wasn't going to last much longer. They rolled to a stop, using every foot of the 7000 available. Without brakes, it was a good thing they had that much runway.

"Pudge," the tail gunner who by some miracle had not even been scratched, helped get Zeamer and the rest of the wounded out of the airplane and started toward the operations hut. He looked up at the wind cone and saw it pointing toward the direction in which the B-17 had landed. Without a second's hesitation, he climbed up to the control tower and to the astonished sergeant on duty yelled, "What the hell is the idea of having that wind sock pointing the wrong way?" The control-tower sergeant took one look at Pudge's face. He knew instinctively that if he told Pudge that his pilot, Captain Zeamer, had landed with the wind, Pudge would certainly sock him. "Okay, Pudge," he said consolingly, "I'll fix it right."

Of course, Pudge was probably a bit excited, perhaps he wasn't thinking too clearly at that moment, and he was wrong about the wind cone, but somehow I liked the subconscious loyalty.

Zeamer and Sarnoski were awarded Congressional Medals of Honor and the rest of the crew Distinguished Service Crosses.

On the 16th, we also flew a detachment of Australian troops and some engineers to Marilinan to start construction of a flying field. These were followed by more troops, more engineers, and more equipment, while work proceeded day and night to speed up the job, install a warning system, and get the airplanes established there before the Nips discovered the place. By the 20th the field at Marilinan proper had been enlarged sufficiently for transport operations and the road constructed to Tsili-Tsili, where we began work on the real operating base.

In the meantime, as I had hoped, the Jap reccos spotted the dust we were raising and the new grass huts we were building around Garoka and Bena Bena and evidently decided we were about to establish forward airdromes there. Beginning on the 14th they attacked both Garoka and the Bena Bena almost daily, burning down the grass huts and bombing the cleared strips, which looked enough like runways to fool them. The natives thought it was all a huge joke and when the Japs put on an attack they would roll around on the ground with laughter and chatter away about how we were "making fool of the Jap man."

About this time I had managed to scrape together about a hundred P-39s, or enough to partially equip five of my nine American fighter squadrons, when out of the blue on June 19th came a wire from Washington saying that until certain changes were incorporated in the P-39s in accordance with an Air Corps Technical Order dated back on March 25th, 1943, we were restricted in the kind of flying maneuvers that could be considered safe. This meant that we could not use them in combat. I immediately wired Arnold that I could not understand why the modifications could not have been made back home before the planes were sent me so that they could go to war on arrival in the theater. I said I had no idea what the modifications were and that no parts or instructions were available if I did find out. I asked Hap to immediately replace all my P-39s with P-38s.

The next day a wire signed Arnold said a new technical order was being issued stating that the P-39s were all right to use in combat. I had hardly finished reading that wire when another from the Air Service Command at Dayton said that as soon as some new part was installed we could let the P-39 go to war again. Luckily one of the newly arrived airplanes had the missing part already installed. We had the Aussie tool shops make up some more and after a week's delay we were back in business again. I was glad the Japs didn't know about our troubles during that time. I was already too weak for comfort without being suddenly deprived of the use of a hundred fighter planes.

On June 20 I recommended the promotion of Colonel Roger Ramey to the grade of brigadier general. He had really justified my decision to let him run the 5th Bomber Command a couple of months previously. General MacArthur approved the recommendation and forwarded it to Washington for confirmation.

That same day the P-47s for the new fighter group began to arrive at Brisbane on the same boat that brought Lieutenant Colonel Neel Kearby, the group commander, and the rest of the squadron personnel. Kearby, a short, slight, keen-eyed, black-haired Texan about thirty-two, looked like money in the bank to me. About two minutes after he had introduced himself he wanted to know who had the highest scores for shooting down Jap aircraft. You felt that he just wanted to know who he had to beat. I told him to take his squadron commanders and report to General Wurtsmith at Port Moresby for a few days to get acquainted and then come back to Brisbane, where we would erect the P-47s as fast as we got them off the boats.

I then went out to Eagle Farms, where the erection was to be done, and found that no droppable fuel tanks had come with the P-47s. Without the extra gas carried in these tanks, the P-47 did not have enough range to get into the war. I wired Arnold to send me some right away, by air if possible. About a week later we received two samples. Neither held enough fuel, they both required too many alterations to install, and they both were difficult to release in an emergency. We designed and built one of our own in two days. It tested satisfactorily from every angle and could be installed in a matter of minutes without making any changes in the airplane. I put the Ford Company of Australia to work making them. We had solved that problem but it would be another month before we could use the P-47s in combat.

In the meantime, everyone in the 5th Air Force, from Whitehead and Wurtsmith down, except the kids in the new group, decided that the P-47 was no good as a combat airplane.

Besides not having enough gas, the rumors said it took too much runway to get off, it had no maneuverability, it would not pull out of a dive, the landing gear was weak, and the engine was unreliable.

I sent for Kearby and told him I expected him to sell the P-47 or go back home. I knew it didn't have enough gas but we would hang some more on somehow and prove it as a combat plane, especially as it was the only fighter that Arnold would give me in any quantities for some time. I told Kearby that, regardless of the fact that everyone in the theater was sold on the P-38, if the P-47 could demonstrate just once that it could perform comparably I believed that the "Jug," as the kids called it, would be looked upon with more favor. I told him that Lieutenant Colonel George Prentice would arrive that afternoon from New Guinea to take command of the new P-38 group which I had formed and had started training out at Amberley Field. He would probably celebrate a little tonight. I told Neel to keep away from Prentice, go to bed early, and the first thing in the morning to hop over to Amberley in his P-47 and challenge Prentice to a mock combat. Neel Kearby was not only a good pilot but he had had several hundred hours' playing with a P-47 and could do better with it than anyone else. Prentice was an excellent P-38 pilot, but for the sake of my sales argument I hoped he wouldn't be feeling in tiptop form when he accepted Neel's challenge.

The "combat" came off as I had hoped. Prentice was surprised at the handling qualities of the P-47 against his P-38 and admitted that Kearby "shot him down in flames" a half dozen times. He still preferred his P-38 but began warning everyone not to sell the P-47 short. At the same time he wanted to go to bed early that night and "have another combat with Neel tomorrow." I interfered at this point and said I didn't want any more of this challenge foolishness by them or anyone else and for both of them to quit that stuff and tend to their jobs of getting a couple of new groups into the war.

On June 30th the Allied SWPA forces landed as scheduled at Nassau Bay, and at both Kiriwina and Woodlark Islands. None of the landings were opposed. Actually we had landed small advance parties several days before the main bodies of infantry and engineers went ashore. Over in SOPAC, landings were made on Rendova, Vangunu, and New Georgia Islands. No Jap aircraft paid any attention to our landings all day, but during the afternoon over in SOPAC a series of attacks were launched against the landing operations by bombers, dive-bombers, fighters, and even floatplanes. The attacks were small piecemeal affairs, most of which came from the Jap fields in Bougainville. One SOPAC vessel was sunk after the cargo was discharged and two others were damaged. SOPAC lost seventeen planes but saved seven of the pilots. The series of attacks cost the Nip 101 aircraft. If their total force had come down all at the same time, the Nip might have really hurt the landing operation, but he was still feeding his planes in fifteen to twenty-five at a time, and with too long an interval between attacks to catch the SOPAC aircraft refueling on the ground.

The 49th Fighter Group kids wanted to know why the Nip didn't come over their way so they could get some of that "easy meat" and get closer to pulling the cork on their magnum of brandy.

Over at Darwin the Japs made their 57th raid since the beginning of the war, with twenty-seven bombers and twenty-one fighters. Three unserviceable B-24s of my new 380th Group were destroyed and six others received some damage. The Aussies put forty-one Spitfire fighters in the air and got eight bombers and two fighters. Six Spitfires were shot down, with three pilots recovered. In spite of the terrific reputation it had in Europe, the Spitfire never did do as well as the American types of fighters in the SWPA after the men once found out how to fight the Nip.

July opened the final struggle for air supremacy over New Guinea and the Solomons. Jap Naval and Army air units based at Rabaul and in Bougainville, constantly augmented in spite of heavy losses from combat and bombing attacks on their airdromes, kept striking at SOPAC forces working their way northward along the Solomons. Supporting the SOPAC drive and at the same time trying to cut down the Jap air strength which threatened our occupation of Kiriwina and Woodlark Islands, the Allied Air Forces of SWPA hammered away at the four main Jap air bases in the Rabaul area at Vunakanau, Lakunai, Rapopo, and Tobera. Wewak was building up rapidly but that area would have to wait until we could establish fighters at Marilinan. We were not strong enough to handle both Wewak and Rabaul, so it seemed wiser to concentrate on the latter target, which helped both SOPAC and ourselves.

On July 1st Fifth Air Force bombers dropped 22 tons of bombs and incendiaries on Lakunai and Rapopo, reporting heavy damage and many fires. That afternoon the Japs lost twenty-two aircraft in a raid over Rendova Island. SOPAC reported losing eight planes in the interception and recovered three of the pilots.

On the 2nd we dropped 33 tons of bombs and incendiaries on Vunakanau and Rapopo, again with excellent results. Again the Nips raided the Rendova area, losing six more aircraft. SOPAC reported three aircraft and pilots missing.

On the night of the 3rd we concentrated on Lakunai, with a nineteen-ton attack. During the day the Japs lost four fighters and a bomber to our P-40s, who suffered no losses, in a fight over Salamaua, while in the Rendova area SOPAC destroyed five Jap fighter planes and lost three of their own in the fight.

Eight Nip bombers plastered our dummy field at Garoka to the great amusement of the natives and to my great satisfaction. Between the beating the Japs were taking all night around Rabaul and all day at the hands of SOPAC fighters in the Solomons and our own fighters around Salamaua, they didn't have much to spare against our airdrome construction in Kiriwina and Woodlark. The continuous attacks on the Bena Bena-Garoka area and the absence of Nip recco planes over Marilinan showed that our deceptive measures were paying off.

In fact, those deceptive measures had worked so well that they stimulated me to write to General Arnold and ask him for some help on another scheme. On June 19th, after discussing the situation in the SWPA, I wrote:

"I have another idea that I would like to sell you hard enough so that maybe you could sell it to someone else. Here it is. Get an old or a new boat fixed up with a painted wooden or other cheap painted deck and a smokestack on the side so that she looks like a carrier, put some dummy aircraft on the deck, and let me have it. I will steam it around the north coast of New Guinea within range of my fighters and let the Nip reconnaissance pick it up, then when Mr. Moto gets a fit and throws every airplane he has in range at it I will take his Air Force apart. Right now I have to go to him with unaccompanied bombers. This scheme will make him come to me, where I can reach him with everything. It should be a natural. The way to do the job right is to make two of these dummy carriers. As soon as the radar picked up the Jap coming in, the crew would get overboard in a launch and let the 'carrier' go steaming along. Maybe it would be a good idea to tie the rudder over and let her circle the way the Jap does. If he really did hit the 'carrier,' let a contact fuze blow her up and announce to the world that we had lost a carrier with heavy loss of life and that the next of kin had been notified. This would make the Jap doubly eager and really convinced when we pushed a second one in front of him. With a minimum loss to ourselves, we ought to be able to clean his whole Air Force out of this theater and then, before he had a chance to replace it, mop up all the shipping within reach in broad daylight, land troops at will anywhere we pleased, and really go places in this war. If you can get Kaiser to spend half a day or so and convert a couple of boats into these decoys of mine, I will bring you home not only the bacon but a couple of Smithfield hams, too."

On July 4, General George Marshall wired General MacArthur giving his approval to the plan. Marshall stated that suitable vessels might be available in our area and suggested that MacArthur's transportation chief see about acquiring them locally.

I was disgusted.

After all the pleas we had made for more shipping, both General Marshall and General Arnold should have known that any vessel already in the theater big enough to cross the Coral Sea was needed to keep the war going. If we had had a ship to spare that was suitable for my scheme, General MacArthur would not have bothered to ask Washington's approval for the idea. He would simply have issued orders to fix the vessel up and told me to go ahead.

MacArthur was as disgusted as I was but was so sympathetic that he said to see if the SOS crowd could fix up a big barge to make it look like a carrier. He knew, however, as well as I did, that the project was now dead. We didn't have any barges that big and, if we did, there would be a lot of reasons why it couldn't be done. I decided that I'd better stick to ideas that I could implement myself.

On the 6th the Japs sent another 27-bomber, 21-fighter show against Darwin. It cost them ten bombers and two fighters. The Aussies lost seven Spitfires.

Over in SOPAC the suckers lost fourteen bombers and ten fighters arguing over Rendova Island. SOPAC losses were zero.

In the meantime nine Jap bombers escorted by ten fighters pasted Garoka again. The natives still thought it funny. So did I, but I kept my fingers crossed as we worked around the clock trying to make an air base at Marilinan.

We were really getting efficient at hauling air freight. Between Port Moresby and Marilinan we escorted the cargo planes with fighters, which also covered them while unloading and escorted them back home. It was quite evident from the start that quick unloading at Marilinan was necessary, as the fighters could only remain overhead for about an hour before the limitations on fuel supply would force them to return to their base.

We started drilling teams, which we labeled Air Freight Forwarding Units, in loading supplies into and out of the DC-3 transports. We set up an old DC-3 wreck on the edge of Seven Mile strip and practiced loading and unloading. It was remarkable how the time studies worked out. For example, at first it took forty minutes to load a jeep into a DC-3 and about the same time to unload it. After a few days of practice, the time was reduced to two and a half minutes to load it and two minutes for the unloading procedure. The trucks bringing supplies from the warehouses to the airplanes were loaded in reverse, so that the cargo fed automatically into the DC-3 and kept the balance in the plane correct. Before very long we were landing, unloading, and clearing Marilinan at the rate of thirty to thirty-five DC-3s an hour.

In order to haul supplies from Marilinan to Tsili-Tsili, at first we flew in jeeps and trailers, but we soon needed something better so we sawed the frames of a couple of two and one-half ton trucks in two, stuffed the pieces into the DC-3s, flew them over the mountains to Marilinan, and then bolted and welded them together. It worked like a charm, so we started converting all our trucks this way ahead of time, sawing them apart and then bolting them together again, so thar they could really be "airborne" equipment. That gang of mine by that time had decided that they could figure out a way to haul anything in a DC-3. In fact, after I told MacArthur about the scheme of cutting trucks in two and welding them back together again, one day at a press conference he said he believed that if he told me to move New York to the West Coast and re-erect it there, the Fifth Air Force would figure out a way to do it.

Our continuous attacks on the Jap airdromes around Rabaul and the steady attrition of Jap aircraft in combat, culminated on the 15th in a definite loss of the initiative on the part of the Nip air force in his area, when their raid of twenty-seven bombers and forty fighters was intercepted over Rendova Island by forty-four SOPAC fighters, who shot down fifteen Jap bombers and thirty fighters with a loss of only three of their own.

With the Jap air strength around Rabaul pretty well beaten down and no attempts being made to build it up, we switched to heavy support of the Aussie drive toward Salamaua and to interfering with the Jap road-construction work from Madang across the Bismarck Range to the

Ramu-Markham Valley trail. Our fighters escorting the bombers on these raids began to really run up their scores.

On July 21st our P-38s shot down eighteen Jap fighters, losing two of our own, while covering an attack of sixty B-25s on Madang. We also took advantage of this raid to fly an advance party of the 35th Fighter Group, with its radar-warning unit and other equipment, into Marilinan. The new field was coming along fast and the Japs still had not discovered it. In fact, they were still sporadically shooting and bombing the dummy grass-hut construction at Garoka and Bena Bena.

Two days later, as our bombers wrecked Jap installations from Madang to Bogadjim with 71 tons of bombs, bombed and strafed Lae and Salamaua, destroyed eleven Jap barges, and damaged eleven more along the New Britain coast, our P-38s shot down thirteen more Nip fighters. One P-38 was reported missing.

On the 26th the Lae-Salamaua area took a plastering of over 125 tons of bombs as our P-38s downed another eleven Jap fighters with no losses to themselves.

That evening our first fighters landed at Tsili-Tsili. One strip had been finished and a stockage of 1500 drums of gasoline flown to it. I told Whitey to call it Marilinan from now on. I still didn't like the other name. In case the Nips should take us out, somebody might throw that Tsili-Tsili thing back at me. My fingers by this time were getting calluses from being crossed so hard, but the Japs still showed no signs of knowing that we were building a field right in their back yard.

On the 28th we ran a test on a B-25 equipped with a 75-mm. cannon, which Arnold had sent over to me to try out. I assigned it to Pappy Gunn, who fell in love with it at first sight. Pappy attached himself to twenty-five B-25s escorted by nine P-38s, who were going on a barge hunt in the Cape Gloucester area. It turned into quite a party for all concerned.

Two Jap destroyers just off Cape Gloucester looked to Pappy as if they were placed there for his especial benefit. Picking out the largest of the two vessels, Pappy scored seven hits with his 75-mm. cannon, but much to his disgust the destroyer didn't even slow down. A 75-mm. gun, which, after all, fires a shell that is only about three inches in diameter, was not enough to worry a destroyer. The two B-25s flying on his wings then told Pappy please to step aside while someone did the job who knew how it should be done. A 1000-pounder skipped into the Jap warship and split her in two. Another 1000-pounder sank the other destroyer, and the B-25s continued along the New Britain coast looking for barges. They found and sank eight barges and two large motor hunches, while the P-38s shot down eight out of fifteen Jap fighters that tried to interfere.

Pappy flew along with the gang, sulking and all mad because they had shown up his pet gun installation. Returning over the Jap airdrome at Cape Gloucester, Pappy looked ahead and saw his chance to redeem himself. Just landing was a Nip two-engined transport airplane. Pappy opened his throttle, pushed ahead of the formation, and fired his two remaining rounds of cannon ammunition at the Jap plane taxiing along the ground. One of the high-explosive shells hit the left engine and the other the pilot's cockpit. The transport literally disintegrated. Pappy reported with great glee when he landed back at Port Moresby, "General, no fooling, as I passed over that Nip plane there were pieces of Jap higher than I was."

We found out afterward that, among the fifteen passengers on that Jap plane, were two generals and three colonels on their way to a staff conference at Wewak.

Kearby's group of P-47s moved up to Port Moresby during the last week of July. They still didn't have the auxiliary droppable fuel tanks, but Ford of Australia promised to deliver me at least two hundred by the second week in August. I told the group they would be limited to local defense of Port Moresby until the tanks arrived. When the kids found out that there hadn't been a Jap airplane around there since the middle of April, I'm afraid that I lost some of my previous good standing with them.

Whitehead reported that as of July 31st over a hundred barges had been sunk and that the kids were beginning to get quite expert at locating them in spite of the most expert camouflage. He predicted good hunting for August if the Japs didn't get discouraged and stop this method of supply.

By the first of August the plan of operations to capture the Lae area had been completed and the liaison groups of the Army, Navy, and Air Force had coordinated the details of each move. During the month of August the Air Force would continue the air blockade of Lae and Salamaua and support the drive of the Aussies and the regiment of the American 41st Division toward Salamaua. Sometime early in September rhe Australian 9th Division under Major General Wooten would land at Hopoi Beach, about ten miles east of Lae, and move west along the coast to capture the place. The day following the Hopoi Beach landing, the American 503rd Parachute Regiment would be dropped at Nadzab, where they would cut the grass on the old strip there, remove any obstructions that the Japs might have installed, and move out to cover the trails coming into the area, while we landed the 7th Australian Division under Major General George Vasey. Vasey would then drive east toward Lae and assist Wooten in wiping out the estimated 10,000 Nips in that place. We were to maintain air control over the amphibious movement of the 9th Division all the way from Buna to Hopoi Beach, transport the 503rd Parachute Regiment and the 7th Division to Nadzab, furnish air support to the ground forces in the drive on Lae, and fly in the food and ammunition required by Vasey's force until Lae was captured and these supplies could be brought in by boat.

I pointed out to General MacArthur that I didn't have enough strength to handle the Jap air forces at both Rabaul and Wewak, so I proposed concentrating everything on Wewak right up to the day of the proposed landing of the Australian 9th Division at Hopoi Beach. I suggested that the landing be made sometime during the first week in September, on a day when the fog in the Vitiaz Straits area was so thick that the Jap airplanes from Rabaul could not get through to interfere with our shipping taking the expedition into Hopoi Beach. At that time of the year a weather condition of this kind normally occurred two or three days a week. The fog would blanket Vitiaz Straits and extend all over New Britain, while the weather from Buna to Lae and on to Wewak would be good. Our long-range weather forecasters should be able to predict such weather at least three days ahead of time and we ought to keep the whole schedule flexible enough to take advantage of nature's help.

The General smiled and remarked that it looked as though the infantry had lost their last prerogative. Up to that time they at least had the right to decide on the date and hour of their attack. Now we were about to turn that privilege over to an Air Force weatherman. He agreed, however, that the scheme made sense, so I put an Australian weather team and an American weather team on the job, with instructions to give the daily forecasts for the weather in the Port Moresby-Lae-Vitiaz Straits area during the first week of September 1943.

At this time it looked as though we would be able to start operating fighters out of Marilinan by the middle of August. so I told Whitehead to concentrate on the Jap barge traffic into Lae from Wewak and Rabaul until we were ready to cover an all-out daylight bombing attack on the Wewak airdromes with fighters based at Marilinan. About the middle of August I wanted him to plan a simultaneous strike on the Jap fields at But, Borum, Dagua, and Wewak itself, using everything that we could spare on the attack.

The barge hunt really got results. During the week ending August 4th, in addition to the two Jap destroyers sunk off Cape Gloucester on July 28th, we had sunk two motor torpedo boats, a patrol vessel, two large motor launches, destroyed ninety-four large motor-driven barges, and badly damaged probably a hundred more. The majority of these barges were capable of transporting a load of around seventy-five tons of supplies, so that the Jap had not only lost his barges but somewhere around 15,000 tons of supplies had been destroyed or prevented from getting to Lae. While the low-altitude boys were shooting up the barges, the escorting P-38 fighters had bagged nineteen Jap aircraft. The week's work had cost us one B-25, and two P-38s had been damaged.

The hunt continued, and during the following ten days the Japs had another 125 to 150 barges destroyed or out of action. The kids were beginning to complain, however, that the Nip was not replacing his barges fast enough to keep them supplied with targets. The hunting was definitely not as good as it had been at first.

On the 14th of August we moved two squadrons of P-39s, thirty-five airplanes in all, of the 35th Fighter Group into Marilinan just before dark. That noon the first Jap reconnaissance airplane to fly over the place since we started work back in July, had circled high overhead for several minutes before heading northwest toward Wewak. As soon as they arrived our fighters gassed up and went on alert, prepared to take on a Jap bombing raid that now was almost certain to come soon. Our deception measures had worked up to then but by this time the Nip must have known that Garoka was a bluff to keep him from discovering Marilinan. We hoped that he didn't know that we had installed our radar, which was good for about fifteen minutes' warning, and that we had thirty-five P-39s on deck to take him.

The next morning at nine o'clock the Japs attacked with twelve bombers escorted by twenty to twenty-five fighters. We shot down eleven of the bombers and three of the fighters. It cost us three P-39s. Two of the pilots parachuted safely. Another P-39 was badly damaged and probably beyond repair. One of the Jap bombers crashed into the chapel on the edge of the airdrome and killed the chaplain and six men. Two of our transports unloading on the ground were destroyed and six more men killed among the ground crews. The battle for control of the air over the Markham Valley had begun.

On the morning of the 16th we sent forty-eight transports, loaded with thirteen fifty-gallon drums of gasoline each, from Port Moresby to Marilinan and escorted them with fifteen P-38s and with thirty-two P-47s from Colonel Neel Kearby's new 348th Fighter Group, who were participating in their first combat mission. The mission met with no interference, so we repeated it in the afternoon. This time the Nips sent sixteen bombers and fifteen fighters down from Wewak to argue the point. We intercepted the Jap force before it reached Marilinan. The P-38s shot down eight fighters and two bombers with no losses to themselves. The P-47s got three bombers and a fighter. One P-47 was reported missing. Pictures taken late that afternoon showed no aircraft, mostly bombers, on Borum, thirty-five fighters at Wewak, sixty mixed fighters and bombers at Dagua, and twenty fighters at But, a total of 225 Jap aircraft for us to get rid of before they took us.

Just before dawn on the morning of August 17th the big take-out of the Wewak airdromes began. Forty-one B-24s and twelve B-17s pasted the Nip airfields at But, Borum, Dagua and Wewak with 200 tons of bombs. Two of our B-24s were reported missing and another B-24 landed on the south coast of New Guinea with four dead crew members. The antiaircraft gunfire over Wewak was reported extremely heavy and accurate. Two hours later thirty-three B-25s with eighty-three P-38s as cover made a simultaneous attack on Borum, Wewak, and Dagua. Sixteen B-25s, scheduled to hit But, had run into bad weather and did not make the rendezvous. Lieutenant Colonel Don Hall, the same big-nosed little blond boy that first used my parafrag bombs at Buna in September 1942, led the B-25 line abreast attack on Borum. Coming in over the tops of the palm trees, Don saw a sight to gladden the heart of a strafer. The Jap bombers, sixty of them, were lined up on either side of the runway with their engines turning over, flying crews on board, and groups of ground crewmen standing by each airplane. The Japs were actually starting to take off and the leading airplane was already halfway down the runway and ready to leave the ground. Off to one side fifty Jap fighters were warming up their engines ready to follow and cover their bombers. Hall signaled to open fire. His first burst blew up the Jap bomber just as it lifted into the air. It crashed immediately, blocking the runway for any further Nip take-offs. The B-25 formation swept over the field like a giant scythe. The double line of Jap bombers was on fire almost immediately from the rain of fifty-caliber incendiaries pouring from over 200 machine guns, antiaircraft defenses were smothered, drums of gasoline by the side of the runway blazed up, and Jap flying crews and ground personnel melted away in the path of our gunfire, in the crackle of a thousand parafrag bombs, and the explosions of their own bomb-laden aircraft. We hit them just in time. Another five minutes and the whole Jap force would have been in the air on the way to take us out at Marilinan.

Wewak suffered the same fate. Thirty Jap fighters were warming up to take off when twelve B-25s caught them by surprise and duplicated the kill at Borum. Only three B-25s attacked Dagua but once again surprise paid dividends and twenty more Jap aircraft were burned and crossed off the Nip list, with at least an equal number damaged.

We found out afterward that the Japs referred to the attack as "the Black Day of August 17th" and that they had lost over 150 aircraft, with practically all the flight crews and around three hundred more ground personnel killed. All our P-38s and strafers returned to their home airdromes.

On the 18th we followed up with another attack in still greater strength. Seventeen B-24s and nine B-17s covered by seventy-four P-38s opened the show by unloading a hundred tons of bombs and incendiaries on the Wewak and Borum airdromes and the big supply installations and antiaircraft defenses in the vicinity of Wewak. Just as the bombs hit, fifty-three B-25 strafers swept simultaneously over the four Jap airdromes. The 3rd Attack Group cleaned up everything left at Borum and Wewak, while the 38th Group took care of Dagua and But. Running out of airdrome targets, the 3rd Group sank four 500-ton cargo vessels anchored off Wewak and set fire to several supply dumps at Borum that had been missed by the heavies.

The heaviest attack, that on Dagua, was led by Major Ralph Cheli of the 38th Group. Several Jap fighters concentrated their attack on him several miles short of the target, setting fire to his airplane while it was still about two miles from the Dagua strip. His speed was sufficient to have enabled him to gain enough altitude to parachute to safety but rather than take a chance on his formation becoming disorganized by leaving it, Cheli kept his position and led his squadron through the minimum-altitude attack. Having completed the most successful attack of the day, which wiped out every Jap plane on the field, Cheli then instructed his wingman to lead the formation back home and said he would try to make a landing at sea. He turned north toward the water which was only a few miles away and made it, but before he could land, one of the gasoline tanks exploded and the airplane plunged into the sea. For this action beyond all call of duty he was awarded the Congressional Medal of Honor posthumously. Between thirty-five and forty Jap fighters intercepted our attack. The P-38s from George Prentice's new group got fourteen of them and the bombers shot down eighteen more.

In addition to Cheli's plane we lost two more B-25s and a P-38.

During the two days' operations we had destroyed on the ground and in the air practically the entire Japanese air force in the Wewak area, had burned up thousands of drums of gasoline and oil, destroyed a number of large supply dumps, and knocked out the greater part of the antiaircraft gun defenses around Wewak. It was doubtful if the Nips could have put over a half dozen aircraft in the air from all four airdromes combined.

Arnold wired enthusiastic congratulations. I sent them on to Whitehead, added some of my own, and told him to keep on hitting those airdromes as long as the Japs persisted in putting airplanes on them. Marilinan had already paid for itself.

11. NADZAB AND LAE
September, 1943

WHEN THE 503rd Parachute Regiment arrived in New Guinea about the middle of August, I was afraid that the news of their presence might leak to the Japs, so we started calling them the Gasmata force, and all conversation, correspondence, and radios referring to our coming operations at Nadzab used the word Gasmata as a code name. In order to make sure that the Japs would believe the story if they did hear about it, I told Whitehead to put the Australian Beaufighter squadron to work on Gasmata and have any reconnaissance airplane going near the place drop a load of bombs on it sometime during the flight. The dummy-carrier scheme had, of course, died as I could not get a vessel to play with, but the idea still appealed to me so I started collecting a dummy invasion fleet. Air Commodore Hewitt, my former Director of Intelligence, Allied Air Forces and now commanding the RAAF units in Eastern New Guinea and Goodenough Island, was given the job of collecting native boats and repairing old Jap barges, dozens of which were scattered along the coast from Buna to Milne Bay. The scheme was to load these barges with dummy figures to look as though they were packed with troops and tow them with motorboats from Goodenough Island in the direction of Gasmata, some morning when the usual Jap reconnaissance plane came over from Rabaul. This time the fighters would not bother the Jap plane but let it report to the boys at Rabaul that the foolish Americans were moving an expedition toward Gasmata in broad daylight. It would look natural, as the speed of the dummy convoy would just about permit it to arrive off the New Britain coast at dawn the following morning, and such landings are traditionally made at first light. I intended to mass all the fighters I could spare to wipe out the Jap air attack which would almost certainly be launched as soon as they got the word from the recco plane. In order to make the "fleet" look real enough to fool the Nips, it would be necessary to have a couple of real destroyers act as escorts, at least until the Jap recco plane had discovered the show, but Admiral Carpender couldn't see it that way. I offered to fill the air with fighters to make sure no one would bother his destroyers, but he said he couldn't spare them and besides he needed them for the Lae operation and could not take a chance on having them put out of action prior to that time. General Blamey had thought my idea was excellent but when I suggested that he let two of his minesweepers act as escorts for my dummy fleet, he too decided that he could not spare his vessels. Hewitt, who in the meantime had been collecting native boats and repairing Jap barges, was quite disappointed when I finally had to tell him to forget the whole thing. I still think it was a good idea but as the time of the Lae operation approached I had plenty of other things to think about, so I just filed it away as another scheme that didn't materialize.

Information indicated that the Nips were repairing their damaged airdromes around Wewak and that some planes had been flown into Borum, so on the 24th, twenty-four B-24s and forty-eight P-38s went out to do something about it. The Borum runway had been repaired but 70 tons of bombs remedied that situation and incidentally destroyed at least five aircraft parked on the edge of the field. The Japs put thirty fighters into the air and lost nineteen of them. We lost a B-24 to antiaircraft gunfire and a P-38 was reported missing. Those thirty Jap fighters were made up of five different types and even included three two-engined night fighters. It looked as though the Japs were finding out what it meant to scrape the bottom of the barrel.

That afternoon the intercepted radio reports indicated that about fifty more Jap fighters were coming into Dagua and But.

The next morning nineteen B-25 strafers of the 38th Group with fifty P-38s of George Prentice's 475th Fighter Group struck the two Jap fields. Fifteen out of twenty airplanes on the ground were burned up by the B-25s, who also shot down three more Jap fighters as they were taking off.

The Japs managed to get between thirty-five and forty planes in the air but our P-38s knocked off thirty-one of them. Four of our P-38s were reported missing and one B-25 was damaged on landing back at the home airdrome.

For the next couple of days we continued our barge hunting, made some small-scale pecking raids on Gasmata, and let the fighters and most of the bombers and strafers rest as we watched the Wewak area. On the 23rd a recco plane reported some activity on the Wewak airdrome, so the next morning thirty-six B-24s smashed the Wewak township, the airdrome, and the supply dumps with 110 tons of bombs, while eleven P-38s that had gone along as escort chased ten uneager Jap fighters that had taken off on the approach of our bombers and headed west to get out of trouble. The P-38s caught up with one of them. That left nine that got away.

It wasn't until the morning of the 29th that Wewak looked worth hitting again. When we heard that another contingent of Jap fighters was coming into the two fields at Borum and Wewak, thirty-five B-24s and forty-four P-38s took off on the attack. Forty Jap fighters intercepted and in a furious fight, during which the Nips broke through the fighter cover and engaged the bombers, twenty-five enemy aircraft were shot down, of which the bombers accounted for eighteen. We had no casualties among the P-38s but one B-24 was shot down, another crash-landed at the home airdrome, and ten more suffered quite a lot of battle damage.

The following day twenty-four unescorted B-24s bombed Dagua and But, with no interference from the Japs. Before the raid, the photos showed a total of only thirty-eight aircraft on all four airdromes which could possibly be serviceable. After the B-24s had finished, it looked as though even that number had been substantially reduced.

The last two weeks in August had been rather hard on the Nip air force in the Wewak area. They had lost over 350 airplanes, two-thirds of which were fighters. In addition to the heavy losses in training flying personnel, it was estimated that at least 1500 ground-maintenance men had been killed during our raids. These would be hard for the Nip to replace.

The barge-hunting business during the last half of August had slackened off a bit, but the strafers had managed to destroy 57 and damage between 60 and 70 others.

Our combat losses during that period had totaled eighteen, of which five were heavy bombers, three were strafers, and ten fighters.

On the 24th General MacArthur and I had flown to Port Moresby, where he expected to stay until the Lae operation was over. Sutherland and most of the GHQ staff had preceded us by a few days and set up shop near the old Government House, where a lot of new temporary buildings had been erected during the previous few weeks. I moved into my own Fifth Air Force Headquarters, where I could eat Sergeant Raymond's food and have Foo, my little Chinese shadow, look after me.

On the 27th, Bob Patterson, the Assistant Secretary of War, and Lieutenant General Bill Knudsen arrived at Port Moresby for a visit, to find out at first hand what was going on, listen to our troubles, and see what could be done to help us out. General MacArthur and all the generals and admirals in New Guinea met the plane and Patterson was hustled off to Government House where he was to have dinner and stay as the General's guest. Knudsen whispered to me that if I could feed him he'd like to get away from all the brass hats and talk airplanes. I told General MacArthur I'd deliver Bill to Government House in time for breakfast the next morning, and Knudsen, Whitehead, and I adjourned to my headquarters. Knudsen said there was a lot of sentiment in Washington in favor of cutting down on the number of types of aircraft and he wanted to know how we felt about discontinuing the production of P-38s. Whitey and I took turns for over an hour explaining that the P-38, with two engines and the capability of flying on one, and with greater range than any other fighter we had, was the ideal airplane for the long over-water and over-jungle operations of our theater. We reminded him that if a pilot went down over Europe the worst that could happen to him was that he would be taken prisoner, with an excellent chance of living through the war. In the SWPA, however, landing in the ocean or in the jungle were both extremely hazardous, with little chance of survival, and if a pilot was taken prisoner by the Jap, torture and death were almost certain.

Knudsen didn't say a word as Whitehead and I pleaded, argued, and stormed at the stupidity of stopping P-38 production. We finally ran out of both breath and arguments. Bill turned to me and, without a trace of a smile and only the shadow of a twinkle in his big blue Danish eyes, said: "George, I gather you want P-38s. Okay, we'll build 'em for you." Then we all laughed and talked about something else until one o'clock in the morning, when I took him down to the 90th Group Headquarters to listen to the final instructions being given for the next morning's raid on Wewak. Bill was tremendously impressed by the teamwork and the thoroughness of the planning and said so in a little talk to the kids.

The next day I flew Patterson and Knudsen over the mountains to Dobodura, pointing out the points along the Kokoda trail where the

Nadzab and Lae

fighting had taken place during the campaign which ended with the capture of Buna. After showing them around the Dobodura airdromes, I drove them over to Buna itself. There were still plenty of signs of the terrific bombing the place had taken nearly a year before. Both our visitors were quite impressed with the bitterness of that struggle.

They left the next morning for Australia, after promising me that they would do everything they could to keep me supplied with B-24 bombers, B-25 strafers, P-38 fighters, and C-47 transports, the four types that I suggested standardizing on for operations in the Pacific.

On the 29th we had a final conference with Major General George Vasey on the airborne show to take out Nadzab. We decided to fly 200 Aussie pioneer troops with light engineer equipment and rubber boats to Marilinan. They were to paddle down the Watut River to the Markham River, and then down the Markham to Nadzab, arriving there about half an hour after the paratroopers landed, and help on preparation of a runway so that we could fly in Vasey's 7th Division. Vasey wanted his troops all flown to Marilinan right away so that they would be that much closer to Nadzab when the time arrived to put them into action. There was a lot of rivalry between Vasey and Wooten, the 9th Division commander, to see which would first enter Lae. Rumor had it that bets totaling twenty cases of whiskey hinged on the outcome. Vasey said he had no doubt that our preparatory strafing attack on Nadzab would remove all the opposition, but he would like to have some artillery on hand to assist the paratroopers with when they landed. I told him to give us one of the guns to play with and we would see if we could drop the main parts by parachute without damaging them. That afternoon, after several practice drops, the scheme looked feasible, so we added an Australian 25-pounder gun battery to the parachute show.

My two weather teams up to this time had agreed that September 4th was the day for the landing of the 9th Division at Hopoi Beach. They both had predicted that the fog over New Britain from Vitiaz Straits to Rabaul would prevent all flying by the Japs in that area until late afternoon. Over New Guinea they had forecast that during that day we would have some haze in the early morning, which would burn off by nine o'clock, and that for the rest of the day there would be blue sky all the way from Port Moresby to Lae and on to Wewak. They predicted that the weather for the 5th would be a repetition of that of the 4th.

On the 29th, however, while the Australian weather forecasters stuck to their original prediction, the American team decided that on both September 4th and 5th the weather would be good between Lae and Rabaul and to play safe we should wait until the 7th for the fog conditions that I wanted.

I told them to get together and settle on something soon, as I would have to tell General MacArthur the date by the evening of September 1st at the latest if we were to start the Lae operation on the 4th. The story got around GHQ that our weather boys were all tangled up, and a lot of the staff that hadn't approved of this method of setting the date of the assault or my scheme of ignoring Rabaul and concentrating on Wewak, began to shake their heads and whisper about the terrible calamity which might take place as a result of listening to my screwy ideas.

I told General MacArthur that on the evening of September 1st I would give him the date and time to start the show. He didn't seem at all worried. I didn't tell him so, but I was beginning to wonder which weather team to shoot.

That afternoon we rehearsed the coming Nadzab operation at full scale over Rorona, the old abandoned airdrome thirty miles up the coast from Port Moresby. The strafers actually fired, preceding the troop carriers loaded with the whole paratroop regiment, with some of the men actually jumping to check the timing of the show. A few minor details were corrected and everyone felt better about looking forward to the real thing.

On the morning of September 1st my weather teams were still hopelessly apart. The Aussies still said the 4th and the Americans stuck to the 7th as the proper date. That evening the Americans said that the day's weather data made the picture look a little different and they were willing to gamble on the 5th instead of the 7th. This sounded better but much to my horror the Aussies now said that the 3rd was the best day and that they were not so sure about the 4th as they had been.

I decided that neither of them knew anything about weather, split the difference between the two forecasts, and told General MacArthur we would be ready to go on the morning of the 4th for the amphibious movement of the 9th Division to Hopoi Beach and about nine o'clock on the morning of the 5th we would be ready to fly the 503rd Parachute Regiment to Nadzab. He passed the word to his staff and told them to start the operation on that schedule.

I discussed the details of the plan for covering and supporting the Nadzab operation and casually mentioned that I would be in one of the bombers to see how things went off. General MacArthur said he didn't think I should go. I said I had obeyed his orders to keep out of combat and that, with Wewak's air force out of the picture and a fog stopping the wild eagles from Rabaul from interfering with the show, I didn't expect any trouble. However, in any case this was to be my big day. If everything went all right, I would still be his Allied Air Force Commander. If the show went sour, I would be what a lot of his staff already thought I was. Furthermore, they were my kids and I was going to see them do their stuff.

The General listened to my tirade and finally said, "You're right, George, we'll both go. They're my kids, too."

"No, that doesn't make sense," I protested. "Why, after living all these years and getting to be the head general of the show, is it necessary for you to risk having some five-dollar-a-month Jap aviator shoot a hole through you?"

General MacArthur looked at me quite seriously and said, "I'm not worried about getting shot. Honestly, the only thing that disturbs me is the possibility that when we hit the rough air over the mountains my stomach might get upset. I'd hate to get sick and disgrace myself in front of the kids."

It was no use. I gave in and arranged for a "brass hat" flight of three B-17s to fly just above and to one side of the troop carriers as they came into Nadzab for the big parachute jump. I decided to put General MacArthur in one bomber, go in another myself, and have the third handy for mutual protection, in case we got hopped. He suggested that we go in the same airplane but I told him that I didn't like to tempt fate by putting too many eggs in one basket. He laughed but agreed.

Just as if we hadn't enough to worry about at this time, the Nips sent a convoy of six large and four small vessels into Wewak and our recco planes reported the Gloucester strip serviceable, with two bombers parked there. We hit Wewak the morning of September 2nd, with sixteen B-25s making a low-level attack which sank one of the small vessels, burned up two large cargo ships, and badly damaged another. The vessels were protected by barrage balloons, with at least one over each ship at altitudes of 500 to 1000 feet. They interfered considerably with the attack, as the B-25s had to change their whole scheme of approach to avoid the balloons. We decided that if the Nips tried the scheme again we would have the fighters clear them out before the B-25s went in. This was the first and only time, however, that we saw this scheme of protection of shipping tried. The B-25s were escorted by twenty-five P-38s, who shot down ten of twenty to thirty Jap fighters which intercepted and on the way home shot down two enemy bombers over Madang. We lost three B-25s and one P-38. The Nip fighters, who incidentally put up the toughest fight we had encountered in some time, were all flying Navy-type planes, which had probably been sent over from Rabaul to cover the shipping. That may have accounted for the sudden fixing up of the airdrome at Cape Gloucester, but assuming the possibility that the Jap suspected our imminent movement on Lae and was preparing the Cape Gloucester field as an operating base against us, we sent twenty-one bombers against it, loaded with 1000- and 2000-pound bombs to dig some nice holes in the runway. All our bombers returned, reporting shooting down seven of twelve intercepting Jap fighters and good results against the airdrome itself.

That evening the Jap radio messages between Rabaul and Wewak started coming through on a different frequency. It turned out to be the same one they had used the previous April at the time of their big daylight air raids on Port Moresby, Oro Bay, and Milne Bay. We never did find out the reason for the change, but coming at this critical time only forty hours from our scheduled landing at Hopoi Beach it worried me considerably. I checked again with my two teams of weather forecasters but got no comfort from them. They still couldn't agree on the proper date for the landing. I told them I could get just as good information by tossing a coin.

On the 3rd we gave the Lae area a farewell slugging with twenty-three heavy bombers, unloading 84 tons of bombs, mainly on the gun defenses, while nine strafers followed up with over 500 fragmentation bombs and 35,000 rounds of machine-gun ammunition. Just to make sure, we hit Cape Gloucester airdrome again with twenty-seven strafers, and that night nine Australian Catalinas bombed the Lakunai and Vunakanau airdromes at Rabaul.

Just before midnight the two weather teams came in to see me and reported that they were now in perfect agreement that the weather on the 4th and 5th was going to be just what I wanted, except that it might clear over the Solomon Sea between Rabaul and New Guinea for a couple of hours between four and six in the afternoon. I felt better, of course, and thanked them as enthusiastically as I could, but by this time I had begun to suspect that all weathermen belonged to one of the most ignorant of professions.

Their final forecast for September 4th turned out to be right on the nose. As eight B-25s strafed and bombed ahead of them, the Australians of Wooten's 9th Division landed at Hopoi Beach, while twenty-four B-24s shackled Lae's defenses with 96 tons of bombs. As the Aussies were landing, a couple of Jap airplanes that had evidently been hidden away at Lae sneaked in under the haze and destroyed one of the small landing barges and damaged another. Our fighter cover, flying above the haze, did not see the Nips, who made just one fast pass at the beach and disappeared over the hills headed north toward Madang.

The landing craft discharged the whole division and its supplies during the morning and headed back toward Buna. They had gotten as far as Morobe Harbor before the fog over the Solomon Sea cleared sufficiently around three o'clock in the afternoon to let about 100 Jap fighters and bombers get through and attempt an attack on our shipping. Sixty-four P-38s and P-47s engaged the Nips, shooting down twenty-three of them while the antiaircraft guns of the convoy got six more. We lost two P-38s and two of the landing craft were so badly damaged that they had to put into Morobe Harbor for repairs.

Colonel Neel Kearby, the commander of the new P-47 Group, the 348th, got two Nips in this fight. Kearby and a wingman spotted a Jap bomber and a fighter flying close together about 4000 feet below them. The two P-47s, with Kearby leading, dived to the attack. Neel lined up the two Nip planes in his sights, fired one long burst, and got them both. A wing tore off the fighter and the bomber exploded before Kearby's astonished wingman could get in a shot.

The seizure of Nadzab by the paratroopers on September 5th went off so well that it is still hard for me to believe that anything could have been so perfect. At the last minute the Australian gunners who were to man the battery of 25-pounders decided to jump with their guns. None of them had ever worn a parachute before but they were so anxious to go that we showed them how to pull the ripcord and let them jump. Even this part of the show went off without a hitch and the guns were ready for action within an hour after they landed.

When we got back to Port Moresby, General MacArthur swore that it was the most perfect example of discipline and training he had ever seen.

By late afternoon the paratroopers had moved out and secured the trail to Lae for a distance of about three miles. The pioneer troops coming down the Markham River in their rubber boats had cleared the Nadzab strip and we were ready to land the troop carriers with the 7th Division at daybreak the next morning. To the east of Lae, Wooten's Australians, in spite of considerable opposition, had advanced to within artillery range of Lae itself. As the kids said, we were now cooking with gas. Not a single Jap airplane had been seen in the area all day.

I wrote to Hap Arnold that night:

"You already know by this time the news on the preliminary moves to take out Lae, but I will tell you about the show on the 5th September, when we took Nadzab with 1700 paratroops and with General MacArthur in a B-17 over the area watching the show and jumping up and down like a kid. I was flying number two in the same flight with him and the operation really was a magnificent spectacle. I truly don't believe that another air force in the world today could have put this over as perfectly as the 5th Air Force did. Three hundred and two airplanes in all, taking off from eight different fields in the Moresby and Dobodura areas, made a rendezvous right on the nose over Marilinan, flying through clouds, passes in the mountains, and over the top. Not a single squadron did any circling or stalling around but all slid into place like clockwork and proceeded on the final flight down the Watut Valley, turned to the right down the Markham, and went directly to the target. Going north down the valley of the Watut from Marilinan, this was the picture: Heading the parade at one thousand feet were six squadrons of B-25 strafers, with the eight .50-caliber guns in the nose and sixty frag bombs in each bomb bay; immediately behind and about five hundred feet above were six A-20s, flying in pairs—three pairs abreast—to lay smoke as the last frag bomb exploded. At about two thousand feet and directly behind the A-20s came ninety-six C-47s carrying paratroops, supplies, and some artillery. The C-47s flew in three columns of three-plane elements, each column carrying a battalion set up for a particular battalion dropping ground. On each side along the column of transports and about one thousand feet above them were the close-cover fighters. Another group of fighters sat at seven thousand feet and, up in the sun, staggered from fifteen to twenty thousand, was still another group. Following the transports came five B-17s, racks loaded with 300-pound packages with parachutes, to be dropped to the paratroopers on call by panel signals as they needed them. This mobile supply unit stayed over Nadzab practically all day serving the paratroops below, dropping a total of fifteen tons of supplies in this manner. Following the echelon to the right and just behind the five supply B-17s was a group of twenty-four B-24s and four B-17s, which left the column just before the junction of the Watut and the Markham to take out the Jap defensive position at Heath's Plantation, about halfway between Nadzab ana Lae. Five weather ships were used prior to and during the show along the route and over the passes, to keep the units straight on weather to be encountered during their nights to the rendezvous. The brass-hat flight of three B-17s above the center of the transport column completed the set-up.

The strafers checked in on the target at exactly the time set, just prior to take-off. They strafed and frag-bombed the whole area in which the jumps were to be made, and then as the last bombs exploded the smoke layers went to work. As the streams of smoke were built up, the three columns of transports slid into place and in one minute and ten seconds from the time the first parachute opened the last of 1700 paratroopers had dropped. During the operation, including the bombing of Heath's, a total of ninety-two tons of high-explosive bombs was dropped, thirty-two tons of fragmentation bombs and 42,580 rounds of caliber .50 and 5180 rounds of caliber .30 ammunition were expended. At the same time nine B-25s and sixteen P-38s attacked the Jap refueling airdrome at Cape Gloucester. One medium bomber and one fighter on the ground were burned and three medium bombers and one fighter destroyed in combat. Two ack-ack positions were put out of action and several supply and fuel dumps set on fire. Between five and a half and six tons of parafrags were dropped and 19,000 rounds of caliber .50 ammunition fired. Simultaneously also, ten Beauforts, five A-20s, and seven P-40s from the R.A.A.F. put the Jap refueling field at Gasmata out of action. No air interception was made by the Japs on any of the three missions. Our only losses were two Beauforts shot down by ack-ack at Gasmata."

All day on the 6th, beginning at first light, we ferried troops of Vasey's 7th Division from Marilinan to Nadzab, while twenty-four heavy bombers plastered Lae with 82 tons of 1000-pounders and forty-eight B-25s dropped 61 tons of bombs and fired 75,000 rounds of ammunition in support of Wooten's advancing troops of the 9th Division.

Late in the afternoon about fifty Jap bombers and thirty-five to forty fighters from Rabaul appeared over Lae, but they abandoned the raid and made a hasty exit when they discovered nearly a hundred of our fighters barring the way. The Nips lost three bombers and six fighters. We lost a P-38. Our fighters made no attempt to follow the Nips home, as they had been ordered to stay with and give close cover to the stream of transports carrying Vasey's troops to Nadzab.

On the 7th, as we continued ferrying the Aussies into Nadzab and slugging away at Lae in support of Wooten's advance, the Japs made their final air-raid attempt against our Lae expedition. Their show ended ingloriously, when twenty-seven Jap bombers approaching Morobe Harbor jettisoned their bombs in the sea and made off toward Rabaul after three of their thirty supporting fighters had been shot down by the fifteen P-38s which intercepted them.

Tokio Radio reported heavy destruction of our shipping in Morobe Harbor and claimed that I had lost over fifty airplanes in the fight, so the returning Jap pilots must have told a pretty convincing story when they got back to Rabaul.

On the 11th, troops of the Australian 3rd Division and the 162nd Regiment of the American 41st Division occupied Salamaua airdrome and by the 13th had mopped up the town itself.

Photo reconnaissance on September 12th disclosed that the Nips had brought some more airplanes into the Wewak area, as there were 107 on the four airdromes that looked serviceable as compared with only eighteen on September 1st. On the morning of the 13th we raided But and Dagua with twenty-one heavy bombers escorted by twenty-five P-38s. Forty to fifty Jap fighters intercepted. Twenty-two of these were destroyed in combat, and in the airdrome attacks another twenty to twenty-five Nip aircraft were written off. That afternoon between eleven and fifteen bombers escorted by ten fighters raided Marilinan. They turned back before reaching the target, when intercepted by ten of our P-38s, who shot down one Jap bomber and one fighter.

The weather stopped us from repeating the attack until the 15th, when twenty-two B-24s escorted by fourteen P-38s raided Wewak and Borum. At least ten airplanes were destroyed on the ground, but the big party took place in the air. Our fighter cover was totally insufficient to stop the fifty to sixty intercepting Jap fighters from attacking the bombers, but the bombers did a phenomenal job of protecting themselves. The P-38s shot down nine Japs, with one P-38 missing. The B-24s shot down forty Jap fighters and all returned to their base, although fifteen of them were pretty well shot up. I had to send to Australia for some more medals.

On September 16th around noon we were still bombing and strafing the Jap defensive positions at Lae when a radio message came through in the clear: *"Only the Fifth Air Force bombers are preventing me from entering Lae.* VASEY."

The bombing stopped. The 7th Division had beaten Wooten's 9th, which had earned the nickname of "The Rats of Tobruk" a couple of years before in North Africa. The "Rats" entered Lae a couple of hours later and settled their bets. The next time we met, Wooten said that he suspected that I had had a bet on Vasey myself and had lifted the bombing on the west side of Lae just in time to make sure I would win.

They were two grand commanders of two grand fighting divisions.

12. MARKHAM VALLEY
September—October, 1943

IN ADDITION to his task of assisting Wooten in the capture of Lae, Vasey had been instructed to secure Nadzab from Jap forces known to be in the Markham Valley, and from other forces which might come down the coastal road from Madang to Bogadjim and then across the Bismarck Range, over the road that the Nips had been working feverishly on for the past four months to give them an overland route to Lae.

Strictly speaking, Vasey could have moved a few patrols ten or fifteen miles up the valley northwest of Nadzab and fulfilled his orders, but Vasey wanted to get rid of Japs and we wanted airdromes as far forward as we could get them. Marilinan and Nadzab would do nicely for operations against Wewak, but 275 miles farther west was Hollandia, which the Japs were making into another big air base. Some day we would have to eliminate that threat. I would then need another forward base for fighters in order to cover my bombers and strafers, as we had done at Wewak, using fighters from Marilinan to cover the show. The Markham Valley, with its wide, flat, well-drained grass plains stretching to the northwest for over a hundred miles, was made to order for quick airdrome construction.

Colonel David W. "Photo" Hutchison, who had been the air task-force commander at Marilinan and had moved over to Nadzab to take charge of air activities there, was told to work the problem out with Vasey. I didn't care how it was done but I wanted a good forward airdrome about a hundred miles farther up the Markham Valley. Photo Hutchison and George Vasey were a natural team. They both knew what I wanted and Vasey not only believed that the air force could perform miracles but that the 7th Australian Division and the Fifth Air Force working together could do anything. By this time those Aussies had already started calling themselves the 7th Airborne Australians.

Hutchison and Vasey got in a light plane, a Piper Cub, and flew up the valley toward a native village about forty-five miles from Nadzab, called Kaiapit, which was held by the Japs and which boasted a landing field used by the Nips as an emergency strip. About eight miles southeast of Kaiapit, near another native settlement named Sangan, the two aerial prospectors noticed a mile-long strip of level land from which the natives had recently burned off the heavy growth of kunai grass. The field was a little bumpy but it would do. They returned to Nadzab and racked up the mission.

The next afternoon, September 17th, thirteen C-47s, led by Major Frank Church, one of Hutchison's "hottest" troop-carrier pilots, landed a company of Australian troops at Sangan. Two of the airplanes were damaged in the landing, one beyond repair, but no one was hurt. The 250 Aussies, without waiting, started on foot for Kaiapit.

They arrived at daybreak on the 19th and, supported by a flight of P-40s that Hutchison had sent up from Marilinan to act as artillery, they took the village. The Japs had some light artillery and machine guns mounted on a hill overlooking the place, but the P-40s took care of that. During the battle the Aussies ran out of ammunition, but they finished the Nips off with their bayonets and hand grenades. The village was completely mopped up by the 20th and, using the old Kaiapit strip on which the natives cut the grass, we brought in food, ammunition, and reinforcements.

In the meantime, the Japs had decided to move a whole division of troops into the Markham Valley from Bogadjim, over their new road across the Bismarck Range. An advance detachment had been ordered to move to Kaiapit. This force of approximately 500 men arrived at the village and, thinking that it was still in Japanese hands, started to march in. The Australian scouts had given a few hours' warning and the ambush was ready.

The battle was short and decisive. The Aussies buried just over 300

Japs. The rest escaped into the jungle. Of the original 250 Aussies who had landed on the 17th at Sangan, seventy-five were dead and fifty were wounded.

On the 22nd Photo Hutch's eighteen C-47s made ninety-nine round trips from Nadzab to Kaiapit, bringing in 2000 troops, artillery, antitank guns, rations, and ammunition for ten days, and evacuated the wounded. A flight of P-39s from Marilinan hovered over the route to insure that there was no interference with the movement.

On the 26th three C-47s, loaded with a small bulldozer, some engineers, and more Australian troops, landed thirty-five miles farther up the valley at the native village of Marawasa, laid out a landing strip, and called for another troop movement. The next day seventy-five planeloads of troops and supplies were landed and we owned another oasis in the valley.

Small detachments of enemy forces were wandering around between Kaiapit and Marawasa, but they were leaving clothes and weapons behind them as they hurriedly worked north to join their main forces and get instructions about what to do in this crazy kind of warfare.

On the 28th, patrols who had moved twenty-five miles farther along the valley to Gusap, cleared the grass for another strip and the following day 350 Aussies were flown in to secure the place. They had a brief encounter with a few Japs just outside the village and we owned another airdrome. This site was about where we wanted it, gravel was plentiful for surfacing a runway, and there was an excellent water supply in the Gusap River close by. We decided this would be the main base and started moving in an airborne engineer battalion, with their equipment and supplies. I formed another air task-force headquarters to operate the projected Gusap air base and put Colonel Don Hutchinson, whom we called "Fighter Hutch" to distinguish him from "Photo Hutch," in command. Fighter Hutch had been Squeeze Wurtsmith's chief of staff of the Fifth Fighter Command. He was a short, slight, brown-eyed, bald-headed veteran of nine years' service, and that nickname, Fighter, really fitted him.

To secure Gusap, more Australians were flown in and, on October 2, the Aussies were in Dumpu, thirty-five miles farther up the valley from Gusap. Vasey hoisted his 7th Division flag there and established his headquarters for the drive which he planned would chase the Nips back over their own road to Bogadjim and Madang. By "getting thar fustest with the mostest," we had stolen the whole valley from the Japs in fifteen days.

Working around the clock, with scarcely time to eat for the past two months, my New Guinea staff was worn down. Freddy Smith was sent to Brisbane with some unexplained fever, Cooper went to Southport, a seacoast resort town fifty miles south of Brisbane, to rest. He had been averaging about two hours' sleep out of the twenty-four for over a month and I was afraid a little more of that would break him. Roger Ramey was so sick that I had to send him to Australia for a leave, and Whitehead was worn down to a shadow, losing weight and unable to sleep. I had gone back to Brisbane the day Lae was captured but flew back to Port Moresby on the 19th, sent Whitehead south for a rest, and took over.

Our reconnaissance kept watching the progress of the Japanese division moving toward the Markham Valley from Bogadjim, and we started daily strafing of everything along the road and bombing the wooden bridges which the Nip engineers seemed to be able to repair almost as fast as we destroyed them. However, we could not allow anything to interfere with the Vasey-Hutchison Markham Valley campaign.

The Jap air reaction during the week following the capture of Lae was puzzling. I did not expect much reaction from Wewak, as we had beaten the place to death, but the Nip had a lot of aircraft around Rabaul and the weather was good enough for flying the route to Lae. I had expected trouble at Nadzab, as we had practically no warning service installed yet. The Nip air force, however, left us alone until the late afternoon of September 20th, when nine Jap bombers escorted by ten fighters made a feeble raid on Nadzab, which killed two men working on the field but did no other damage. A flight of P-38s got three of the Nip fighters. That night one bomber dropped a few bombs near Marilinan and two others raided Port Moresby, also without effect.

When the alarm sounded that night as the Jap bomber approached Nadzab, on what later turned out to be an attack on Marilinan, Photo Hutchison and three members of his staff were playing bridge by lantern light in a tent near the Nadzab strip. Hutch had just bid a grand slam, which had been doubled. The opening lead had been made and the dummy hand laid down just as the siren blew. Someone moved to turn off the lights. Hutchison would have none of it. He was going to play that hand, war or no war. Pulling his pistol from his shoulder holster he laid it on the table and announced that the siren meant that he had five minutes before the bombs began to drop, that he could play the hand in less time than that and he intended to play it. The hand was played. Hutch made his bid. The lights were extinguished and everyone ran for a slit trench. A few minutes later the all-clear signal was sounded and his opponents tried to explain to Hutch that the hand should not count as they had not had their minds on the game. It was no use. Hutchison insisted that his mind had been on the game all the time and it was their own fault if they couldn't concentrate with nothing more serious than an air raid going on.

On the 21st, twenty to twenty-five Jap bombers and fighters appeared over Hopoi Beach; there they turned and headed back toward Rabaul, where sixteen of our P-40s intercepted. We got nine Nip aircraft. They shot down one P-40.

I still couldn't figure it out. Of course, by this time I had over 300 fighters in New Guinea, but we were covering an area as big as the whole northeastern part of the United States north and east of a line from Washington to Buffalo. We had to look after Kiriwina and Woodlark Islands, cover the shipping at Milne Bay, Port Moresby, and Oro Bay, convoy the shipping from Buna to Lae, escort the transports from Port Moresby to Nadzab, from Nadzab up the Markham Valley, and protect the bombers hitting the Bogadjim road. There was still plenty of room left in the air, if the Japs wanted to use it.

On the 22nd, the 20th Brigade of Wooten's 9th Division landed two miles north of Finschaven without opposition. After feeble resistance, the Jap garrison withdrew to their defenses in the town. Late in the afternoon the Japs from Rabaul timed a big air attack just right for us. I had four squadrons of fighters overhead, covering the unloading of Wooten's troops and supplies. Four other squadrons had just arrived to take over the patrol, when the Nip force of fifty-five airplanes, twenty bombers and thirty-five fighters arrived on the scene.

With over a hundred of my fighters opposing, it was just too bad for the Japs. We shot down forty of the Japs' aircraft and several others were listed as probably destroyed. A destroyer up near Vitiaz Straits on antisubmarine patrol duty reported that only five Jap planes passed by him on the way back. We had three planes shot down and rescued one of the pilots.

Our recco planes reported that the Nips seemed to be coming to life again in the Wewak area. On the 26th eleven B-24s escorted by twenty-seven P-38s laid 33 tons of "eggs" on But and Dagua. Fifteen Jap fighters took off to argue about the matter. The P-38s got nine of them. All our planes returned.

That afternoon General Summerville, the top War Department Supply Chief, and Major General Echols, Arnold's Supply man, arrived in Port Moresby from Washington to see what we needed and what they could do about it. We told them plenty. They both promised to try but reminded us that Europe had the first call. They got a good picture of the conditions that we were up against and actually did get a lot of equipment sent out to us that we had been unable to get before their visit.

The next day I flew them over Marilinan and Nadzab and landed at the old Jap field at Lae. They were tremendously impressed by the destruction in the town, where not a building remained intact and the defensive positions had been blasted out with hundreds of heavy bombs. When they counted over 150 wrecked Jap aircraft scattered around the perimeter of the Lae airdrome and in a big junk pile on the edge of the field, they both admitted that for the first time they realized that we had a real war on our hands.

While we were inspecting the airdrome, the Aussie troops on guard there picked up three Japs who had been hiding among the wrecked airplanes and living on food stolen from the Australian ration dumps. When we saw them, the three Nips were sitting apprehensively in a bomb crater with two Australian soldiers guarding them and looking as if they wished the prisoners would try to escape. They were waiting for an Aussie intelligence officer to come from headquarters and take charge of them. The Japs had evidently done pretty well at foraging. They looked much fatter than most of the others who were still being buried all around the town.

We then flew down to Dobodura. Summerville stayed with the SOS crowd at Oro Bay, while Echols and I went over to the 3rd Attack Group headquarters for the night. Colonel Don Hall had gone back to the States and the new commander was a big, good-looking ex-Notre Dame football player, named Lieutenant Colonel John P. (Jock) Henebry, who had flown a wing position alongside Ed Larner at the Bismarck Sea Battle and participated in practically every big strike the 3rd Attack Group had made for the past year. You only had to talk with Jock for five minutes and you knew that he had what it took to lead any organization in peace or war. Jock fixed us up for the night and then invited us over to the opening of the group officers' club, which he claimed was the best thing this side of Hollywood. After a shower and clean clothes, we followed Jock along a winding path lit with colored electric lights, by the side of a stream, and suddenly turned into a little clearing, at the end of which stood a huge bungalow type of palm-leaf-and-bamboo structure, with a twelve-foot-long neon sign which blazed on and off with more colored lights, spelling the name "Tropical Paradise." Where those rascals got that neon sign I had no idea and I didn't see any reason for asking. Bursting with pride, Jock took us into the club, showed us around, and introduced us to the crowd gathered for the grand opening.

I turned Echols over to Lee Van Atta, the INS correspondent who had attached himself to the 3rd Attack Group and was flying with them almost every day, to get material for his stories. They were getting a tremendous play back home. Lee promptly gathered in Major Richard H. Ellis, Captain Chuck Howe, and Captain Ken Rosebush, three of the real "hot rocks" of the group and experts of the skip-bombing technique that had already made the group sensationally famous at the business of sinking Jap shipping. Echols said later that they backed him into a corner and practically talked him to death, but that he was willing to repeat the evening all over again any time it could be arranged. Those kids had so much fire and enthusiasm you couldn't help getting that way yourself if you listened to them very long.

There was plenty of good Australian beer and the dinner was excellent. I could have asked questions about that, too, but I didn't. I didn't have to. The Liberty boat crews who were constantly coming into Oro Bay and Buna lived exceedingly well, and it wasn't long before some of my enterprising aviators found that a case of cheap Australian gin, which could be smuggled up on a newly delivered airplane, made wonderful trading material, particularly when the other party, with access to a well-stocked larder, wanted some gin but had none.

The next morning ninety-seven B-25s and eighteen B-24s escorted by 128 fighters raided the Wewak airdrome, the supply installations in Wewak itself, and the shipping in the harbor. Forty Jap planes were destroyed on the ground and eight of the ten Nip fighters that got into the air were shot down. Huge fires were set in the town and several gasoline and oil dumps blazed up as the strafers swept over the area. In the harbor we made a real killing, sinking every ship in sight. The bag included a 7000-ton cargo vessel, three 2000-ton tankers, nine 300- to 500-ton luggers, six large power barges, and two motor patrol craft. We lost two B-25s.

I took Echols down to the interrogation of the 3rd Attack Group crews which had participated in the strike. He told me afterward that he had never seen such spirit and morale and that he no longer wondered why I bragged about my kids.

The low-altitude pictures taken during the Wewak attack had disclosed a number of huge ammunition dumps and thousands of drums of gasoline, hidden under the trees and camouflaged so well that the high-altitude photographs had not picked them up. On the 28th forty B-24s with twenty-nine P-38s practically blew up and burned Wewak out. The big ammunition dump exploded with so much force that it turned two of the B-24s over on their backs, although they were flying at 12,000 feet. Black smoke from burning fuel billowed up higher than the bombers, followed by flames that were visible for fifty miles. The fighter escort shot down eight out of thirty Nip fighters who intercepted. We had no losses.

We had a lucky break on the Bogadjim road. Eight B-24s tried to bomb the biggest bridge along the route, but missed. The bombs, however, struck the side of the mountain rising abruptly from the river and caused a huge landslide, which blocked the road for nearly a mile and dammed up the river to a height of thirty feet above the water. The river rose rapidly and, on September 30th, our recco planes reported that the dam had finally burst, washing out several miles of road and taking out the large bridge we were trying to hit and two other good-sized bridges farther downstream.

On that same day, with Whitehead and Cooper back on the job and rested, I went back to Australia and took a few days off myself.

I came back to the office on October 3rd and found a tall, slim, good-looking, burning-eyed sergeant named Ernest Moser waiting for me with a note from Jock Henebry, the commander of the 3rd Attack Group.

Some months before Moser had started out with Jock as his aerial engineer and top-turret gunner. In a combat shortly afterward the co-pilot had been badly wounded and the sergeant had taken his place. When the co-pilot went to the hospital, Jock had put Moser on the crew in his place. It seemed that prior to his enlistment in the Air Force the sergeant had about 1300 hours' flying time as a private pilot, so the B-25 didn't faze him a bit. He had now participated in fourteen combat missions, during which Jock said the kid had done most of the flying as well as the take-offs and landings.

Some of my hawk-eyed inspectors had found out about this unauthorized flying on the part of the lad who had not been officially rated as a pilot by the Air Corps Training Center back home and had ordered Jock to pull him out of the copilot's seat.

The youngster wanted to fly in combat. In fact, as I listened to him, pleading with a catch in his throat I knew he wanted it harder than anything else in the world. He didn't care whether he wore wings or not. I could demote him to a private if I saw fit, but, "Please, General, let me fly combat." Jock had said practically the same thing in his letter.

I sent the sergeant out to Amberley Field and told the chief test pilot out there to check his proficiency on all the types of aircraft at the depot.

The next day I was told he had flown every type of single-engined and two-engined airplane we had and that the rest crew rated Moser as a superior pilot. Although it was contrary to the rules, I assigned him as a co-pilot to a troop-carrier squadron to keep his hand in, put in an application to GHQ in Brisbane to make him a second lieutenant, and wrote General Arnold asking him to waive the rules as he had for Pappy Gunn and give the kid an official pilot's rating.

It took quite a while to break through the various walls of rules, regulations, and objections to such unorthodox procedure, but about six weeks later I was able to pin the second lieutenant's bars and a pair of silver wings on ex-Sergeant Ernest Moser and send him away happy, on his way to New Guinea and combat duty.

After our devastating raids on the Wewak airdromes during August and September, the Japs gradually filled in the bomb craters on the runways but showed little inclination to replace their airplane losses there. Instead they began to base more and more aircraft at Wakde Island, Hollandia, and Tadji, occasionally using the Wewak airdromes as forward refueling points for raids on Nadzab and Finschaven. For the first ten days in October our photographs showed only about seven to ten serviceable aircraft on each of the four strips. We didn't trust the Nip, however, so our reconnaissance aircraft kept the area under constant surveillance and to make doubly sure we ran a patrol of a flight of fighters up there once in a while to see if any of the Nips could be induced to come up and accept the challenge.

Colonel Neel Kearby, the commander of the 348th P-47 Group, decided that this was a good way for him and members of his group headquarters staff to keep their hands in. Kearby had already established himself as a superb shot and combat leader, even in the short time he had been in the theater, but on October 11 he set a real mark for the boys to shoot at. With three other pilots, Major Raymond K. Gallagher, Captain John T. Moore, and Captain William D. Dunham, he was making a sweep over Wewak on that date when he sighted a single Jap plane below him. Kearby, followed closely by his "Three Musketeers," promptly dived on the Nip and shot him down in flames. As the quartet pulled up they sighted a Jap formation of 33 bombers and 12 fighters dead ahead. Without hesitating, Kearby gave the signal for an attack. Plunging into the midst of the enemy he quickly shot down three planes, then reversed his course and knocked down two more that were on the tail of one of the P-47s. Kearby then pulled off to one side and, seeing one of his formation threatened by a Jap fighter diving from above, made a final pass, practically cutting the Nip airplane in two with one burst from his eight fifty-caliber guns. With his plane getting low on fuel, he then reformed his flight and led it safely home. The other three members of his flight had gotten two more definite victories and another Jap plane went down on fire, but everyone had been too busy to notice whether or not it had crashed. It was listed as a probable. The whole Jap formation had turned and was headed west as our fighter patrol left the area.

General MacArthur and I had flown to Port Moresby the day before to be on hand for a big strike I was planning against Rabaul on the 12th. Kearby came over to my headquarters soon after he landed and told us the story of the fight. I remarked to General MacArthur that the record number of official victories in a single fight so far was five, credited to one of the Navy pilots, and that he had been awarded a Congressional Medal of Honor for the action. I added that as soon as I could get witnesses' statements from the other three pilots and see the combat camera-gun pictures, if the evidence proved that Kearby had gotten five or more Nips, I wanted to recommend him for the same decoration. The General said he would approve it and send it to Washington recommending the award. We both congratulated Kearby and I left for Fighter Command Headquarters to get the evidence.

The testimony of the rest of his flight, and the pictures, confirmed the first six victories beyond a doubt. At the time Kearby got the seventh, however, the other three pilots had been so absorbed in extricating themselves from combat that they had not watched his last fight and, much to my disgust, the camera gun had run out of film just as it showed Kearby's tracer bullets begin to hit the nose of the Jap plane. I wrathfully wanted to know why the photographic people hadn't loaded enough film, but they apologetically explained that this was the first time anyone had ever used that much. They hadn't realized that enough film to record seven separate victories was necessary but from now on they would see that Kearby had enough for ten Nips.

I put in the recommendation for a Medal of Honor for six victories in one combat, which, after all, was a record unequaled up to that time by any aviator of any nation in the whole history of air combat. Those six victories ran Kearby's total to fourteen.

13. TAKING OUT RABAUL
October—November, 1943

FOLLOWING the agreements reached in Washington during March 1943, between representatives of SWPA and SOPAC, the Joint Chiefs of Staff had instructed MacArthur to capture Lae and to secure control of the Vitiaz Straits by taking Finschaven and Cape Gloucester. In the meantime, Halsey was to push north through the Solomons and into Bougainville Island. In order to help Halsey's advance, the Allied Air Forces of SWPA were to keep Rabaul beaten down as much as possible commensurate with supporting the New Guinea campaign. Prior to our operation to capture Cape Gloucester, Halsey was to get airdromes established in Bougainville so that his SOPAC aircraft could take over the responsibility for Rabaul. The setting of the dates was left to MacArthur.

During September, at a number of conferences between the staffs of the two theaters, the schedule of these operations was fixed. Halsey was to speed up his airdrome construction in New Georgia so that by October 15th his bombers and fighters could neutralize the Jap airdromes on Bougainville Island and deny the Buin-Faisi anchorage to Jap shipping. On November 1st, following a preliminary operation to seize Treasury Island about forty miles south of the Buin-Faisi anchorage on October

27th, he would land at Empress Augusta Bay on the west coast of Bougainville and build an airdrome from which his aircraft could take over the neutralization of Rabaul.

My job was to pound away at the Rabaul area, beginning about the 12th of October, to get rid of the Jap air force there, destroy the supplies in the town, which were estimated at over 300,000 tons, and by sinking the shipping in the harbor make the place untenable for Jap vessels.

On the 15th of December, MacArthur's ground forces were to land on the south coast of New Britain near Gasmata, and on December 26th, the 1st Marine Division, which was then in the Melbourne area getting rid of the malaria acquired in the Guadalcanal campaign, would go ashore at Cape Gloucester under operational control of General Krueger, MacArthur's 6th Army Commander.

On October 12th we hit Rabaul with the biggest attack so far in the Pacific. A total of 349 aircraft took part, including eighty-seven heavy bombers, 114 B-25 strafers, twelve Australian Beaufighters, and 125 P-38s, with some miscellaneous weather and photographic aircraft. Everything that I owned that was in commission, and could fly that far, was on the raid.

The B-25s and the Beaufighters opened the attack with simultaneous low-altitude sweeps of the Jap airdromes at Vunakanau, Rapopo, and Tobera and achieved complete surprise, destroying at least 100 airplanes on the ground and badly damaging another fifty-one. Huge fires blazed up from burning fuel dumps, and heavy explosions at Vunakanau indicated that the Japs would have to restock their bomb and ammunition dumps at that base.

The B-24s attacked the shipping in Rabaul Harbor, sinking three merchant vessels of between 5000 and 7500 tons each, forty-three small merchant ships of 100- to 500-ton size, and seventy small harbor craft. In addition, direct hits were scored, badly damaging five more medium-sized cargo vessels, a destroyer tender, three destroyers, and a tanker.

Photographs taken the 11th had shown 128 bombers and 145 fighters on the four airdromes in the Rabaul area, but the Nips managed to put only thirty to thirty-five fighters in the air against our raid. The P-38s swarmed all over them and shot down twenty-six.

Our losses were two B-24s, one B-25, and one Beaufighter, all to antiaircraft gunfire from the ground.

By the time I got through listening to the interrogation of the crews I was proud of the Fifth Air Force. Whitehead had done a great job with them. His planning had been faultless and his detailed instructions had been carried out to the letter. As he said, this was the first team. Their timing, precision, and bombing had been excellent and, in spite of extremely heavy antiaircraft fire, every airplane but one had attacked its assigned target. During the approach to Rabaul Harbor and while still about three miles from the town, one of the B-24s had been knocked almost over on its back by an antiaircraft shell which had burst just under a wing. In order to gain control, the pilot ordered the four-ton load of bombs released. As they hit the ground, a huge explosion occurred and great clouds of smoke billowed up over a huge area. A direct hit had been inadvertently scored on a big Jap fuel dump that we hadn't known even existed. Even Lady Luck was working for us that day.

The weather stopped us from repeating the attack for nearly a week, but while Port Moresby was fogged in, the weather between Rabaul and New Guinea cleared on the 15th enough to let thirty Jap dive-bombers with sixty-five fighters attack our shipping at Oro Bay. The shipping suffered no damage but the Nips ran into considerable trouble. The raid was intercepted by seventy-five of our fighters from Dobodura. When the smoke cleared away the Japs had lost twenty-seven dive-bombers and twenty fighters. We lost a P-38 but recovered the pilot, a thin little black-mustached lad named Captain Tommy McGuire, who had knocked down three Nip planes before one of his engines was set on fire and he had to take to his parachute.

Incidentally in this flight, Lieutenant LeRoy R. Donnell shot down the 300th enemy plane for the 49th Fighter Group. The kids brought out the magnum of cognac that evening and patted it for luck as they chalked up the score.

On the 16th the weather still wouldn't let us at Rabaul but it was good toward Wewak, so we sent the 3rd Attack Group and an escort of forty P-38s up there to make certain the Nips realized that we were in earnest about not wanting them to build up again in that area. The recco boys had said that it looked as though the Nips had brought in about fifty fighters in the last day or two. They were right.

Between forty and fifty Nips contested our raid. We shot down thirty-three and destroyed four more on the ground. All our planes returned.

We repeated the attack the next day, getting fifteen more planes on the ground, but only five Japs took off to intercept. The kids got four of the five.

In the meantime, the Nips came down from Rabaul with another formation of around fifty fighters and dive-bombers to repeat the attack on Oro Bay. Forty-six of our fighters intercepted. The Nip dive-bombers immediately dropped their bombs in the sea and fought back ferociously.

The Japs lost twenty-four aircraft. We lost five. At the same time one of our patrols of four P-39s over Finschaven ran into six Jap bombers and twelve fighters and shot down four of the bombers and two fighters, without loss to themselves.

We tried for Rabaul again on the 18th. Off Gasmata the weather shut down to a ceiling of about 200 feet and the heavy bombers and the fighters turned back. Lieutenant Colonel True, leading fifty-four B-25s, pretended that he didn't hear the fighters say that they were leaving and kept on flying just over the waves until they hit the New Britain coast. How True navigated I don't know but he hit Tobera airdrome on the nose and practically ruined the place. Thirty-nine airplanes were destroyed on the ground and the Japs suffered a heavy loss of life, as mechanics and crews were clustered around their airplanes warming up the engines and were killed in their tracks. As the group passed over Tobera they broke into clear weather. Just ahead of them was a freighter escorted by a corvette. Both were sunk. Fifty fighters, who roared down from Lakunai and Vunakanau, tried to interfere but found the tight defensive formations of the B-25s flying right on the deck a difficult problem to solve. The Japs lost thirty-two more planes in the air combat. Three of our B-25s failed to return.

When True returned, I called him over to headquarters and bawled him out for disobeying our standing instructions that bombers were also to turn back if the fighters had to. I told him that if he repeated the offense I would chase him home and then gave him a Distinguished Service Cross. After congratulating him on the success of his mission, I said, "Now that it's all over, tell me, True, didn't you hear the P-38s say that they were going home?" The rascal looked me in the eye, grinned, and said, "General, I didn't hear a word."

He was lying and he knew that I knew it but he was sticking to his story. I said, "Okay, we'll forget it, but don't do it again." He said he wouldn't, saluted, and left. I had a suspicion he didn't have a single regret, in fact, was rather proud of himself. As a matter of fact, I was kind of proud of him, too.

While the weather kept us grounded at Port Moresby the Japs raided Finschaven weakly on the 19th with fourteen bombers and again on the 20th with thirty planes, doing minor damage only. We managed to put over one raid of twenty-five B-25s on Wewak on the 22nd and sank two 500- to 1000-ton cargo vessels, destroyed twenty barges, and burned up fifteen Jap aircraft on the ground without any losses to ourselves.

On the 23rd the weather boys gave us a break. Forty-five B-24s and forty-seven P-38s plastered Rapopo, destroying twenty planes on the ground and shooting down twenty Jap fighters of the fifty that intercepted. All our bombers returned, but we lost two P-38s.

The next day sixty-two B-25s escorted by fifty-four P-38s swept over Vunakanau, Rapopo, and Tobera, getting forty-five more Nip planes on the ground. The P-38s shot down thirty-five Jap fighters and the B-24s got eight out of the fifty-five to sixty that intercepted. The afternoon reconnaissance reported the Japs were still flying heavy air reinforcements in to Lakunai and Vunakanau.

On the 25th sixty-one B-24s and fifty P-38s worked over Lakunai and Vunakanau, destroying twenty-one planes on the ground. Between sixty and seventy Nip fighters intercepted. We shot down thirty-seven. We lost a B-24.

For the next three days the weatherman refused to play ball with us. Then, about the time I decided to calm down and wait for the sun to come out, I got a wire from General Arnold in answer to one of mine begging him to hurry up replacements of my P-38s; he said that on account of the serious situation in England he could not give me any more of that type or any more P-38 pilots for the next two months. With the constant losses and wear and tear during the past month, my available number had steadily gone down from the 125 with which we covered the October 12 strike on Rabaul to about 50. A few more missions and I would not be able to give enough support for a decent daylight raid without exposing my bombers to inordinate losses. Furthermore, my bomber losses were not being replaced fast enough to keep up with the attrition we were suffering. My bomber availability was also going down fast. I sent Arnold another wire, pointing out what he was doing to me and yelling for help.

On the 29th we went back at Rabaul. The bad weather had helped our fighter maintenance and we managed to send fifty-three P-38s to Rabaul escorting thirty-seven B-24s. We burned up twenty Jap planes on the ground and it cost the Nip twenty-five more, shot down in combat. We lost no aircraft, but a lot of them were pretty badly shot up. I decided to put on the next big show on the 31st, the day before Halsey was due to land at Empress Augusta Bay.

Our photo reconnaissance of the 29th showed that the total serviceable Jap aircraft in the Rabaul-Kavieng-Bougainville area numbered only fifty fighters and twenty-five bombers. We didn't know it at the time, but during the next two days 115 more fighters and twenty-five additional bombers were flown in to the Rabaul area and another forty-five fighters and fifteen more bombers were at Truk, ready to fly south.

The weather stopped us from hitting Rabaul on the 31st of October and again on November 1st. The only ray of sunshine was a wire from Arnold which said he would give me forty P-38s sometime before January 1st instead of the 100 I was originally scheduled to get.

Halsey landed the 3rd Marine Division at Empress Augusta Bay that day as scheduled. No opposition to the landing was encountered.

Rabaul, November 2, 1943, was the title of another big event in the history of the Fifth Air Force, destined to be compared with the Bismarck Sea Battle and the take-out of the Jap air force at Wewak on August 17th of that same year. It was not a big show as far as numbers went. Seventy-five B-25 strafers and fifty-seven P-38s did the job, but what a job! Nine B-25 squadrons and six P-38 fighter squadrons, all at about half strength, were represented.

Two fighter squadrons swept in ahead to cover four B-25 squadrons, led by Major Ben Fridge, who were to neutralize the antiaircraft positions around the harbor with machine-gun attack and phosphorus smoke bombs. Lt. Col. Jock Henebry led the other five B-25 strafer squadrons in to Rabaul Harbor to take out the shipping, covered by the remaining fighters, led by Captain Jerry Johnson, with Captain Dick Bong as his deputy.

The phosphorus bomb attack was a distinct success, creating a thick wall of smoke, which blanketed practically all the shore antiaircraft gun stations, and in addition setting most of the town on fire.

Driving through and over the smoke, Henebry's attack swept across the harbor at mast height with a rain of machine-gun fire and heavy bombs that sank or damaged every ship in its path, in spite of an intense concentration of fire from the eighteen naval vessels and twenty merchant ships under assault. Of the thirty-eight vessels in Rabaul Harbor that day, thirty received direct hits in the toughest, hardest-fought engagement of the war. The list included one heavy cruiser, one destroyer tender, one submarine tender, three destroyers, two naval auxiliary craft, three minesweepers, sixteen merchant vessels, two tankers, and a tug.

We had expected to be opposed by perhaps fifty to sixty Jap fighters at the most, but between 125 and 150 fighters engaged the P-38s and broke through to attack the B-25s during their strafing and skip-bombing of the shipping. These were no amateurs, either. They put up the toughest fight the Fifth Air Force encountered in the whole war. The P-38s definitely destroyed forty-one Jap fighters and claimed thirteen more as probables. The B-25s got twenty-seven definites and ten probables. In addition, after putting down his phosphorus smoke cover, Fridge's group destroyed ten floatplanes sitting in the harbor and burned up seven more Jap fighters on the ground at Lakunai airdrome, making our total score for the day eighty-five definitely destroyed and twenty-three probables.

We had six B-25s shot down in the harbor. Three were reported missing, although we picked up most of their crews later on.

One of the B-25s shot down was piloted by Squadron Leader Major Raymond Wilkins. The citation for the Congressional Medal of Honor, awarded him posthumously, tells the story:

"Major Raymond H. Wilkins, 0-429531, Air Corps, United States Army. For conspicuous gallantry and intrepidity above and beyond the call of duty in action with the enemy near Rabaul, New Britain, on 2 November 1943. Leading his squadron in an attack on shipping on Simpson Harbor, during which intense antiaircraft fire was expected, Major Wilkins briefed his squadron so that his airplane would be in the position of greatest risk. His squadron was the last of three in the group to enter the target area. Smoke from bombs dropped by preceding aircraft necessitated a last-second revision or tactics on his part, which still enabled his squadron to strike vital shipping targets but forced it to approach through concentrated fire, and increased the danger of Major Wilkins' left-flank position. His airplane was hit almost immediately, the right wing damaged, and control rendered extremely difficult. Although he could have withdrawn he held fast and led his squadron in to the attack. He strafed a group of small harbor vessels, and then, at low level, attacked an enemy destroyer. His thousand-pound bomb struck squarely amidships, causing the vessel to explode. Although antiaircraft fire from this vessel had seriously damaged his left vertical stabilizer, he refused to deviate from the course. From below mast-head height he attacked a transport of some nine thousand tons, scoring a hit which engulfed the ship in flames. Bombs expended, he began to withdraw his squadron. A heavy cruiser barred the path. Unhesitatingly, to neutralize the cruiser's guns and attract their fire, he went in for a strafing run. His damaged stabilizer was completely shot off. To avoid swerving into his wing planes he had to turn so as to expose the belly and full wing surfaces of his plane to the enemy fire; it caught and crumpled his left wing. Now past control the bomber crashed into the sea. In this fierce engagement Major Wilkins destroyed two enemy vessels, and his heroic self-sacrifice made possible the safe withdrawal of the remaining planes of his squadron."

Jock Henebry's airplane, riddled with bullet and shell holes and with one engine gone, landed in the water just short of Kiriwina. Jock and his whole crew were rescued that day. One P-38 was shot down at Rabaul and eight others were reported missing. We picked up four of the pilots. One B-25 and three P-38s cracked up landing back at Dobodura after the fight.

In the space of twelve minutes we had destroyed or damaged 114,000 tons of Japanese shipping, shot down or destroyed on the ground eighty-five Nip airplanes, and burned out half the town of Rabaul, with a loss of supplies to the enemy estimated at 300,000 tons.

Never in the long history of warfare had so much destruction been wrought upon the forces of a belligerent nation so swiftly and at such little cost to the victor.

During the early hours of November 2nd the Japs had sent a task force of twelve naval vessels, including four cruisers, out of Rabaul toward Bougainville. In a night engagement Halsey had chased the Nips back, sinking a cruiser and a destroyer and damaging two other cruisers and another destroyer, with minor damage to his own force. About eight o'clock that morning approximately seventy mixed bombers and fighters had attempted to attack the landing operations at Empress Augusta Bay. The Nips lost twenty-two airplanes and withdrew. Forty-five minutes later thirty Jap planes were seen over Buka at the north end of Bougainville, but no attack was made and Halsey's fighter cover made no more contacts all day.

Our attack on Rabaul which came later in the day evidently spoiled any ideas the Nips had of seriously interfering with Halsey's show, for on the 3rd one force of six and later one of twenty airplanes were seen in the Empress Augusta Bay area, but they did not attack, while on the 4th the standing air patrols reported no enemy contacts of any kind.

On the morning of November 4th one of our reconnaissance planes sighted a Jap fleet of five heavy cruisers, three light cruisers, five destroyers, four freighters, and two corvettes, about 100 miles north of Kavieng.

Halsey decided to attack the fleet on the morning of the 5th, when it would have arrived in Rabaul Harbor at the rate of speed reported by the reconnaissance aircraft. He radioed General MacArthur and asked that I follow up his attack, which would be by aircraft from all five of his carriers, the Saratoga, Princeton, Essex, Bunker Hill, and Independence. Halsey said he would hit the shipping and he asked that we strike the Jap airdromes and the town, with a maximum effort. I told the General that my maximum effort would be pretty low until I got some replacements and repaired all the shot-up airplanes which had gotten way ahead of the repair and maintenance crews during the past couple of weeks. The strafers came out of the November 2nd attack pretty well shot up so that I could not count on any of them. I said I thought I could promise about twenty-five B-24s and fifty P-38s for an attack on Rabaul on the 5th, arriving over the town about twelve-thirty. If Halsey was to attack ahead of me there would be no Japs on the airdromes. They would be out fighting with his carrier aircraft or on the way to attack the carriers, so that I considered the town the proper target for our attack. There was still a section of it on the west side of the harbor that had a lot of warehouses that were probably loaded with supplies.

General MacArthur told me to go ahead on that basis. After an exchange of radios with Halsey, our attack was set for twelve-thirty, while the carrier strike was scheduled an hour earlier.

At eleven-thirty on the 5th, Halsey's carrier boys hit Rabaul Harbor with ninety-seven planes, including twenty-three torpedo bombers, twenty-two dive-bombers, and fifty-two fighters, damaging six cruisers and a destroyer and shooting down twenty-seven Jap fighters out of between seventy and one hundred that intercepted. The Navy lost eight planes and two others crash-landed on return.

We did a little better than I expected with our maximum effort. A total of ninety-four aircraft, including twenty-seven B-24s and sixty-seven P-38s, struck the Rabaul dock area at twelve-twenty-five, with excellent effect. It was a good thing we had decided to hit the town, for the Jap airdromes were completely bare. Apparently the Nips were out hunting for Halsey's carriers. About fifteen uneager Jap fighters were aloft but only two of them got close enough to get themselves shot down. We lost a P-38.

On the 7th the weather cleared just enough to let twenty-five B-24s and sixty-four P-38s get as far as Rapopo, which received about 90 tons of bombs intended for Rabaul. We destroyed thirteen Jap planes on the ground and shot down twenty-three out of fifty fighters that intercepted. We lost five P-38s.

Halsey radioed that he was planning on striking Rabaul shipping again on November 11th, using all five carriers, and asked us for all-out efforts to neutralize the enemy air in that area on November 9th and 10th. General MacArthur replied that, weather permitting, we would hit the Rabaul airdromes on the 9th and 10th as requested and, in addition, we would follow up Halsey's attack on the 11th. I told Whitehead about the job we had been given and said to have all available strafers arrive on the scene about an hour after the carrier aircraft had hit Rabaul, and to skip-bomb all the vessels the Navy chased out of the harbor away from the shore antiaircraft batteries. I told him to have the heavies hit the town and burn down what was left of it.

With New Guinea fogged in on the 9th, all I could put on the Rabaul airdromes were twenty Australian Beauforts from their field on Goodenough Island just at dusk. They attacked Vunakanau and Rapopo and reported several fires from burning aircraft and fuel dumps. That night we got nine B-24s to Lakunai, but the weather was so poor that the results were unobserved.

Weather again interfered all day on the 10th, although we did get thirteen B-24s on Lakunai and four Beauforts on Vunakanau during the night. A number of fires were reported, but it was impossible to tell how much damage had really been caused.

All our plans for the Rabaul attack, in conjunction with Halsey's show on the 11th, were washed out by the weather. We started jumping the heavies off at seven in the morning, but by eight-thirty everything had to come back. Two of our airplanes evidently got mixed up in the storm area and failed to return.

Halsey's carriers were on the other side of the frontal activity and made Rabaul all right, sinking a destroyer and damaging two cruisers and three destroyers as well as shooting down twenty-four intercepting Jap fighters.

I would like to have gotten in on that show as we had managed to get quite a lot of aircraft repaired and ready for a big strike.

About this time the Torokina field at Empress Augusta Bay was completed sufficiently to base SOPAC fighters there to cover strikes by the B-24s of the 13th Air Force from the field at Munda on New Georgia. By agreement between General MacArthur and Admiral Halsey, SOPAC took over the responsibility for continuing the neutralization of Rabaul, leaving me free to devote more attention to Wewak, which was beginning to build up again. That threat could not be allowed to get too serious, especially with operations coming up soon to capture Gasmata and Cape Gloucester. As Kavieng was within fighter distance from the new field we were building at Finschaven, I was still responsible for reconnaissance and neutralization of the Kavieng airdrome and shipping.

Colonel Fighter Hutchinson had driven the airdrome construction at Gusap so well that by this time we had moved the 49th Fighter Group and a squadron of strafers from the 3rd Attack Group up there and were ready for business against Wewak. There were, of course, no roads into the area from the direction of Lae or Nadzab, so that the whole Gusap base as well as the needs of Vasey's 7th Division had to be supplied by air haul from our new supply base at Lae. In spite of the fact that we had flown in and were maintaining a total force in the Markham Valley of over 20,000 men, were still maintaining about 2000 at Marilinan, and still had to haul supplies by air to Nadzab, we had managed to stock Gusap with over 5000 fifty-gallon drums of gasoline and 1000 tons of bombs.

The Japs evidently began to figure that our land and air forces in the Markham Valley were a threat not only to their troops on the Bogadjim Road, who were being steadily pushed back over the mountains, but also to the maintenance of the Wewak area itself. On November 6, fifteen bombers and sixteen fighters had raided Gusap,. Dumpu, and Nadzab. On the 7th, nine bombers and ten fighters attacked Nadzab airdrome, destroying four P-39s and damaging twelve transports. We got an interception on this raid, however, and shot down seven Nip bombers and seven fighters.

On the 9th of November a direct telephone line between Brisbane and Washington was inaugurated. General Barney Giles, Arnold's Chief of Staff, talked with me for a few minutes and mentioned that my son Bill, who had recently graduated from Officers Candidate School and been given a commission as Second Lieutenant in the Air Corps, was on his way to the Pacific. I told him not to send Bill to me as it would not be fair to either of us. I said to send him to the Thirteenth Air Force and let him work for General Miff Harmon. Barney said he would fix the orders up and then he put Alice Kenney, my wife, on the line. The connection was not too good and I had trouble understanding her excited soprano voice, but we both got quite a kick out of the conversation. Arnold had very thoughtfully brought her in for the occasion, warning her that the very existence of the newly opened line was a deep military secret. She told me afterward that she was so impressed she didn't dare mention it even in her diary.

On the 9th the Japs lost twenty-two out of forty to fifty airplanes that tried to raid the Markham Valley.

This quieted the Nip down until November 15th, when twenty-one bombers and thirty fighters from Wewak attacked Gusap, destroying some of our precious fuel that we had flown in and damaging six of our planes on the ground. The Nip paid for his raid, however. It cost him four bombers and sixteen fighters. We lost two fighters.

As the reconnaissance showed that the Wewak area was now low on airplanes again and the Japs were building more runways at Hollandia, we started the softening up of Gasmata to get rid of the Jap defenses there before our troops went in to capture the place. The date was still not settled, as it depended on the availability of shipping and the completion of our new airdrome at Finschaven, but the tentative schedule called for the landing around December 1st.

Between the 20th and the 26th we just about wrecked everything around Gasmata with a daily series of attacks that dropped a total of 440 tons of bombs. The Japs made no interceptions and, except for an occasional antiaircraft shell hole, the airplanes made the attacks with very little damage. The kids started referring to the missions as "milk runs."

In the meantime, I had been up at Port Moresby working on the coordination of our efforts with those of General Krueger's 6th Army troops who were assigned to the Gasmata operation. When reports from people who knew the locality showed that it was a poor place to build an airdrome, I got cold on the whole plan. Another bad thing about it was that Rabaul was closer to Gasmata than my nearest airdrome at Dobodura, so that the Japs could always make an early-morning attack on Gasmata and get away before I could get any fighters over there to stop them. The Navy wanted it as a PT-boat base, to keep down the Jap barge traffic along the south coast of New Britain. The Army wanted to go there because the plans were all drawn up. There may have been some other reason, but I didn't know of any.

I asked for a conference at General MacArthur's office with the General, Admiral Carpender, and General Krueger, and when MacArthur said okay, I flew back to Brisbane and presented my story. I said I didn't need any airdrome on the south coast of New Britain if I could have one at Gloucester and suggested that Arawe, about fifty miles west of Gloucester, had a harbor good enough for a PT base and, in addition, had only about 200 Nips garrisoning it against over 3000 at Gasmata. A regiment could handle Arawe, while it would take a whole division to capture Gasmata. We wouldn't have enough shipping to take care of hauling a division before Christmas, but we could ship a regiment with what we had, almost any time. I suggested that I keep on bombing Gasmata to make the Japs think we were serious about it and, a couple of days before we landed at Arawe, slug that place hard and then let the troops go in. Following this, I would plaster Gloucester with everything I had, right up to the hour the First Marine Division was due to go ashore there.

Both Carpender and Krueger opposed my plan but the next day General MacArthur made the decision adopting my scheme completely, set the date of the capture of Arawe for December 15th, and scheduled the landing at Cape Gloucester for December 26th.

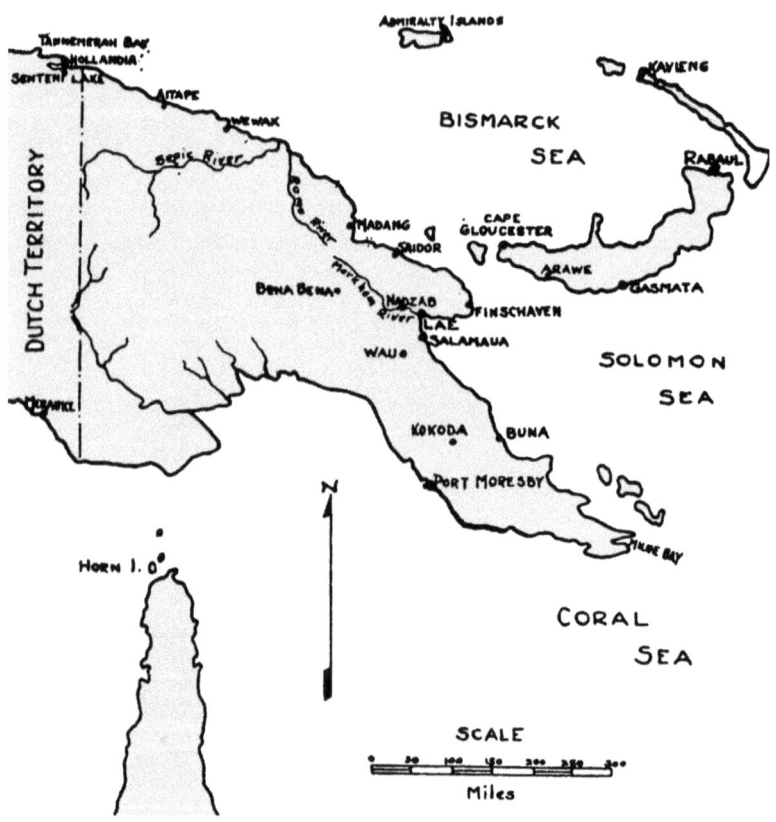

14. CAPE GLOUCESTER
December, 1943

W<small>E KEPT</small> working away at a reduced scale on Gasmata, but on the 29th I told General MacArthur that I'd like to start working over Gloucester, as our information indicated that Japs had a garrison there of probably around 5000 troops, who were building a lot of defensive works and preparing artillery positions which would have to be taken out if we were to go ashore without suffering heavy casualties. I said I wanted to see those Marines go ashore with their rifles on their backs and that I believed the bombs could make it possible if I put enough of them in there. He said to go ahead. We started working that day with a preliminary load of forty-six tons of bombs and wound up the month of November with another sixty-eight tons.

Between the 1st and the 12th of December, Gloucester really took a beating. Over 1100 tons of bombs were poured into the target area during twelve straight days of attacks that plastered the Jap air facilities, camps, artillery and machine-gun positions, and supply dumps. Dozens of enemy barges endeavoring to make the run from Rabaul were destroyed, and a constant watch was kept over the whole Bismarck Sea area and the New Britain trails leading into Cape Gloucester for Jap attempts to supply or reinforce their garrison.

From eight o'clock in the morning to six o'clock in the evening of December 13, we ran almost continuous attacks on Gasmata to give the Japs a last-minute impression that we were about to take the place. Two hundred and forty-eight tons of bombs and over 100,000 rounds of ammunition were expended by sixty-two B-24s and forty-eight B-25s that took part in the raids.

On the 14th, in the heaviest raid in the Pacific war up to that time, we unloaded 433 tons of bombs on Arawe, and the next morning the 112th Cavalry Regiment went ashore without opposition and occupied the village. The only casualties were suffered by a diversionary force of 150 men in rubber boats, which was supposed to go ashore just before dawn at a beach about ten miles east of Arawe to fool the Japs into thinking that was the main effort. The idea was that the Nip, in order to combat the rubber-boat expedition, would pull his troops out of Arawe and leave it undefended when the main force of the 112th Cavalry went ashore.

Several things went wrong. In the first place, by the time we got through with bombing Arawe, there were no Japs there. Then, instead of landing just before dawn, the rubber-boat expedition did not arrive off their landing area until two hours after daylight. The Japs had the beach covered by a few machine guns which opened fire on the diversionary force as soon as it got within range, killing twenty-three men, wounding several more, and breaking up the show. The boats that were not sunk beat a hasty retreat and were picked up by one of our destroyers.

This minor disaster must have upset the rest of the troops considerably, as they shot at every airplane I sent over all day. Liaison planes, reconnaissance planes, fighters, and strafers, sent over on call to help them, were all treated alike. Reports coming in from Arawe sounded as though the whole Jap air force was sitting overhead ready to destroy the 112th Cavalry. It was another excellent example of the lack of training of our ground troops in the recognition of their own types of aircraft, plus the normal trigger-happiness to be expected of insufficiently trained soldiers going into combat for the first time.

The Jap air reaction to our landing at Arawe was unaccountably slow and generally quite haphazard. About ten o'clock on the morning of the landing, ten bombers and twenty fighters came down from Rabaul, but on sighting our cover of sixteen P-38s they turned around and went back home.

On the 16th six different raids were made during the day by small formations of seven to twelve bombers with from fifteen to thirty fighter escorts. All were chased away before they reached Arawe, except one attack by fifteen Jap fighters, which broke through our fighter cover and strafed the beach area, causing a few casualties. The day's work cost the Nip eight bombers and seven fighters. We had no losses.

On the 17th twenty Jap dive-bombers and fifteen fighters ran into our patrol of eight P-47s just east of Arawe. All our fighters returned, but the Japs were driven off with a loss of nine dive-bombers and two fighters.

Another raid of twenty-five to thirty Jap fighters on the 18th turned back when they had lost three planes at the hands of our patrol of sixteen P-38s. On the 20th the Nips tried three more times to break through without success. Our P-38 and P-47 patrols, although outnumbered in every case, shot down twelve Jap dive-bombers and two fighters out of a total of thirty-five dive-bombers and forty-five fighters that took part in the raids. Again we suffered no losses.

The final raids took place on the 21st when, during the day, five different raids were made by a total of about thirty-five dive-bombers and around seventy fighters. The Japs lost fifteen dive-bombers and four fighters. We lost a P-47. One of the enemy attacks got through to Arawe Harbor and caused some minor damage and inflicted a few casualties among the crews unloading supplies. From that day on, except for an occasional reconnaissance plane, the Jap air force left Arawe in peace.

With the loss of one fighter airplane during that week we had destroyed forty-three enemy bombers and eighteen fighters.

On the 13th General George C. Marshall, Admiral Cook, Major General Tom Handy of the War Department General Staff, and Brigadier General Hansell of Arnold's staff arrived at Port Moresby to look over the theater and confer with General MacArthur, who was then over on Goodenough Island at General Krueger's Sixth Army Headquarters watching the Arawe show develop.

The next morning I flew the party over Marilinan, Nadzab, Lae, down the coast over Buna and Dobodura, and finally landed them at the field on Goodenough Island.

General Marshall was much impressed by the amount of work we had done by airborne means at Marilinan. As we circled Nadzab, where the bulldozers and trucks were working for all they were worth, building strips and roads and hauling supplies from the constantly landing transport planes, the General remarked that we were lucky to have a road in from the coast. I retorted that we not only didn't have a road but there probably wouldn't be one for another couple of weeks, although we had owned Nadzab for nearly three months.

"Those trucks down there," said General Marshall. "How did those get in?"

"We flew them in," I said. "I don't mean the jeeps. I meant those two-and-a-half-ton trucks," retorted the General. "You can't get those in a C-47." "Oh, those?" I pretended to yawn. "We just sawed the frames in two, stuffed the pieces in the C-47s, and then welded them together after we got them up here."

General Marshall got a great kick out of the story, especially when he discovered that it was true, and for years the episode was one of his favorite after-dinner anecdotes.

That afternoon General Marshall briefed us on the situation in Europe and at home and General MacArthur described the situation in the Southwest Pacific and discussed our plans for the future. He emphasized the need for keeping up the strength of my air units as a prerequisite for any offensive operation.

Admiral Cook spoke of the necessity for a big main effort west across the Central Pacific, to clear the Japs out of the Marshall Islands, the Carolines, and the Marianas, followed by jumps to Formosa, the Ryukus, and then to Japan. That was the plan the Navy believed should be implemented as first priority.

I said I didn't see any reason for fooling with the Central Pacific Islands, as they were too far away from our line of advance along the north New Guinea coast, through the Halmaheras to the Philippines, and then to Japan via Formosa and the Ryukus, for the Jap air force to bother us. I believed that if the Navy were based on the Admiralties and Palau, they could counter any action of the Japanese fleet and support the movement of General MacArthur's forces along the line I had indicated, and we could all be in Japan quicker than by any other method.

General Marshall looked as though he agreed with me. I knew that General MacArthur already did, but I was also certain that Admiral Cook was definitely in favor of the Central Pacific plan, even though it meant having two commanders in the Pacific instead of one. To me the difficulties already experienced, trying to coordinate the efforts of two independent commanders whose only superiors were the Joint Chiefs of Staff back in Washington, had already pointed out the desirability of putting our whole Pacific effort under one control.

The next day the meeting continued while we listened to the news that everything was going along all right with the Arawe operation and Krueger and I outlined our plans for the Gloucester show the day after Christmas.

Cape Gloucester 251

On the 16th we all returned to Port Moresby and General Marshall and his party left early the following morning for the United States via Guadalcanal, where they were to stop for a conference with Admiral Halsey and Miff Harmon.

With the capture of Arawe, we again began pounding away at Cape Gloucester. For eleven straight days, averaging around 260 tons of bombs a day from December 15th to the 25th, the "milk run" went on. Following each day's bombing, photographs taken from a few hundred feet altitude were examined, and bombs were delivered the next day on anything that looked like a target. By the time the day of the landing arrived we had slugged Cape Gloucester with over 4000 tons of bombs and several million rounds of ammunition. The kids had coined the battle cry of "Let's Gloucesterize the place," which became standard Fifth Air Force jargon for the duration of the war.

On the morning of December 26th the 1st Marine Division landed on both sides of Cape Gloucester with rifles on their backs. No opposition was encountered at either landing beach and, outside of a few complaints about having to bury a lot of dead Japs and repair the roads and bridges that we had torn up, the Marines seemed quite happy about the support they had been given. While the landing operation was going on, we made sure that the Jap ground troops would not interfere with the show by unloading another 422 tons of demolition and phosphorus bombs on the Borgen Bay area, a few miles to the east of the main landing point.

During the late morning two small formations of Jap aircraft had approached the Cape Gloucester area but had shied off on the approach of our covering fighters. Between five and six-thirty in the evening, however, the Nip air effort from Rabaul went into high gear. Approximately forty-five dive-bombers and forty fighters pressed home the attack on the shipping unloading at the beach and the destroyers standing just offshore. The destroyer Bronson was sunk, the Shore badly damaged, and two other destroyers received minor damages.

We intercepted with thirty-three P-38s and forty-seven P-47s, who shot down thirty-seven Jap dive-bombers and twenty-four of their fighters, for a total of sixty-one out of the eighty-five Nip planes which took part in the fight. The antiaircraft fire from the Naval vessels and the landing craft accounted for five more Jap airplanes.

Our losses were three P-47s, two P-38s, and two B-25s. We recovered two of the P-47 pilots, one of the P-38 lads, and the complete crew of one of the B-25s. That crew was headed by Colonel Clint True, the commander of the 38th Group, whom I had reprimanded and then decorated a few weeks before for going to Rabaul without fighter cover.

True said he was shot down by antiaircraft fire from one of our own landing craft. He made a water landing some miles offshore north of Cape Gloucester and paddled ashore that night. As he approached the beach he was challenged by the Marines and asked for the password. True didn't know it, but his remarks in response were so forceful and indicated such an excellent command of the finer aspects of profane eloquence that identity as an American was quickly established. Two days later when I asked about him I was told he was out leading his group on a barge hunt along the north coast of New Guinea.

The day that General Marshall left for the United States General MacArthur told me he was sending me to Washington again about the first of January to take up my personnel and supply problems directly with Arnold and that I would attend a conference in Pearl Harbor on January 23, 1944, at which representatives of Central Pacific, South Pacific, and our own theater would discuss and recommend to the Joint Chiefs of Staff our ideas on how to go ahead with the war. I was to attend as his Air chief. Sutherland and Chamberlin would join me in Hawaii.

Before leaving for Washington I attended several top-level conferences called by General MacArthur to discuss plans for our operations after the Gloucester operation. There was the matter of the Joint Chiefs of Staff directive to capture Kavieng, Manus Island in the Admiralties, and Hansa Bay on the north coast of New Guinea, and the question of what to do about Major General George Vasey, the Australian 7th Division commander, who seemed willing to try to capture all of New Guinea if the Fifth Air Force would give him a little help once in a while. He was already beginning to chase a whole Jap division back across the Bismarck Range toward Bogadjim and at the same time pushing northwest from Dumpu. General Blamey was afraid Vasey was getting overextended and wanted to limit his mission to the movement on Bogadjim.

Admiral Thomas Kinkaid, who had replaced Admiral Carpender in command of the 7th Fleet, and his smart amphibious commander, Admiral Dan Barbey, represented the Navy; Krueger, or his operations man, Colonel Edelman, the Sixth Army; and Sutherland and Chamberlin, the General Staff.

It was decided that on January 2, 1944, the American 32nd Division, which had recovered from the malaria that most of them contracted during the Buna campaign, would occupy Saidor, about halfway up the New Guinea coast, between Finschaven and Madang, and build an

Cape Gloucester

airdrome there to cover our movement later on to capture Hansa Bay and Manus Island. After securing the airdrome area around Saidor, the 32nd Division was to push west along the coast to join Vasey's troops somewhere in the vicinity of Bogadjim. Vasey was to stop his progress beyond Dumpu and concentrate on driving the Nips back on Bogadjim along the road they had built over the Bismarck Range.

General MacArthur had hoped to conduct the Kavieng, Manus Island, Hansa Bay captures as one simultaneous operation, but he announced that due to the shortage of amphibious equipment that idea was now dead. As there might be enough equipment to do two of the tasks at once, the staffs were ordered to work up a plan to pick up Kavieng and Hansa Bay the same day, to be followed by the capture of Manus Island by an expedition starting from Finschaven. I told General MacArthur that the Air Force would be ready to support the Hansa Bay job as soon as the engineers would produce the airdromes at Saidor and Gloucester for three groups of fighters and two groups of B-25s and fix up Nadzab so that I could move my two heavy-bomber groups up there from Port Moresby.

The engineers estimated that meant sometime after April 1st. General MacArthur immediately said that the target date for Kavieng and Hansa Bay would be April 1st and told us all to get our staffs going on that basis.

Halsey's carrier aircraft were to support the Kavieng landing and I was responsible for covering the Hansa Bay show. Who would cover the Manus expedition was left up to a conference between Halsey and MacArthur, scheduled for sometime in January. I told Whitehead, if I had not returned by that time, to represent me at this conference and argue for our covering the expedition out of Finschaven during the day that it left. At daybreak when the convoy would be due just off the coast of Manus, the carrier airplanes should take on the job. My fighters, operating out of either Saidor or Gloucester, would be too far away to remain over Manus long enough to do any good as cover.

On the 27th Neel Kearby's 348th Group P-47s made a remarkable interception. The group picked up a Jap formation about halfway between Rabaul and Arawe and, in a combat lasting about ten minutes, shot down twenty-eight out of the thirty-seven Nips they encountered. All our P-47s returned. Since Kearby had brought the Group to New Guinea on August 16, 1943, they had engaged in thirty-seven combats and had destroyed 162 Japanese planes for a loss in combat of only two of their own.

15. THE SECOND TRIP TO WASHINGTON

January, 1944

I LEFT Brisbane for the United States on December 29th, taking with me Colonel Gene Beebe, my Chief of Operations. We landed at San Francisco New Year's Eve and reported to General Arnold at the Pentagon Building in Washington on January 2nd.

Everyone in Washington seemed to know what I had come back for—more airplanes and more people to fly and maintain them. From General Arnold on down they were quite sympathetic and willing to help, but it was emphasized over and over that the European part of the war had first priority on both things I was asking for. There were, however, several ways of increasing the striking power of the air forces in the Southwest Pacific without detracting from the show on the other side of the Atlantic.

With the capture of Kavieng or the Admiralties, the Japanese forces in New Britain, New Ireland, and Bougainville would be isolated and left to die on the vine. The South Pacific area would no longer have a combat mission and accordingly would not need the 13th Air Force. I proposed that it be turned over to me. If that were done, I would move the two heavy-bombardment groups into the Admiralties as soon as we got airdrome space, maintain the air blockade of New Ireland and New Britain, and start bombing the island fortresses of Truk and Woleai in the Caroline Islands, to assist the Central Pacific forces in their drive to the west. The two fighter groups and the B-25 Group would be installed on the north coast of New Guinea to add to the punch that would be needed to help MacArthur's ground forces forward.

As the disposition of the Thirteenth Air Force when the SOPAC area folded up had not yet been settled, my proposition offered a solution which General Arnold and General Marshall both accepted and said they would recommend to the Joint Chiefs of Staff. General Arnold said he was sending Major General St. Clair Streett out to me, who could take command of the Thirteenth when it came under my control. I asked him to let me have Streett right away, so that he could have some preliminary duty with me and learn the business of fighting Japs before he took over command of an air force. I did not believe the Thirteenth could be released under the present schedule before May, and three or four months of experience would certainly help Streett to do his job later on. Arnold agreed. Major General Hubert Harmon, who then commanded the Thirteenth, was to go back to the United States at the time of the transfer, as Hap had some job that he wanted to put him on.

The B-24 production was in excess of European requirements, so I got Arnold to let me turn my 22nd Medium Bombardment Group, then equipped with B-25s, into a heavy group. As I was short B-25s to equip my other medium-bombardment groups, I proposed to use the 22nd Group's airplanes to help out that situation. After a little argument, that was also conceded.

The A-20 light-bomber production also was more than taking care of European needs, so it was agreed to equip my 312th Light Group with A-20s, to take the place of the P-40 fighter planes they had been using, and let me turn over the P-40s thus made surplus to the Australians, who needed them to keep their fighter units in aircraft.

As long as the subject of the RAAF had come up, I proposed that, as we had an apparent surplus of B-24s pouring out of Ford's Willow Run plant, I would like to have the RAAF organize seven heavy-bomber squadrons as fast as we could give them the bombers and they could train the crews. I offered to help out by lending them six B-24s immediately and helping the Aussies on the training end as well. Arnold agreed to let them have twelve B-24s in May 1944 and send them six a month, beginning in July, 1944. That would allow me to take the 380th Heavy Group out of Darwin, from where it was then bombing targets in the Netherlands East Indies, and move it north to increase the punch of the

Fifth Air Force. Australian B-24s, as fast as they could go into combat, would replace the American B-24s in that work. It was also agreed that my 380th Group would be increased to a strength of forty-eight planes, instead of the thirty-six that they then had.

I pleaded all over again with Hap to give me P-38s. Arnold said he didn't know who he would steal them from, but he would ship fifty to me on an aircraft carrier in about two weeks. For the next two or three days he told everyone that I had wept real tears so copiously all over his office that his own eyes were beginning to water, so that he had given me my P-38s to save both of us from making a scene. I had put up a good argument, but the tears were Hap's invention.

I wanted some B-29s, which would be coming available in a few months, and explained that, with their heavy bomb load of 20,000 pounds, as compared with the 8000-pound capacity of the B-24, and with double the range, I could destroy the oil refineries of the Netherlands East Indies and the Japs would be unable to keep the war going. Arnold would not make any promises but said that if I had a runway big enough to take them by July he might let me have fifty B-29s at that time. I sent word immediately to Bertrandais to start things going to give me a 10,000-foot runway at Darwin, with parking area for a hundred B-29s.

People were another problem. Due to the growing shortage of manpower I got nowhere. I suggested that if they would send me a thousand women I would put them to work as telephone and teletype operators, file clerks, typists, and messengers around my headquarters and in the Service Command. This would let me release at least that many men to build up combat units. There were plenty of WACs available and Arnold decided that if I could get General MacArthur's approval I could have the women.

On January 5th General Marshall sent for me. I went to his office and found General Arnold and General Eisenhower there. General "Ike" had just flown in from England and his presence was the most closely guarded secret in Washington. I had known him for nearly twelve years, ever since we used to sit at the riding ring at Fort Myer outside of Washington, watching our children taking riding instruction and waiting to drive them home for dinner when it was over. I wanted to spend a lot of time discussing the war situation with him, but outside of the conference that day we never did get a chance to talk things over until after the end of the war.

He and Spaatz were worried about having control of the air when the time came to go ashore in France. Arnold had bragged about how I had gained such control in the Pacific every time we had gone forward, so I was supposed to tell Eisenhower how to do the job in Europe. I explained our low-altitude tactics and the necessity of always having the preponderance of machine-gun firepower in air combat whenever we got into a fight. I also stressed the importance of staying on one target until it was knocked out before switching to another and, whenever possible, of attacking in superior force to strike targets that the enemy had to defend if he expected to stay in business in that locality. I suggested, however, that the two wars were entirely different and being fought against opponents who were also quite different in their methods, their military intelligence and capacity, and their psychology. I therefore could not presume to tell either Eisenhower or Spaatz how to fight their war.

Eisenhower wanted to know if I could spare one of my bright young leaders for a while, to go to England and describe our methods to Spaatz and his staff, with a view to putting into effect any of our tactics which might apply to the European theater.

I told him that Freddy Smith was his best bet. Freddy was getting worn down by the tropics and I had decided to send him back to the United States for a rest in a few weeks, anyhow. If he could have a month at home to get back in shape and I could have him back in six months, I would be glad to loan him out and I thought he might be of considerable help. At any rate, he was the only one I could spare. Actually if I were not afraid that he would soon burn out completely, I wouldn't be letting him go.

Eisenhower said he'd be glad to have him under those conditions. I wired General MacArthur my recommendation to send Smith back home immediately and the next day I learned that Freddy would be on his way back in a few days.

On the 15th I saw the President for about two hours and a half. He wanted to know all about our operations since the time I had last seen him in March 1943. He got out his maps to follow the geography as I talked and showed a surprising knowledge of the names of obscure native villages, rivers, and islands all over the theater. I think that the only name he didn't recognize was Gusap. It was no wonder. It was not shown on any of his maps. He asked at least a hundred questions about the details of our seizure of Marilinan, Nadzab, and the various places up the Markham Valley, and chuckled over some of the rather unorthodox things we had done. He particularly enjoyed the story of an Air Force weatherman deciding when the ground forces should land at Lae. When the President wanted to know how I was making out this time getting more airplanes, I said, "Not too well." He laughed and replied, "Yes, it's too bad. I understand all you could get out of them was a couple more groups and another whole air force." We both laughed.

I believe that he was not only interested but enjoyed the whole discussion. I know I did.

When I went back to say goodbye to Arnold I ran into Edwin McArthur, who had gone home the previous November after doing such a grand job of entertaining at every Air Force field in New Guinea. I told him to get his accordion and come back to the Pacific with me. He said he would be glad to if I could get him orders. I did and the next morning he got aboard as I took off for the return trip to Australia. We arrived at Brisbane on the morning of the 19th. I had left Colonel Beebe in Washington to take care of a few details of the agreements arrived at during my trip and had told him to meet me in Hawaii on January 25th.

I reported to General MacArthur the results of my trip to Washington and was told that he wanted me to leave on the 23rd for Hawaii for the big Pacific conference. Sutherland and Chamberlin would leave at the same time, as well as Admiral Kinkaid with some of his staff. We were to have a preliminary conference at Noumea with representatives of SOPAC before going on for the big get-together at Pearl Harbor. The General congratulated me on what I had accomplished in Washington and then told me that he had just gotten word that the British had awarded me the decoration of Knight Commander of the Military Order of the British Empire, one of their highest decorations, which carries with it the title of Sir Somebody if the recipient is a British subject. I thanked him as I knew that he had recommended to Blamey sometime ago that the Australians give suitable recognition for my services in the Buna Campaign. Whitehead, Colonel Paul Prentiss, who had run the troop carriers during that show, and Squeeze Wurtsmith and Freddy Smith had been made Companions of the same order.

The next morning a radio from Washington stated that Colonel Neel Kearby's Congressional Medal of Honor had been approved. Neel happened to be in my office when it came in. I immediately rushed him up to General MacArthur's office and had the decoration ceremony done by the Old Man himself. The General did the thing up right and so overwhelmed Neel that he wanted to go right back to New Guinea and knock down some more Japs to prove that he was as good as MacArthur had said he was.

Neel, who had scored his first victory on September 4, 1943, during the Lae operation, was now credited with nineteen kills and tied with

The Second Trip to Washington

Captain Dick Bong. Bong had gone home on leave, after getting his score to nineteen, and was due back in a couple of days. I told Kearby not to engage in a race with that little Norwegian lad. Bong didn't care who was high man. He would never be in a race and I didn't want Kearby to press his luck and take too many chances for the sake of having his name first on the scoreboard. I told him to be satisfied from now on to dive through a Jap formation, shoot one plane down, and come on home. In that way he would live forever, but if he kept coming back to get the second one and the third one he would be asking for trouble. Kearby agreed that it was good advice but that he would like to get an even fifty before he went home. I said I wished he would try to settle for one Nip per week.

Lieutenant Colonel Tommy Lynch, who had led the P-38s in their first fight over Buna in December, 1942, had also just returned and was anxious to go after the top score. Tommy had led the pack with sixteen victories when he went home on leave but now he needed three to tie Bong and Kearby. I told him to take it easy, too, and not to get in a race with anybody. He left for New Guinea, promising to behave himself.

Bong, Kearby, and Lynch were the big three in the Southwest Pacific as 1944 began. Major George Welch, with sixteen to his credit, had gone home in November 1943. Captain Daniel Roberts, who had knocked down fifteen Nips in eleven weeks, was killed over Finschaven on November 9th and Major Edward Cragg, also credited with a score of 15, was shot down in flames over Gloucester on December 26 of that same year.

The nearest fighter pilots to the top three were now Major Gerald Johnson, Major Tommy McGuire, and Captain James Watkins, all crack shots, who were getting Nips at the rate of two or three in a combat and at this time were in a triple tie at eleven victories.

On the 23rd we flew over to Noumea and spent the rest of the day discussing the coming operations to capture Kavieng and Manus Island. Halsey was in the United States but Admiral "Mick" Carney, his Chief of Staff, told us that Halsey wanted to take Emirau Island to the north of Kavieng instead of Kavieng. No decision could be reached on that point as the final decision would have to be made by MacArthur and when we had seen him last he wanted to go into Kavieng. I suggested that the two B-24 groups and the two P-38 groups of the 13th Air Force be moved over to Dobodura, and that we combine them with the heavy bombers and P-38s of the Fifth Air Force in a series of big strikes on Rabaul to clean out the Jap Air Force there, which photographic reconnaissance indicated consisted of about 250 aircraft. I believed it was essential to get rid of this threat before we went into Kavieng with a big amphibious expedition, regardless of the number of carrier aircraft Halsey had to cover the landing. Captain Morehouse, representing Admiral Fitch, the SOPAC air commander, agreed to write his boss about the scheme and recommend it. After discussing other details of the Kavieng operation to get SWPA representatives familiar with Halsey's plan, we decided to go to Pearl Harbor the next day.

I arrived in Pearl Harbor the next afternoon and General Richardson, the ground-force commander in Hawaii, put me up. That evening we discussed the advisability of pooling all the forces in the Pacific and driving along the north coast of New Guinea to the Philippines, then to Formosa and on to Japan, bypassing the whole bunch of islands in the Central Pacific. I felt there was already enough power in the Pacific to go fast if we drove ahead on just one axis instead of having two drives, one by MacArthur, along the New Guinea-Philippines axis, and another by Nimitz, west through all the islands that the Nips held. Richardson agreed with me and said he would back the idea if it came up during the conference. I made a date with Admiral Jack Towers, the oldest airman on Nimitz's staff and a great guy whom I had known for years, for the next morning to sound him out on the idea.

The next day Towers and I discussed the scheme of pooling our combined efforts. He was for it, too, and said that as soon as he could get hold of Carney he would advocate it and believed that Carney would go along. Towers said Admiral Sherman, the Chief of Operations for Central Pacific, was already on our side. We saw Admiral Calhoun, the Central Pacific logistics man, and he declared he liked the scheme much better from the supply and construction viewpoints than going through the Jap-held island groups. Kinkaid decided to support the idea, and Sutherland and Chamberlin also were lined up to talk for the scheme, which Towers said he would bring up on the 27th for discussion.

At Nimitz's office we were briefed on the latest war news and the progress of the amphibious movement then on the high seas and scheduled to capture Majuro in the Marshall Islands on January 31st. Nimitz, in greeting us and speaking of the task ahead of us, said that he believed we would have to get into China through the Philippines, open China's ports, and drive the Japs from North China, Manchuria, and Korea, as well as bomb Japan itself before we could count on defeating the Japanese.

The next morning after the briefing on the war situation in Nimitz's office, Towers said Carney was also on our side. Halsey had not arrived and was not expected for several days, so it was decided to start the conference the next day without him. Sutherland, Chamberlin, and I agreed to present a tentative schedule, if Manus and Kavieng were taken on April 1st, to go into Hansa Bay during the first half of May, Hollandia on July 1st, Palau in July, and then jump to the Halmaheras and on to Mindanao as soon as possible after that.

On the 27th we had a regular love feast. Admiral McMorris, Nimitz's Chief of Staff, argued for the importance of capturing the Carolines and the Marshalls, but everyone else was for pooling everything along the New Guinea-Philippines axis. Admiral Sherman and Sutherland were to go to Washington to present the case to the Joint Chiefs of Staff for approval. The meeting finished with everyone feeling good and ready to work together and get the war over.

On the 28th the conference wound up with a presentation of the details of the Marshall Islands landings, beginning on January 31st, to be followed by an invasion of Eniwetok Island on February 19th, after which the amphibious equipment and fleet support were to be available for the Kavieng-Manus and Hansa Bay operations. We all had lunch with Nimitz, who spoke about now seeing an end to the war. I think it looked clearer to everyone there. I returned to Brisbane, arriving there late the night of the 30th.

I saw General MacArthur the next morning and gave him the story of the Pearl Harbor conference. He was extremely pleased with the way things had worked out, but said he was not going to get too excited until he got a report from Sutherland as to whether or not the Joint Chiefs of Staff would agree with the plan. I remarked that it looked to me as though, with the Pacific commanders all agreeing, the scheme was in the bag. MacArthur said he hoped so, but we'd better wait and see what happened in Washington.

16. LOS NEGROS

February, 1944

With Halsey scheduled to occupy Green Island, a little over a hundred miles east of Rabaul, on February 15th and build a fighter base there, it was up to us to make sure that the Kavieng airdromes were knocked out prior to the amphibious movement. SOPAC aircraft were to make sure that Rabaul's airdromes were similarly taken care of. In addition, while the four Wewak airdromes averaged only twenty airplanes apiece, they had to be watched and kept beaten down, and if we were to go into Manus Island, it was about time to start softening up that place. Saidor and Cape Gloucester were the two places I was depending on as bases for the fighters to cover the expedition from Finschaven to Manus, but airdrome construction was far behind schedule already on account of heavy rains that had turned both airdrome sites into mudholes. I decided it was time to make certain about getting rid of any possible threat from the Wewak area.

We bombed the airdromes at Wewak itself and Borum, the two nearest to us, for three successive missions in an endeavor ro persuade the Japs that they were safe at the other two strips at But and Dagua. We got rid of about thirty-five Nip planes in these attacks and on February 3rd put fifty-eight heavy bombers on Wewak and Borum again to crater the strips with 2000-pounders. Right behind the heavies and just over the tree tops came sixty-two B-25s escorted by sixty-six fighters, which kept on going and swept over But and Dagua. The strategy worked. The Nip behaved perfectly and had his planes lined up nicely, most of them with the engines turning over, crews in their seats and mechanics standing by, when the storm hit them. At But and Dagua alone, sixty Nip planes were destroyed on the ground while the fighters shot down another sixteen in air combat. We had no losses.

As I was shuffling papers in my office in. Brisbane the morning of February 8th, I received a message to meet General MacArthur at Amberley Field in an hour. On arrival I found my crew lined up in front of my airplane, with cameramen and war correspondents standing by. The General got out of his car, told me to stand in front of my crew, and then pinned the Distinguished Service Medal on my shirt while the cameras clicked and the correspondents asked MacArthur for a statement and copies of the citation.

His aide handed out this statement:

GENERAL HEADQUARTERS
SOUTHWEST PACIFIC AREA

PRESS RELEASE

In speaking of General Kenney, General MacArthur said:

"General Kenney is one of the world's outstanding air leaders. His resourcefulness, his ingenuity, his aggressiveness, and his loyalty have made his services invaluable. He has air vision, by which I mean an understanding of the almost limitless potentialities of air development. It is still in its infancy and fifty years from now the world will look back on today with something akin to amazement at its air antiquity. The effect on war and indeed the impact on civilization as a whole of this new element has yet to be comprehended. Its destructive qualities are becoming understood and applied but its logistic possibilities are as yet only mildly realized or utilized. The whole history of man has shown his avidity for speed. The surface of the earth and the surface of the sea, which satisfied him for so long, no longer suffice. He accordingly has taken to the air. The imaginative boldness with which General Kenney approaches this great subject is only one of the qualities which has so greatly endeared him to me. No living man will probably contribute more to the air age which is now upon us."

The citation accompanying the award said:

"For exceptionally meritorious and distinguished service to the Government in a position of great responsibility from August 4, 1942 to September 1, 1943. As commander of the Allied Air Forces, General Kenney revitalized the air arm in the Southwest Pacific Area. Initially his dynamic leadership made inadequate resources effective out of all proportion to their size and, as the force was built up, his constant bold extension of activity kept all elements of his command at a high pitch of aggressive effort. His brilliant tactical conceptions were largely responsible for the defeat of the Japanese attempt to capture Port Moresby, for the transportation and support of the ground forces which drove the enemy out of Papua, and for the relentless reduction of enemy air and naval strength on the north coast of New Guinea and in the Bismarck Archipelago. During this period General Kenney wrested the command of the air from the enemy, thus creating in the Southwest Pacific area a situation favorable for large scale coordinated offensive operations."

As we rode back to the office together I asked MacArthur why he had finally decided to present a decoration publicly, when he had consistently refused to do it a dozen different times when I had asked him to in the case of some of my outstanding medal winners. He laughed, said he didn't know himself, but that he wouldn't do it again even for me. I found myself getting rather fond of Douglas MacArthur.

With Wewak quieted down for a while I told Whitehead to concentrate on Kavieng until the 15th, with Los Negros, the island next to Manus in the Admiralty Group, as a secondary target in case the weather interfered with striking Kavieng. On Los Negros the Nips had built an excellent coral runway 4000 feet long, which could be extended easily to 7000 feet, if necessary. I said I didn't want that airdrome cratered but to have the Japs cleaned completely off the island. I asked him to report the results to me every day by radio. To make sure of his information I told him to have air reconnaissance made and photographs taken at 200-feet altitude. We might get a chance to finesse.

The weather did interfere for several days. On the 10th five heavy bombers managed to get through to Los Negros Island. On the 11th we got a good strike on Kavieng by fifty heavies and a congratulatory wire from Halsey on the results. The 13th saw thirty-five more B-24s hitting Kavieng, while seventy-seven B-25 strafers worked over Los Negros Island. The weather cleared all over the area the next day. Forty-two heavies slugged the two airdromes at Kavieng with 147 tons of bombs, destroying the last two airplanes left, rendering one of the runways unserviceable, burning several fuel tanks, and knocking out antiaircraft gun positions. At the same time, eighty-four B-25s unloaded 87 tons of bombs on the stores and personnel areas of Los Negros and strafed the island from end to end.

The morning reconnaissance reports had indicated that the Japs were fixing up a strip at Tadji about seventy-five miles west of Wewak, so twenty-four strafers from our new airdrome at Gusap made an attack. Seven Jap planes were burned on the ground and the Nip construction problems were complicated by a dozen large bomb craters in the runway. To make the job complete the gang sank a freighter and a large lugger just offshore and, from the looks of the heavy black smoke that came from both vessels, the Nips at Tadji would have to wait for some other vessels to bring them fuel and oil. The accompanying fighters shot down two Jap planes that were trying to get out of the area. One of them was credited to Lieutenant Donald Mentin. It was the 400th official victory for the 49th Fighter Group. Only one hundred to go now.

One A-20 pilot was not ready to take off when his flight left. He finally got away but was unable to join his formation, so he decided to put on a one-man attack on Wewak and Borum, which he did. He got away with it. After he had spent his whole load of bombs and ammunition strafing and bombing both fields for nearly half an hour, four Jap fighters appeared on the scene, but they did not follow as our hero hastily decamped. The Nips probably were afraid of an ambush or some other dirty American trick.

On the morning of the 15th Colonel Cain, my intelligence officer, came into the office with a tall, spare, nice-looking master sergeant. Shaved to the blood, his little black mustache neatly trimmed, clothes neatly pressed, he looked every inch a soldier as he stood there erect, waiting for Cain to introduce him.

"General," said the colonel, "this is Sergeant Gordon Manuel, who was shot down over New Britain last May and who came out just a few days ago."

I got up, went over and shook hands, and said, "Sit down, Sergeant, and tell me all about it." For nearly two hours I listened to his story, and what a story it was!

On May 21, 1943, the airplane on which Sergeant Manuel was the bombardier was to make a night attack on Vunakanau airdrome. It was a B-17, belonging to the 64th Squadron of the 43rd Bombardment Group. The sergeant seemed to want to be sure that I knew that part of the story. Just before reaching the target, a Jap night fighter had set fire to the airplane and the crew had been ordered to jump. The co-pilot and Sergeant Manuel had been the only two to get away before the plane had exploded. It was night and the two landed in the water some distance apart. They never saw each other again. He heard later that the co-pilot had made shore so exhausted that he lay down in the middle of the road along the beach where a couple of Japs found him; they took him to Rabaul where he was executed.

Manuel, with one broken leg and the other with several pieces of Jap shrapnel in it, managed to reach a coral reef about 300 yards off the coast, rested there for several hours, and then swam ashore. As soon as he saw the road, he figured that he'd better hide, so he crossed the road and lay in the bush by the side of a stream, covering himself with leaves. For four days he remained there, tending his wounds and the broken leg, with nothing but water and snails he found in the stream to keep him alive. The next day he hobbled inland along a path a short distance and ran into a couple of natives, who fed him, looked after him, and hid him when the Japs came looking for him, until his legs had healed so that he could do a better job of looking after himself.

In the meantime, he had learned pidgin, the mixture of English and native words that is the common language of all the tribes, and made friends with the people. He told them that the war was going well for the Allies, the Japs were bound to lose, and that it was time for them to get on the winning side. He must have put over a good story. The natives built him a special hut, put in furniture that they brought from an old wrecked Australian plantation house, and fed him like a king. He even acquired an orderly, who was at the same time a native police boy in the service of the Japanese. This boy, named Robin, used to go into Rabaul for Manuel to get him quinine for his malaria, cigarettes, and, occasionally, even a bottle or two of Japanese beer.

Manuel then organized the natives as lookouts all over the eastern half of the island to watch for our kids when they crashed or parachuted. As soon as possible these natives would get to the spot and guide the aviators to some place where they would be safe from the Nip patrols. Then they would be taken to Manuel, who would pass them along to an Australian spotting station which we had recently established about twenty-five miles from where Manuel had come ashore originally. The Aussies had a radio and would call for a submarine to come in at night to bring out our kids.

Besides organizing and operating this escape chain, Manuel with his little band of followers gathered a lot of information on Jap installations which he thought would be useful to us. He located beach defenses, gun emplacements, Jap camps, depots, food stores, and even infiltrated his agents into Rabaul itself, where he gathered a lot of useful data on the harbor defenses.

He had several opportunities to get out himself on a submarine, but there were always another two or three kids that he first had to rescue, or some other section of the island needed to be organized. He settled tribal quarrels, made peace between irate husbands and erring wives, and became known all over the island as the real boss man who could settle any problem the natives brought to him. After about eight months he decided things were running satisfactorily and that he should bring out the information he had gathered. A submarine brought him and three officers he had rescued to Cape Gloucester, and from there he had been flown to Brisbane. Now he wanted to go back with some arms and ammunition to organize the natives to fight the Japanese. He was quite sure that in a few months he could drive all of them back into Rabaul itself and, with a little help, maybe even capture the place for us.

I told him that we were not much interested in Rabaul any more, that we had the place practically surrounded already and were on our way to the Philippines. Anyhow, I wanted him to go back home to Hodgdon, Maine, where he came from, take a couple of months' leave, and then I'd have a job for him. He admitted that a trip back home sounded pretty good. He mentioned a girl named Mary.

I got out orders for his leave, recommended him for a commission as second lieutenant in the Air Corps, and sent him home. He was given his commission a couple of months later, but the doctors decided that, as he had had five separate attacks of malaria while on New Britain, he should not go back to the tropics for a while. He was given duty in the United States and, in spite of his continual applications to be sent back to the Fifth Air Force, remained there until the end of the war.

The final Kavieng attack to support Halsey's landing came on the 15th as he was making an unopposed occupation of Green Island. Forty B-24s ruined the remaining Jap airdromes, silenced three heavy antiaircraft positions, and set fires that were still blazing two hours later. Sixty B-25s set fire to the town area, destroyed the airplane and engine-repair facilities, burned or sank ten floatplanes in the harbor, sank dozens of barges and harbor craft, and blew up a huge ammunition dump, which released debris bringing down five of our strafers. Three of the crews were picked up in one of the most striking rescues of the war, when one of the Navy Catalina flying boats assigned to the Fifth Air Force for air-sea rescue service picked up all fifteen men in Kavieng Harbor itself, while under fire from the Jap shore batteries. The pilot, Lieutenant Gordon, landed, picked up three men, saw two more clinging to a piece of debris, landed again, kept seeing more survivors, and kept on landing, until he had gone into that hornets' nest seven times. Two B-25s, which stayed behind to cover him by strafing the Nip guns on the shore, ran low on fuel and had to land on the field at Cape Gloucester which was still under construction. One B-25 was wrecked beyond repair. The other stayed there for about a week, until enough runway had been built to get the airplane off the ground and back to its home station.

The Catalina pilot had a rough time getting off the water after each landing. As he kept picking up survivors he kept adding weight, and each time he came back into the harbor he collected some more bullet holes, and finally more water leaking into the hull added still more weight. It was a masterful exhibition of courage and airmanship. I submitted his name for a Medal of Honor and recommended the rest of the crew for Distinguished Service Crosses.

Late that afternoon we sighted a convoy of fourteen small cargo vessels and patrol craft off Mussau Island, about 120 miles northwest of Kavieng. For the next five days we worked on this convoy and three others of five to seven ships apiece. It was the last effort the Japs made to run shipping into Rabaul or Kavieng. We had sunk a total of twenty-six out of the thirty-two ships in the four convoys when on the 21st I got word that some of Halsey's destroyers were coming into the Kavieng area to make a raid. We had a lot of airplanes in the air on the way to finish off the remaining Jap ships, but to avoid any chance of our bombers operating against our own destroyers, I called the airplanes back. The destroyers got most of the remaining Nip vessels. We heard afterward that only two of the original thirty-two got back to Truk.

We went back to the job of depopulating Los Negros Island and throwing a few bombs at Manus Island every once in a while just for luck.

On the evening of the 23rd the daily reconnaissance report indicated that the Jap might be withdrawing his troops from Los Negros back to Manus. There was nothing for him to stay for. The airdrome installations had been taken out, the last airplane destroyed, and his fuel supplies burned. Even his antiaircraft guns had been knocked out of action.

The message of the evening of the 24th confirmed my estimate. It said that the reconnaissance plane had flown at low altitude all over the island for half an hour. No one had fired a shot at it. There was still a heap of dirt in front of the Jap field hospital door that had been piled there two days before by the bombing. There had been no washing on the lines for three days. In short, Los Negros was ripe for the picking.

I went upstairs to General MacArthur's office and proposed that we seize the place immediately with a few hundred troops and some engineers, who would quickly put the airdrome in shape so that if necessary we could reinforce the place by air. Kinkaid had a lot of destroyers at Milne Bay and we could use them for a fast express run as the Japs had done to us all through the Buna Campaign. We could load a couple of hundred of General Swift's crack 1st Cavalry Division on each destroyer, run up there during the night, and unload and seize the place at daybreak. I could have fighters overhead and bombers to knock out the Japs if they did try to stop us from stealing Los Negros from under their noses. If the weather should stop me from supporting the show, it would also prevent the Jap air force from interfering. We need not take any real chances. On arrival off the island, if the Nips did too much shooting, we could always call it an armed reconnaissance and back out. On the other hand, if we got ashore and could stick, we could forget all about Kavieng and maybe even Hansa Bay. Manus was the key spot controlling the whole Bismarck Sea. That coral strip on Los Negros, a little over 200 miles from Kavieng, was the most important piece of real estate in the theater. The whole Jap force in New Britain and New Ireland could be cut off and left to die on the vine. With Manus in our hands, we could jump the next show up the New Guinea coast to Tadji or maybe even to Hollandia.

The General listened a while, paced back and forth as I kept talking, nodded occasionally, then suddenly stopped and said, "That will put the cork in the bottle. Let's get Chamberlin and Kinkaid up here." I hunted up Chamberlin and, while we were waiting for Kinkaid, Chamberlin bought the idea. When the Admiral arrived, it didn't take long to get things moving. Kinkaid sent word to Barbey to assemble at Milne Bay three APDs, or destroyers, fitted to carry troops, eight regular destroyers, and the cruiser Phoenix, as Kinkaid's flagship. MacArthur decided he would go along with Kinkaid.

I sent a courier to Whitehead with a letter outlining the whole plan. I told him to keep on hitting Los Negros, paying special attention to any gun positions covering that Jap airdrome there, and to play safe with Wewak by combing over all four airdromes the afternoon of the 28th and the first thing on the morning of the 29th, which was the date of the actual landing. Whitehead was to work up the plan of air support and cover with Krueger, who was at 6th Army Headquarters at Finschaven.

Krueger was notified ". . . to prepare plans for an immediate reconnaissance in force of Los Negros Island in the vicinity of Momote airstrip, with the object of remaining in occupancy in case the area is found to be inadequately defended by the enemy; or in case of heavy resistance to withdraw after all possible reconnaissance had been accomplished." Low-altitude photographs the next day showed that practically all Jap artillery on Los Negros had been moved out of the gun positions. Whitehead's estimate of the number of Japs on the island was "not over 300 combat troops."

On the 26th Whitehead's courier letter said he and Krueger spent most of the preceding day at Finschaven working out the details of the little surprise party for the Japs. Krueger decided to move an engineer battalion and some additional troops up by ship if the reconnaissance force managed to stick. The reinforcements and the engineers, totaling about 2700 men, would arrive on March 2nd at daybreak.

General MacArthur and I flew to New Guinea on the 27th. He continued on to Milne Bay to board the Phoenix with Kinkaid, and I went on to our advanced headquarters at Nadzab to go over the last-minute changes which Jap reaction or the weather might make advisable. That day the airdromes at Hollandia showed a count of over 270 enemy planes. I told Whitey to make a night attack with a full heavy group the night of the 28th-29th, in order to cool down any enthusiasm the Nips might have to take off for Los Negros, in case their reconnaissance should pick up our movement after it passed Finschaven.

Whitehead's planning as usual covered every detail. He had even ordered eight old B-17s, fixed up to drop supplies by parachute, into Finschaven, where they were to be at Krueger's disposal if an emergency arose.

Both Whitehead and Cooper were worried about the show and my having gotten myself and MacArthur out on a limb. I argued them into a state of partial reassurance, but had to start all over again when on the morning of the 28th a report, from some scouts that Krueger had sent to Los Negros the night before, stated, "The place is lousy with Japs." I told Whitey and Coop that the scouts had seen only one spot on the south end of the island where we would expect the Japs to be, to keep away from the airdrome and camp areas which were being bombed. Moreover, if there were as many as twenty-five Japs in those woods at night, the scouts would think that the place really was "lousy with Japs." I told them that I was sure the scheme would work, that in a few days we would be landing airplanes at Los Negros, and that the next stop would be Hollandia. They shut up, but were still worried.

On February 29th the landing was made on Los Negros by two squadrons of the 5th Cavalry Regiment and the airdrome secured. Very slight opposition was encountered as the troops went ashore. General MacArthur wandered up and down the strip, pacing its length and width, and dug into the coral surfacing to see how good it was, while his anguished aide, Colonel Lehrbas, tried to get him to hurry his investigations and get out of range of occasional sniper fire that was beginning to develop. The General told Brigadier General Chase to stay and hold the place. The cork was now in the bottle.

The weather had interfered considerably. Our fighters had been unable to get through, but it had also stopped the Nips if they had had any idea of doing anything. Three B-24s and fifteen B-25s, flying all the way from Nadzab just over the wave tops, had managed to find Manus and had assisted Chase considerably in occupying the airdrome area. That night and all the next day the Japs infiltrated through the woods and made desperate suicide charges to drive our troops off the island, but Chase hung on stubbornly and wiped out every attempt to stop him from carrying out his mission of holding on. When the rest of the 1st Cavalry Division began arriving the morning of March 2nd, the crisis was over. We owned Los Negros, and the capture of Manus Island and the rest of the Admiralties was a mop-up job.

The day before the Los Negros landing, Dick Bong and Tommy Lynch, who had decided to pair up when they were out hunting Jap planes, were sent to the Wewak airdrome to intercept a Jap transport plane. We had picked up and decoded a radio message, which gave the time of arrival at Wewak as six o'clock that evening and suggested that some very important people were on board. By the time we got the information it was almost too late for an interception, but Bong and Lynch hurriedly took off and flew wide-open all the way, arriving over Wewak about two minutes before six. The inconsiderate Jap, however, was ahead of schedule, had landed, and was taxiing down the runway. Lynch dived to the attack but found that, in his hurry to get away, his gunsights had not been installed. He called to Bong to take the Nip. Bong fired one burst, the plane was enveloped in flames for a second or two, and then it blew up. No one was seen to leave it before or after it was attacked. The two kids then machine-gunned a party of at least a hundred Japs, who had evidently come down to greet the visitors. Subsequent frantic radio messages passing between Wewak and Tokio indicated that the victims were a major general, a brigadier, and a whole staff of high-ranking officers.

Bong and Lynch came over to my office at Nadzab and told me the story. I wanted to let Bong have credit for an airplane shot down in combat, but, of course, if it was on the ground, it couldn't count on the victory scoreboard. I asked Bong if he were sure that the Jap transport was actually on the ground when he hit it. Couldn't it have been just an inch or so above the runway? Maybe the plane's wheels had touched and it had bounced back into the air momentarily? Everyone knew the Jap was a poor pilot. Lynch stood there, grinning, and said that whatever Bong decided was okay with him. Bong listened as seriously as if it were a problem in mathematics and then looked up and said simply, earnestly, and without a trace of disappointment, "General, he was on the ground, all right. He had even stopped rolling."

Everyone was watching the scores of these two and Neel Kearby. Lynch had gotten two more, bringing his score to eighteen. Bong and Kearby were still tied. Bong had shot down a Jap bomber a few days after returning from his leave in the United States, breaking the old tie of nineteen that he had shared with Kearby. Neel had been terribly unhappy all that day, but the next morning, with a wingman for a witness, he had gotten two Jap fighters. When he landed back at his home airdrome, however, he was greeted with the news that Bong had shot down a Nip that same morning. The score was still tied, but it was now twenty-one. Each one of the three added a victory a few days later. Twenty-two didn't look so far away now from Eddie Rickenbacker's old World War I record of twenty-six.

On March 4th Kearby decided to break the tie. With his headquarters flight of Major Sam Blair and Captain Dunham, Neel headed for the Wewak area and trouble. Sighting a formation of fifteen Jap aircraft, he signaled for the attack. A quick burst and a Nip went spinning to earth in flames. Kearby was back in a tie. Now for the lead position. He turned back into the Jap formation and, with a beautiful long-range shot from a sixty-degree angle from the rear, got his second victim with a single burst. As he pulled away, three Jap fighters closed in from above and behind. Dunham got one. Banks got another. The third Jap poured a burst into Neel's cockpit from close range. The P-47, with twenty-two little Jap flags painted on its side, plunged straight down. It never came out of the dive and no parachute opened as it kept going until it disappeared in the jungle bordering on the Jap airdrome at Dagua.

Somehow I felt a little glad that Neel never knew that Bong had also added two to his score that morning in a fight over Los Negros.

The score remained tied at twenty-four until March 9th. Bong and Lynch each got a Jap plane over Tadji. Those two were all there were, so looking for more trouble they spied a Jap corvette just offshore, heading toward Hollandia. Burst after burst from the fifty-calibers of each plane poured into the Jap vessel, which responded with its own deck defense guns. Suddenly Bong saw Lynch pull around and head for shore. One of his engines was smoking. Then it burst into flames. Lynch jumped but he was too close to the ground for his parachute to have time to open. Dick flew around for several minutes to see if he could discover any indication that by some miracle Tommy had escaped. There wasn't a chance and Dick knew it. He flew back and reported. I sent him to Australia that evening to ferry a new airplane back to his squadron and told the depot commander that if the airplane was ready to fly before another ten days, I would demote him at least two grades.

Arnold radioed me that, now that Major General Streett had arrived in my theater, he wanted Whitehead. I wrote a two-page reply saying that Whitey was essential, that it was inconceivable that he should be ordered away, he was not war-weary, he did not want to move, that I had no intention of releasing him and would protest any attempt to take him away from this theater. General MacArthur said if I heard any more about the subject to let him know and he would start something.

The reply came back saying that Arnold wanted Brigadier General Don Wilson, my Chief of Staff for Allied Air Forces. I replied that I would not let him go unless he were given a promotion. A couple of days later they did promote him and I had to make good. Colonel Gene Beebe went into Wilson's place and I sent Gene's name in for promotion to brigadier general.

Right in the middle of my troubles, while trying to keep Arnold from stealing Whitehead, when I would have much rather had him take my right arm, Squeeze Wurtsmith came in with a new problem. He said, "General, you'll have to help out a couple of my kids. I'm afraid if you don't, they are in for some trouble."

It seems that about a week previously a couple of former pilots of the 49th Fighter Group reported in to him at Nadzab. I had sent them home for combat fatigue about three months before. One had a Nip to his credit and the other had two official victories. Both had well over 300 combat hours. Squeeze said he recognized them and when they asked if there were any P-38s for them to fly and could they start flying again right away, he said, "Sure," and told them to report to their old squadron. They went off as happy as a couple of larks. During the next two or three days each one of them added another Nip to his score up over Wewak. Then their story leaked out. When they had returned to the United States, they were sent to a rehabilitation or rest center somewhere in Florida. After a few weeks of doing nothing but rest, they got bored and finally decided that they "had just thought they were tired," so without saying a word to anyone they simply packed up and left. Two weeks later they had thumbed their way to Brisbane. Their names, of course, did not appear on the cargo or crew lists of any of the planes they hitchhiked their way on. Our pilots would cover up for each other in a case like this, any time any place, and you couldn't stop it, no matter how many orders you put out. As a matter of fact, I liked the spirit so much that I wouldn't try to stop it.

The only trouble was that, when it came to drawing pay, they had no orders—they were AWOL, deserters, in fact, they had broken enough rules to qualify for a nice court-martial.

I told Squeeze to tell them I'd take care of them and dropped a note to General Bevans, the Air Forces Personnel Officer in Washington, asking him to get out some orders to make their trip legal and reassign them to me.

The details of how they managed to go from Florida to New Guinea in wartime without getting caught would have made a good story. You couldn't lose when your youngsters had a spirit like that.

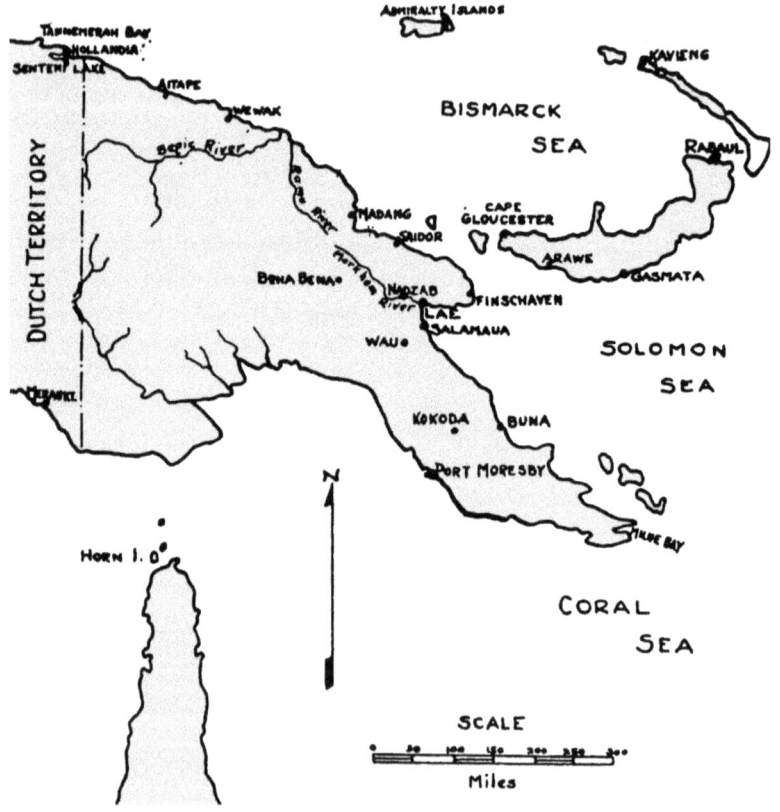

17. THE HOLLANDIA OPERATION
March—April, 1944

When General MacArthur came back from the Los Negros landing, I discussed with him the proposition of bypassing Hansa Bay and jumping all the way to Hollandia, with a preliminary operation to seize Tadji for an advanced fighter base to cover the troops and the shipping as they landed at Hollandia itself. He was all for it, as soon as Halsey finished the Kavieng operation.

Halsey came over on March 3rd and we spent a couple of days discussing the next moves. MacArthur asked Halsey to advance the day of the Kavieng show to March 15th. Halsey said he could not do it that soon but could occupy Emirau Island on that date. The General said no, that he wanted Halsey to go ahead with the original plan as soon as possible.

Everyone agreed on the substitution of Hollandia for Hansa Bay. The date agreed on as feasible was between the 15th and 24th of April. My job was to take out the Jap air strength from Wewak to Hollandia and keep it beaten down so that the carriers escorting the expedition could handle any attempted air attack against the convoy.

We sent a radio to the Joint Chiefs of Staff stating that MacArthur and Halsey were in complete agreement on the substitution of Hollandia for Hansa Bay and asked for a directive confirming our agreement. We talked on the trans-Pacific phone to Sutherland in Washington about the message we were sending. Sutherland said the Los Negros show took everyone in Washington by surprise and he thought the Joint Chiefs would issue the directive when they met on the 8th.

On March 12th we got the Joint Chiefs of Staff directive we had been waiting for since the Pearl Harbor conference. It also contained the answer to our radio regarding the next move.

The decision was announced that the best approach to the Luzon-Formosa-China area was by way of the Marianas-Carolines-Palau-Mindanao route.

We were not to bother with Kavieng but to isolate the Rabaul-Kavieng area with minimum forces. This sounded as though Halsey would occupy Emirau, after all.

We were told to expedite the clean-up of the Admiralties and its development as a base for: (1) SWPA air forces, to complete the neutralization of Kavieng and Rabaul and assist Central Pacific in neutralizing Truk and Palau; (2) Nimitz's air forces, to assist in the neutralization of Truk and Palau; (3) units of the United States Fleet.

We were to capture Hollandia with a target date of April 15 and, after that, do whatever was feasible in preparation for the support of Nimitz's capture of Palau the 15th of September, and for our own movement into Mindanao the 15th of November.

After that the route and the dates were indefinite, except that Nimitz would work up a plan to go to Formosa February 15, 1945, and MacArthur would work up his plan to go into Luzon on that date. It was hinted that one of these plans, only, would be implemented at a later date.

Nimitz was to occupy the southern Marianas about June 15th and build bases for the B-29s for operations against the Japanese homeland.

Nimitz was also to use his carriers to provide cover for our occupation of Hollandia and other operations in the Southwest Pacific Area.

I was dumfounded when I read the dispatch. I couldn't figure out why the Joint Chiefs had reversed a recommendation made by the three commanders in the Pacific. The unanimity at the Pearl Harbor conference evidently had not made much impression anywhere else. I had almost laughed at MacArthur when he suggested not getting too excited about that conference until we got the Joint Chiefs' directive. He had certainly been right.

The suggestion that we bypass the main Philippine island of Luzon and go to Formosa before liberating the Filipinos was unthinkable to me.

The Hollandia Operation

That people belonged to us and had fought loyally on our side. If we blockaded them and left a half million or so Japs in their country to steal their food and subject them to misery and starvation, I felt we would not be keeping faith with them. The Japs certainly would live off the country and let the Filipinos starve rather than go hungry themselves. From a purely military view we would have a large land-mass, with a lot of airfields and a lot of airplanes, right in our midst; it would offer much more of a problem than the bypassing of some small island or an isolated place like Wewak on the edge of a comparatively uninhabited place like New Guinea. However, as MacArthur observed, we had Hollandia to think about right now and could worry about the future later.

A stirring story came in that day from Major General Swift, the commander of the 1st Cavalry Division.

In the last stages of the mopping up of Los Negros Island, just at daybreak one morning, about 200 surviving Jap troops made a suicide charge on an American position. Before attacking, for some unknown reason, one of them had sneaked forward in the dark and set up a phonograph. The playing of the record was evidently the signal for the charge. The tune was "Deep in the Heart of Texas."

Sergeant McGill, with twenty-one years' service in the 1st Cavalry Division, was sitting behind a machine gun in a little sandbagged position out ahead of his company. The commander yelled to him to withdraw. The sergeant refused to budge. "No yellow-bellied Jap is going to sing about Texas," he called back. "I'm from Texas." After he ran out of machine-gun ammunition, he clubbed his rifle and fought until he fell with a dozen bayonet holes in him. A hundred and thirty-five Japs were credited to Sergeant McGill that day.

They picked him up and carried him to the field hospital, where he lived for two days. Just before the sergeant died, General Swift went down to see him. The General, who had known McGill ever since he had joined the Army, could not prevent the tears from coming into his eyes, but the old sergeant said, "Don't you worry about it, sir. I've been training in the army for twenty-one years, looking for a scrap like this. And after all, I can't think of a better way of going out than taking one hundred and thirty-five Nips along with me."

A six-day series of smashing attacks finished off Wewak as an airdrome area as far as the Japs were concerned. From the 11th of March to the 16th we delivered over 1600 tons of bombs and poured nearly a million rounds of ammunition into the final clean-up of the But, Dagua, Wewak, and Borum strips. On the 16th there was no target left. The runways were full of craters fifty to seventy-five feet across and twenty to thirty feet deep; hundreds of wrecked and burned-out aircraft littered the airdromes and the dispersal bays. Even the trees in the vicinity looked like gaunt skeletons, with their tops and branches gone and showing great gashes and scars from machine-gun fire and bomb fragments. During that six days we had shot down sixty-eight defending Jap fighters that intercepted us and destroyed around a hundred aircraft on the ground. The Japs didn't even try to fill the bomb craters. They had given up on Wewak.

Hollandia now became the big Jap air base to succeed Wewak. Here, beyond fighter range even of P-38s operated from my closest airdrome at Gusap, the Nips had built three airdromes on the north side of Sentani Lake, about fifteen miles inland from Humboldt Bay on which was located the port of Hollandia, a former Dutch trading post, which had a population of perhaps two thousand natives and a dozen whites and boasted a stone pier and several substantial warehouses. A road good enough for light motor traffic ran into the Sentani Lake area from a small village called Pim, also on Humboldt Bay and about five miles east of Hollandia, to which it was connected by another road similar in construction to the one leading to Sentani Lake and the three airdromes.

Sure that they were safe enough from our attacks, the Japs built up their air strength and parked airplanes almost wingtip to wingtip, with big gasoline and ammunition dumps on the edges of the fields. Tokio Radio even boasted of the big air force they were building up at Hollandia in preparation for an early offensive that was to sweep the arrogant Americans out of the New Guinea skies.

During February, fifty-eight new P-38s had arrived, with extra tanks in the wings that gave them sufficient range for us to escort our bombers all the way from Gusap to Hollandia and back with enough reserve to allow us a full hour of air combat over the target. With over a hundred Jap fighters certain to intercept us as soon as we got anywhere near Hollandia, I needed more than fifty-eight fighters to protect the bombers and strafers on a big-scale clean-up type of operation. I wanted another party like the "Black Day of August 17, 1943," which had broken the Jap air force in our first big attack on Wewak.

I told Bertrandais to make up the extra wing tanks for another seventy-five of my old P-38s as fast as possible. He promised to have them ready and installed by March 25th. I then told Whitehead not to let any of our P-38s, whether they were the long-range ones or not, go any farther than Tadji. At Tadji, even if they were in combat, no P-38 was to remain any more than fifteen minutes before turning around and heading for home. I wanted the Nips to keep on thinking they were safe so that they would bring into Hollandia all the planes they had room for. If they became alarmed for any reason, they would start dispersing their aircraft so that they could not be taken out so easily during a bombing and strafing attack.

To make sure that the Japs would be certain that I did not dare to make a daylight attack with unescorted bombers, I ordered Whitehead to make single-plane night attacks on the Sentani area, beginning about the middle of March and not worry about hitting anything. If the crews could not see the ground on account of the cloud cover, they were to drop the bombs, anyhow, when the navigators figured they were about over the spot.

While Tokio Radio jeered at us nightly for our futile bombing, and claimed that we were splashing the waters of Sentani Lake and messing up the scenery all over the place without hitting anything worth while, Bertrandais worked his depot twenty-four hours a day remodeling our old P-38s, and the tighter pilots cursed because I wouldn't let them go hunting past Tadji. In the meantime, the reconnaissance photographs showed the Hollandia air force steadily growing. Better yet, the Nip was still parking his airplanes wingtip to wingtip and even in double lines alongside his runways.

While we had driven the Jap air force out of the battered Wewak area, there were still around 100,000 Japanese troops there that had to be supplied and kept up to strength. Some information came in that the Nips would try to run in a convoy of reinforcements and supplies sometime around the 18th or 19th of March. On the night of March 17th two B-24s of the old 63rd Squadron of the 43rd Bombardment Group, equipped with the new radar bombsight, took off on an armed reconnaissance of the waters north of New Guinea between Wewak and Hollandia.

About one o'clock the next morning they spotted a destroyer about 100 miles northwest of Wewak, steaming west at high speed. They straddled it with two 1000-pound bombs, slowed it down to a crawl, and then pushed on west to look for the rest of the convoy. An hour and a half later they located eight vessels about sixty miles northeast of Hollandia, heading southeast. This was it. They launched their second attack, sinking one freighter and leaving another listing and dead in the water. Getting low on fuel, they headed back home. En route they again sighted the destroyer they had bombed earlier. It had moved a few miles while they had been gone, but dropping their last bombs they got a couple of close misses that brought the Jap vessel to a dead stop.

The next morning a group of heavy bombers picked up the convoy which now consisted of six ships, sank two vessels, and crippled another with direct hits. The strafers, eighty of them from the 3rd, 38th, and 345th Groups, then went in for the kill. There were three vessels in the area when they got there. In a few minutes the strafing and skip-bombing attack had gone, leaving behind nothing but a few pieces of floating debris and hundreds of dead Japs bobbing up and down in life jackets amid the oil-soaked flotsam.

That night the destroyer damaged in the attack the night before was found and sunk by another pair of radar-equipped bombers, and the Japs had made their last effort in the war to supply or reinforce Wewak. Two of the nine vessels involved staggered back to Hollandia.

One of the airplanes which took part in the low-altitude attack on the convoy came back with a strange souvenir of the battle. The strafer was an A-20 flown by Lieutenant Robert Vukelic. When he landed and inspected his airplane, he found a piece of paper lodged in his port-engine cowling, blown there by the blast of an explosion as he zoomed over the merchant ship. The paper was part of the ship's records. It revealed that the name of the vessel was the Taiyei Maru, a 3231-ton Japanese freighter. The register, dated November 30, 1936, had been issued to The Imperial Japanese Marine Corporation by the British Corporation Register of Shipping and Aircraft and certified the vessel as "fit to carry dry and perishable cargoes."

Halsey's troops landed unopposed on Emirau Island, seventy-five miles northwest of Kavieng, on March 20th.

Admiral Nimitz and Sherman came to Brisbane on the 25th for a final conference in General MacArthur's office on arrangements for the cooperation of the carrier force with our Hollandia show. Nimitz was emphatic about the danger to his carrier force if the Jap air strength at Hollandia was not greatly reduced by the time he was called on to support our landing. His carriers were to make a series of raids on Palau, Yap, Ulithi, and Woleai, and then be off Hollandia at the time of our landing to give air support to Krueger's troops going ashore. One division was scheduled to land in Humboldt Bay at Pim and drive west to the airdrome area, while the other was to land at Tannemerah Bay and move east to the same objective. At the same time, one regiment was scheduled to land near Aitape, seize Tadji, and fix up the Jap airdrome there as soon as possible for our fighters.

Nimitz said he would start the Palau strikes between April 1st and 3rd with eleven carriers and that this number, less losses, would be what we would have at Hollandia on the 20th; the carriers could only stay around until the afternoon of the second day following the landing. I said that if I didn't have an airdrome ready for a group of fighters at Tadji or Hollandia by that time I could not cover the unloading of the supply vessels in Humboldt or Tannemerah Bays and requested that some carriers stick around a little longer. Kinkaid was to get some small carriers loaned him to cover the Aitape show. It was agreed that he could keep them for eight days after the landing, in order to fill in on fighter cover at Hollandia in case I didn't get a forward airdrome by that time.

Nimitz kept bringing up the threat of the Jap air force at Hollandia and said that he didn't want to send his carriers to Hollandia with two or three hundred Jap airplanes in there at the time he arrived. I promised to have them rubbed out by April 5th. Everyone except MacArthur looked skeptical, especially when I said I did not expect I would start to hit Hollandia before March 30, when my long-range P-38s would be in position. I warned everyone, however, that it would take my whole striking force and that I could not take on any other jobs for the heavy bombers. The bombers of the 13th Air Force, working out of Los Negros, would have to take care of any operations desired in the Marianas or Carolines.

The Joint Chiefs' plan concerning Formosa and the Philippines was mentioned during the course of the discussion. Nimitz said he believed the occupation of Luzon was an essential preliminary to the movement against Formosa and that Formosa could not be attacked by any expedition springing from the Marianas-Palau line. General MacArthur agreed and said that the idea of a direct attack on Formosa, bypassing the Philippines, was not only unsound strategically but basically immoral, because it would abandon all the American prisoners that the Japs held in the Philippines and seventeen million Filipinos to the enemy.

Nimitz left on the 27th. He had hardly gone before the Central Pacific liaison officer asked me to start a four-day bombing show against Woleai. I told him I did not have the bombers available without detracting from the Hollandia strike. Chamberlin came to me with the same story that afternoon, as he had been approached by the liaison officer. I gave him the same answer.

Major General Kuter from Arnold's office visited us at this time and gave us some interesting information.

He and a lot of the Air Corps officers in Washington believed that B-29s, operating from Guam, Saipan, and Tinian against Japan proper, could knock the Japs out of the war before we could capture Luzon.

Kuter said he and Arnold and, he believed, the rest of the Joint Chiefs were in favor of bypassing the Philippines in February 1945 with a direct assault on Formosa.

In spite of the fact that Washington knew that I was ready to operate B-29s out of my new airdrome at Darwin against the oil refineries at Balikpapan, which my information showed was producing most of the aviation fuel for the Japanese, the first 100 B-29s were going to the China theater, where half of them would haul gasoline from India across the Himalayas to the fields in China from which the other half would operate. The next groups would go to the Marianas when the Navy captured them and had airdromes built to receive them. Kuter said Washington's information was that Palembang, which could be reached from India, was a bigger producer of aviation fuel than Balikpapan, which was considered of much less importance.

I remarked that I hoped no one expected me to cheer either the decisions or the beliefs that he had voiced.

On March 29th the 307th Bombardment Group, "The Long Rangers," of the Thirteenth Air Force, earned a presidential citation by an outstanding performance against the Eton airdrome, the key air base in the Truk Islands.

Twenty B-24s made the attack, flying 850 miles from their take-off point and breaking through two heavy weather fronts to get at their target. Forty-nine Jap planes were destroyed on the ground, twenty-one hangars, shop buildings, and warehouses were destroyed or severely damaged, and thirty-seven direct hits scored on the concrete runway. The 307th Group formation was continuously attacked by seventy-five Jap fighters for forty-three minutes in one of the bitterest fights of the war. Thirty-one Nips were shot down, twelve probably destroyed, and ten damaged.

We lost three B-24s.

The plan for the attack on Hollandia, at first, was for a big low-level surprise affair, like that on Wewak in August 1943. Then the photographs began to show what looked suspiciously like a massing of antiaircraft machine-gun batteries, at both ends of the valley in which the three airdromes were located. An attacking force would have to climb over some low hills at each end of this valley several miles from the airdromes themselves. The amount of visual warning to the defenders on the ground might not be much, but it might result in our losing the first wave of strafers and maybe the second wave, too.

After a long discussion with Whitehead, Cooper, and Crabb, the Fifth Bomber Commander, I told them to hit first with all available B-24s carrying a maximum load of fragmentation bombs and dropping from an altitude of ten to twelve thousand feet. If we put sixty B-24s over the target they could carry 200 frags apiece, and 12,000 frags would blanket the place like rain, killing Japs, knocking out machine guns, destroying airplanes, cutting open gasoline drums, and messing the place up in general. On the first day's attack I wanted planes destroyed, of course, but the primary targets were the antiaircraft machine-gun positions shown on the photographs and the fuel dumps. The second day's attack should also be by the heavies, with the airplanes themselves the primary target. The third attack, for a final clean-up, should be opened by the heavies and followed by all available strafers mopping up everything not already destroyed. Each operation would be covered by all the long-range fighters we could muster.

In the meantime, I said to keep on with small night attacks to lull the Nips into the belief that we did not dare to hit them in the daytime. The photos of the three Hollandia airdromes taken on the morning of March 30th showed nearly 300 aircraft on the ground and sixteen others that had evidently been destroyed or badly damaged by our previous night's attacks.

That morning the first big daylight strike on Hollandia took place. Sixty-five B-24s, dropping over 14,000 23-pound fragmentation bombs, destroyed twenty-five planes and badly damaged another sixty-seven on the ground. Fuel dumps burned all over the lot, with heavy black smoke rising to 10,000 feet and visible 150 miles away. The photos showed that we did quite effective work on the antiaircraft machine-gun positions I was worried about. The heavy antiaircraft gunfire was described as slight and inaccurate. Eighty P-38s took on the forty Jap fighters who intercepted. Our score was ten definite and seven probable. We had no losses.

The 31st, we hit the Hollandia airdromes again. This time sixty-eight B-24s and seventy P-38s participated. The photos showed the total destroyed definitely on the ground as a result of the two days' operations was now 138 aircraft, with another forty-two damaged. From the intense smoke following the bombing of dumps we estimated that the Nips had lost another large increment of his fuel stocks. About half the bombers this time carried heavy bombs, which dug craters in all three runways and made them at least temporarily unserviceable. Thirty Jap fighters intercepted briefly and headed west after losing fourteen planes and probably losing eight others. We lost one P-38.

On April 3rd we completed the destruction of all the airplanes based at Hollandia. The photographs indicated a total number of 288 wrecked and burned-out aircraft. Sixty-three B-24s opened the attack, dropping 492 1000-pound bombs on the antiaircraft positions. One hundred and seventy-one B-25 and A-20 strafers then came in over the tree tops and swept up the pieces of what had once been a Jap air force. Seventy-six P-38 fighters escorted the heavy bombers. Thirty Jap fighters intercepted. The P-38s got twenty-four of them and the bombers got two more. We had one P-38 shot down. All the rest of our aircraft returned.

Congratulations came in from Admiral Nimitz, who, after all, had a right to be interested, along with a nice message from General MacArthur, which he sent to me uncoded so that even the Japs could read it.

The heavy-bomber strikes on Hollandia had been led by Colonel Carl Brandt, the new commander of the 90th Bombardment Group. Carl had reported to me in Brisbane in May 1943, burning with enthusiasm to fly B-24s in combat. He had worked for me in the Air Corps Materiel Division in Dayton back in 1941 and, in addition to being one of the best bomber pilots I had ever seen, he knew all about the B-24 from his experience as Air Corps representative at the Consolidated Aircraft plant at San Diego and later on at Willow Run, Michigan, where the Ford Motor Company was producing the airplane.

I told Brandt I knew how he felt but I had another job for him. I had started a supply and maintenance depot at Port Moresby and I wanted someone to run it and straighten things out. Aviation supplies of all kinds had been coming in by boat for months and had simply been dumped all over the place. Everything had been dispersed so well that no one now knew where half of it was. It was his job to find the stuff, store it under cover, catalogue it, and have it ready for issue. In addition, he would have a constant stream of crated fighter airplanes coming in by ship, to unload and erect, and bombers with bullet and shell holes in them which had to be patched.

Carl looked a bit disappointed. I said that when he got the depot running properly and had trained a successor capable of taking over, I would consider letting him go to war.

After I had returned from my last trip to Washington, I called him in and told him that I was sending him to the 90th Bombardment Group for duty as a pilot of a B-24. Colonel Art Rogers, then commanding the group, had done an exceptionally good job for nearly a year but was worn out and I intended to send him home in a few weeks. Rogers had flown over 500 combat hours already and 300 was about the number that the average man could take without cracking. Three hundred hours of flying, during which you live in constant expectation of having someone shoot at you, is a lot of time. Carl ranked everyone in the 90th Group but I told him that my groups were commanded by the best leaders. If a youngster had the gift and the ability to lead, I would give him the job and promote him afterward.

While I had every confidence that Brandt would be the best man to lead that outfit, the kids didn't know it yet, so he would have to prove it to them. Normally when the group commander went home or was killed, the best-qualified squadron commander took over and before I would deviate from this procedure I would have to have a good reason for it. Otherwise the youngsters might think there was no future in pushing a B-24 around. They wouldn't have the same incentive to become star performers and the morale would begin to drop. It was up to Brandt to fly with the kids, show them first that he was a better bomber pilot than any of them and then that he was the kind of pilot they would like to have lead them—the kind they had more confidence in than anyone else.

Carl took my advice literally. For a whole month he flew formation with youngsters, who were not too good at it. He taught them how to maintain their airplanes, how to get the most out of their engines and, without their really suspecting it, imparted to them a world of knowledge that he had gained in long years of flying experience.

One day the four squadron commanders came in to see me. One of them acted as spokesman. He said they understood that Colonel Rogers was going home. I nodded. The lad continued.

"General, we know the custom is for a squadron commander to succeed to command of the group. We've all been working toward that goal and any one of us can run the 90th Group for you and do a good job, but Colonel Brandt is the best man in the 90th and we would like to fly behind him any time, any place." The others all said those were their sentiments, too.

I picked up the phone, called Art Rogers, and asked him to come over to my office. The next call was to Brandt. I told him I was sending Rogers home that evening and that Colonel Carl Brandt was the new commanding officer of the 90th Heavy Bombardment Group, effective at once.

"Now go on back and congratulate your new boss," I said. The kids grinned, saluted, and hurried off.

By the time Rogers arrived I had his medals all laid out for the decoration ceremony—he had been awarded everything except the Medal of Honor, a seat on a plane going back home that night, and a place set at my own headquarters mess, where Sergeant Raymond would give him a demonstration of how good ordinary rations can taste when a master chef is at work.

Rogers wouldn't admit it, but he was really tired and glad to go home for a while.

We all realized that making the long jump to Hollandia was quite a gamble, but if that operation went through all right we should keep moving as fast as possible before the Japs had a chance to dig in anywhere. Besides, we needed airdromes beyond Hollandia if we were to keep rolling. The Nips had a good one on Wakde Island and had built more on Maffin Bay around Sarmi. The main thing we were worried about was that the Japs would get an idea that we were going to make our next landing at Hollandia and would start reinforcing their troops already there and pushing a lot more aircraft into Wakde and Sarmi.

Every information indicated that the Japs thought we were aiming for some operation along the coast between Madang and Wewak. They had moved a whole division out of Wewak to Hansa Bay and were evidently getting ready to contest any landing in that area.

To make sure that they kept on thinking that way, General MacArthur told us to make some air attacks in the Hansa Bay area, drop some parachutes, and simulate taking night photographs with flares. The Navy was to shoot up the shore every once in a while and a few rubber boats would be spotted at selected localities between Wewak and Madang, to create the deception that scouting parties had landed and gone inland.

I also asked General MacArthur to let me start planning for a paratroop operation against Selaroe, an island two hundred miles north of Darwin. The idea would be to seize the Jap airdrome there, put in a couple of squadrons of fighters, and extend the field so that it would take bombers. Then we would be close enough to Balikpapan to hit the vital oil refineries there. I suggested that if the plan leaked and the Japs got wind of it, they might move more forces into the Netherlands East Indies instead of into New Guinea. Moreover, if they thought there was a threat to Selaroe, it might stop them from sending aircraft from that area to New Guinea to replace losses suffered at Hollandia.

The General said to go ahead on the planning but not to help the news leak out, as he might want to carry that operation out if anything went wrong at Hollandia or the Japs decided to put too much strength on the north coast of Dutch New Guinea for us to handle.

Our low-altitude photographs of the proposed landing area in Tannemerah Bay came in, showing the ground running steeply up from the only beach in the bay. Furthermore, the beach was small and, if the troops had any difficulty making their way inland toward Sentani Lake, the supplies would get stacked on the shore so that a single bomb would take them out. It didn't look to me like a good place to put a whole division of infantry, with their artillery and supplies of ammunition and food. I suggested to Chamberlin that Tannemerah Bay be ignored and that the 24th Division, scheduled to land there, jump into Maffin Bay at Sarmi or capture Wakde.

Chamberlin would not buy my idea, at all. We had a big meeting on the 9th, presided over by General MacArthur, with Admiral Kinkaid and myself and representatives of the GHQ staff and the 6th Army. I brought up the subject of Tannemerah Bay and said the photos indicated to me that it was no place for a division to go ashore. If they had to go in there, a battalion would be enough, as there was nothing to indicate that the Japs had anything in that locality. The 6th Army crowd said they were sure their plan was all right, that anyhow they could not change the plan at this late date. The idea that, in spite of the fact that we had ten days before the Hollandia operation took place, their plan was that inflexible did not make sense to me and I told them so. The decision was made, however, that the 6th Army had the mission to capture Hollandia and their plan would be followed. Accordingly, the whole 24th Division, plus six battalions of artillery with tanks, engineers, and service troops, was set up to go into Tannemerah Bay, while the 41st Division went into Humboldt Bay.

After the conference the General told me he would go ashore with the 24th Division and, if it didn't look good, he'd stop the landing and send the troops back to Humboldt Bay to help out the 41st.

In conference with SOPAC representatives, it was agreed that the two heavy-bombardment groups of the 13th Air Force, the 5th and the 307th, would move from Munda in the Solomons to Momote, the field on Los Negros, by May 15th, when they would come under command of Major General St. Clair (Bill) Streett, who had been gaining experience for the past month, operating the field at Saidor. It was still not much more than an emergency field but we did not need it so much now that we had captured the Admiralties and were making the next jump to Hollandia.

The two fighter groups and the group of B-25s of the Thirteenth Air Force would move into Cape Gloucester and a second airdrome was to be built on Los Negros as soon as airdrome space was available. Marine and New Zealand and Naval air squadrons would operate from Emirau and Green Islands, Empress Augusta Bay on Bougainville Island, and Treasury Island. The whole Thirteenth Air Force was to be turned over to me about June 15th.

It was about time to start the final preparation attacks on the landing areas around Hollandia and Tadji, if we were going to continue to have our troops go ashore against negligible opposition. On April 6th 210 bombers struck the Hollandia area, hitting buildings, shipping, personnel, and stores of supplies. A cargo vessel at the dock was burned, the town area partially destroyed, and heavy fires set in large fuel dumps near the town. Sixty-five P-38s went along as escort but there was no interception. We had no losses. During the day 126 of our fighters, carrying bombs, struck all Jap-held spots along the coast from Madang to Tadji, where they destroyed two serviceable Jap planes on the ground and set fire to buildings and fuel dumps.

On the 10th Hansa Bay took a beating when a 227-ton bombing attack was inflicted on the Nips to persuade them that we would land there soon. The following day we repeated the attack with another 251 tons of bombs.

On the 12th we returned to Hollandia, hitting the place with 322 tons of bombs, destroying the rest of the town, sinking a freighter in the harbor, and burning two other vessels. One hundred and eighty-eight bombers and sixty-seven fighters took part in the raid. Twenty Jap fighters, evidently from Wakde, intercepted. Our P-38s shot down eight of them and probably got another. We lost a B-24. Two of the Nips' planes that were definitely destroyed in combat and the one probably shot down were added to the score of Captain Richard I. Bong, making his score twenty-seven, one more than Eddie Rickenbacker's high score in World War I. The next day Eddie wired me that he was sending me a special delivery "letter" that he wanted me to deliver.

I made Bong a major and sent him back to the United States to take a course in gunnery. It was a funny thing, but he was not a good shot. He got his Nips because he was good enough as an aviator to place himself within a few yards of the enemy plane so that he couldn't miss. Whenever he tried a long-range shot, he missed. He had said many times that he wished he could take a good gunnery course. I figured this was a good time to let him go home, see his girl back in Poplar, Wisconsin, and

The Hollandia Operation

be acclaimed as America's leading air victor of World War II. I didn't want that pitcher to go to well too often.

That day the Australian 7th Division reached the coast at Bogadjim and joined elements of the American 32nd Division which had advanced west along the coast from Saidor. The Aussies had accounted for about 10,000 Japs in their drive from the Markham-Ramu Valley across the mountains to the coast.

On the 15th we "softened up" the Tadji area again. One hundred and eighty tons of bombs was the prescription administered by 131 bombers in a series of attacks lasting for nearly three hours. The bombing and strafing results were reported excellent.

April 16th went down in the records of the Fifth Air Force as Black Sunday. We took a beating. It was not administered by the Jap. The weather did us wrong. Forty-five B-24s, forty-nine B-25s, thirty-seven A-20s, and forty fighters raided the Tadji area and put the final touch on the place. On the way home, however, they found that the low clouds and fog that had been hanging offshore for the previous three days, had moved inland and blocked them off from the home airdromes. That night there was gloom in the Fifth Air Force. Of 170 airplanes that had left on the strike seventy were unaccounted for. By noon the next day we had recovered thirty-six which had made emergency landings away from home and remained there until the weather cleared. Others were gradually accounted for, but the final cost was thirty-two kids and thirty-one airplanes. It was the worst blow I took in the whole war.

I had landed at Nadzab the afternoon of the 17th, leaving General MacArthur at Port Moresby, where he was having a final meeting with his staff before taking off with Admiral Barbey on the Hollandia landing. The situation was definitely improved over that of the day before but we were still missing too many crews and airplanes. There were plenty of grim faces around Nadzab. That evening we got a chance to smile, however, when one of the Catalina rescue planes brought in the crew of an A-20 which had made a forced landing on the north coast of New Guinea, about thirty miles west of Madang, nosing up the plane and breaking one of the propellers and a wingtip. Flying right on the deck for nearly half an hour on the return flight from the attack of the 16th, the crew had been lost but were quite sure they were over territory controlled by Allied forces when they spotted the little stretch of beach. They were already on their reserve fuel supply, so they decided to land, even though the spot didn't look good enough to insure that they could get away without damaging the airplane.

After getting down without hurting anyone, they calmly went to work building a shelter for the night, got out the rubber boat, and caught a few fish to supplement their emergency rations. These they broiled over a fire which they kept going to attract any plane out searching for them and, after a good meal, the lads turned in for the night. The next morning, while they were enjoying a surf bath or sunning themselves on the beach, with their newly laundered clothing drying on the lines near by, a Catalina suddenly landed and taxied ashore. The nose gunner of the rescue plane, looking very businesslike behind his pair of caliber fifties, yelled to them to hurry up and get aboard. The castaways called back, "Don't be in such a rush, we got to get dressed and get our things out of our plane. Come on in and have a swim. The water's fine."

When the Catalina crew profanely but briefly explained that there were about twenty armed Japs less than half a mile away, coming along the beach in their direction, the picture changed instantly as the Robinson Crusoes hurriedly grabbed their clothes and clambered aboard the rescue plane. The somewhat crestfallen A-20 lads recovered some of their lost face when they watched the machine-gunning of the intruding Jap party by the Catalina on its way home. By the time they landed back at Nadzab the rescued were taking credit for decoying the Nip troops to the scene in order to furnish a good target for the rescuers and make them feel that they had earned their day's pay.

From the 17th to the 21st we concentrated on the coast defenses of Hollandia. Final reports indicated nothing left to shoot at. During the last two days there had not even been any antiaircraft gunfire. No enemy aircraft had been seen in the air. No attempt had been made to repair the damage to the strips in the Sentani Lake area, which were still littered with the debris of over 300 wrecked and burned-out aircraft.

18. HOLLANDIA TO NOEMFOOR
May—June, 1944

ON THE morning of April 22nd, the 41st Division landed one regiment at Tadji simultaneously with the rest of the division at Humboldt Bay. The 24th Division, in the meantime, started unloading at Tannemerah Bay.

At Tadji something went wrong with the navigation and the boats beached three miles from the spot chosen by high-powered planners after months of study of the problem. They found a much better beach, no opposition, and a good road to the backyard of the Jap camp. The place was captured with little effort, most of the Nips leaving their breakfast in their hurry to get away.

At Humboldt Bay the 41st Division went ashore with no opposition and made good going but the troops started burning a Jap dump which had some ammunition in it. The fire spread to their own dumps but before they could put it out, just after dark, a lone Jap plane attracted by the fire came over, dropped some bombs, and eleven landing-craft loads of food and ammunition—the whole of the unloading for the day—went up in smoke.

At Tannemerah Bay two battalions of infantry and a battery of artillery also found the place undefended when they landed with practically all the food for the whole division. It was then decided that, as the beach was too small and the slope too precipitous, they couldn't get any more troops in and worse luck they couldn't get the food back on the ships. General MacArthur arrived about this time and told Eichelberger, who was in command, to move the rest of the show over to Humboldt Bay.

Nimitz's carriers, just down from a highly successful raid on Palau, were off the coast at daybreak that morning and launched their planes to support the Hollandia landings. They flew around looking for targets around the landing beaches but found nothing to bomb so they took a few shots at the Hollandia strips just for luck, re-wrecking the wrecks there, and plastered the Wakde and Sarmi fields, where they shot down twenty-three Jap fighters and destroyed a lot of aircraft on the ground. There was nothing else for them to do, so that afternoon the carriers pulled out. Kinkaid still had a few baby carriers which had covered the Tadji landing. These stayed around for several days, covering the shipping in Humboldt Bay, and furnished support to our ground troops, which pushed inland rapidly and captured the Hollandia, Cyclops, and Sentani airdromes on the 26th. The Tadji strip had been repaired in the meantime and two Aussie P-40 squadrons were in there supporting the ground forces and assisting in covering the shipping between Tadji and Hollandia.

The lack of opposition at Hollandia puzzled us until later, when interrogation of the natives and a few prisoners told an eloquent story. Our big attacks on the airdromes around Sentani Lake on March 30th, 31st, and April 3rd had caught the Nips off base. Besides destroying all the airplanes and burning up practically all their gasoline, over 2000 ground crewmen and service troops had been killed and the antiaircraft guns destroyed or knocked out of action. The subsequent attacks on Hollandia itself and the shore defenses had been so heavy, particularly the attack of April 17 by over 250 bombers, that the Nips on that date began to flee into the hills, leaving their equipment behind, abandoning their artillery, and leaving the whole area undefended.

From the 23rd to the 28th our air efforts were almost entirely confined to hauling food and ammunition to the troops in the Hollandia area. After they had burned their food up on the day of the landing, we dropped their rations by parachute for a couple of days until the ships brought in another supply. In the meantime, Eichelberger's tanks and trucks had broken the road all apart between the village of Pim on the bay and the Sentani Lake area, where his troops had mopped up the few Nips in the vicinity. Now they were isolated. There was plenty of food only twenty miles away, but even caterpillar tractors got stuck trying to get through. It was surprising how fast those troops repaired the Sentani runways and cleared away the debris so that we could haul food to them by transport airplane. We dropped it from bombers for the first day but after that we landed it. It was eight days before the road was repaired enough to get food in to those troops by road.

When the ground forces finished the count of wrecked and burned-out Jap aircraft on the Hollandia airdromes, the total was 380. In addition, beginning with our initial attack on March 30, we had shot down seventy-two more. To occupy this base which Tokio Rose had boasted was so safe had cost the dumb Nip 452 airplanes.

Just before Bong left for home he told me that the airplane he had listed as probable in the combat of April 12 near Hollandia, had gone into the water in Tannemerah Bay. The reason it had been called a probable was that no one except Bong had seen it go down. Under the rules he could not get credit for it. He showed me on the photograph exactly where the plane went into the water and said that it was a single-seater fighter of the type we called an Oscar, that he had hit the left wing, the pilot, and the engine, but that the plane had not burned. When we had secured the Hollandia area I got a diver to go down where Bong had said the airplane went in. The diver located it almost instantly and we pulled it up. It was an Oscar, the left wing had eleven bullet holes in it, the pilot had been hit in the head and neck, two cylinders in the engine were knocked out, and there was no sign of fire. I put out an order giving Bong official credit for the victory—his 28th.

On the 25th I flew back to Port Moresby for a conference with General MacArthur, who had just returned from the Hollandia operation, and the GHQ staff, about speeding up our schedule of advance west along the New Guinea coast. The supply people argued that there was not enough shipping to move any farther. I wanted to seize Wakde and the coast opposite Wakde around Maffin Bay as soon as I could get a couple of fighter groups into the Hollandia area to cover the amphibious movement. They had too many troops in Hollandia, anyhow, and it would not be much more difficult to feed them at Wakde than where they were.

We needed airdromes. I was abandoning the Dobodura area and Port Moresby as operating bases. The Kiriwina, Woodlark, and Goodenough Island airdromes and the fields around Milne Bay had ceased to be of any value since the capture of the Admiralties. Most of my bases were now out of range of the war. The Hollandia area would not take care of very much, as the space was limited. Wakde would help, but I needed four or five more good-sized fields to bring the Fifth Air Force and the Thirteenth into play. The area around Maffin Bay looked like a possibility for airdrome development, but I wanted to keep on going and take over Biak where the Nips already had three coral runways. Coral was good stuff. Where there was plenty of coral, the engineers could give us a field in a matter of a few days. In the case of ordinary sand-clay soils, the period would be a month or even two.

The discussion—and argument—was settled by General MacArthur, who set May 15 as a target date for the seizure of Wakde and Sarmi on Maffin Bay and wired General Krueger and Admiral Barbey to start figuring on that date and to send representatives to Port Moresby right away to make the detailed plans and arrangements. He wouldn't commit himself on Biak until we saw how the Wakde-Maffin Bay thing worked out, but I wasn't worried. His staff might want to go slow but MacArthur believed in moving fast when he was winning. We were definitely on the move. The Philippines didn't look anywhere near as far away as they had just a few months before.

The General decided to move his advanced headquarters to Hollandia as soon as practicable. I told Whitehead to fix me up some buildings, as I intended to move my whole headquarters up from Brisbane as soon as I could get the communications in and enough buildings erected to let me operate. I told Whitey that he would make his next move into Biak after we captured that place.

After giving instructions to Whitehead to Gloucesterize Wakde and the Maffin Bay area, I flew back to Brisbane with General MacArthur on April 29th.

During the low-altitude strafing and bombing attacks of the Maffin Bay area, the close-up visual and photographic evidence began to show that it was extremely doubtful if we would have any luck getting airdromes there without so much engineer effort that our movement toward the Philippines would be delayed for months. Biak looked better than ever as an objective to follow quickly after the Wakde show. I talked it over with General MacArthur. He called a staff meeting to discuss our next moves and it wasn't long before May 27th was being mentioned as the date to pick up Biak. I said that the coral strip at Wakde could be ready to handle plenty of fighters in a week after the ground troops captured the place, if they would put enough engineers and equipment ashore to really go after it. Therefore I would be ready to cover a landing at Biak easily by the 27th if Wakde was taken on the 17th, which was the date so far agreed on. The staff, except for some dubious remarks by the supply section, offered no objection. I reminded them that the Air Force had been doing most of the hauling of food and ammunition so far and I could keep on helping out, although I did think that sooner or later I would like to have them ship my gasoline and bombs. At that time we were flying gasoline into Hollandia and using the Jap bombs that we had captured there, in order to keep the war going.

General MacArthur told Sutherland and me to go to Krueger's Sixth Army Headquarters at Finschaven and have a conference there, at which GHQ, the Sixth Army, the 7th Fleet, and the Allied Air Forces would be represented, to set up the date for the Biak show.

The conference took all day on the 9th of May and was a bit stormy in spots. The big objection came from the supply people. In addition to the lack of shipping, insufficient time to load, and all the other arguments, the Sixth Army supply staff said the Biak operation was impractical as there was no way of getting stuff ashore. I remarked that the Japs had put between eight and ten thousand troops in there, were operating aircraft from three airdromes on the island, and it seemed funny for us to confess that we were not as good as the Nip.

The Navy had an Admiral Fechtler there, an extremely capable amphibious man, who was going to handle the landing job for the Navy. I asked him if he was worried about his ability to get supplies ashore. He emphatically said he was not. That really broke up the argument and when I reminded Krueger that General MacArthur had called this conference to decide how to capture Biak on the 27th, not when to do it, the die was cast.

Right up to the morning of the 17th, Whitehead "Gloucesterized" Wakde, the Jap shore defenses, camps, supply establishments, and antiaircraft positions around Maffin Bay, and the airdromes on Biak. In the meantime, my 380th Heavy Bombardment Group, commanded by Major Brissey, a real fireball who had succeeded Colonel Bill Miller when Miller became Chief of Staff of the 5th Bomber Command, operating under the Aussies from Darwin, kept beating down the Jap airdromes in Dutch New Guinea, west and south of Biak, around Nabire, Fakfak, Babo, and Manokwari. I still wanted to keep the same pattern of our moves up to that date: (1) gain complete air control over the landing area; (2) destroy the shore defenses that might bring fire to bear against our troops while landing; and (3) maintain cover over the amphibious expedition en route, during the unloading operation, and on the way back. If we could do those things, we could boast that "the troops were going ashore with their rifles on their backs."

From May 1st to the 17th, 1500 tons of bombs and half a million rounds of machine-gun ammunition reduced Wakde and Sarmi to ruins and incidentally destroyed twenty-seven Jap fighters and twenty-five of their bombers on the airdromes there. No Jap air opposition had been offered, probably because in the same period we had visited Biak and Noemfoor with practically the same treatment. During the first Biak raids the Japs had sent up from fifteen to twenty fighters to oppose us, but in each case it had cost them around three-quarters of their aircraft and by the 10th no further air opposition was encountered. The Jap air units were withdrawing to the new airdromes at Galela, Lolobata, Miti, and Kaoe in the Halmaheras, where it looked as though the Nip intended to build up another big base like Hollandia out of range of our combined fighter-bomber strafer attacks.

On the 17th our troops landed unopposed at Toem, near Sarmi on Maffin Bay, opposite Wakde Island, and the following morning Brigadier General Jens Doe took his 163rd Regiment of the 41st Division ashore at Wakde itself. The landing there was not opposed initially but immediately afterward it met determined opposition by a stubborn bunch of Jap marines. The occupation was completed by the afternoon of the 19th. One prisoner was captured who told our interrogators that the Nips had known for several days that we were contemplating the capture of Wakde and that, two days before we landed, all Jap service and air troops had been evacuated to Sarmi and a crack marine detachment of about 800 troops moved in. That was what had caused us the trouble. It looked as though something had leaked. The Nip may have cracked one of our less secure codes or maybe he had tapped some of our telephone lines around Hollandia. We never did find out. These things did happen throughout the war on both sides, although we were way ahead of the Japs in that respect.

Colonel Jack Murtha, who at that time was Whitehead's Chief of Operations, got permission to fly over Wakde on the 19th to observe and report how the show was progressing. He was flying a B-25 in which a 75-mm. cannon was mounted. The ground troops were being held up at one point by two Jap coconut-log bunkers, so Jack decided to help out. He fired a few rounds, scoring direct hits on both Jap positions which he destroyed, and our doughboys rushed in and finished the job.

In anticipation of making quick use of the Wakde field, we had a detachment of mechanics and communications and maintenance men go ashore with some of their equipment immediately after the infantry had landed. The unexpected resistance brought everyone into the fighting before the party was over.

On the afternoon of the 19th Sergeant Vincente E. Barbosa put on a one-man war that earned him the Distinguished Service Cross for extraordinary heroism in action. When forty-six Jap troops attacked his squadron's position Sergeant Barbosa was the first to open fire, killing a Nip officer and two soldiers. An enemy bullet removed the end of his trigger finger and damaged the clip of his gun. The sergeant removed the clip and inserted another but, after firing a couple of rounds, the gun jammed. Observing an enemy soldier only twenty yards away preparing to open fire with a light machine gun, Sergeant Barbosa charged from the cover of his foxhole, clubbed the Nip to death with his jammed gun, picked up the enemy machine gun, and brought it back with him to his foxhole. Then with a carbine dropped by one of his men who had been wounded, he killed two more Jap soldiers. By this time, with everyone in the squadron area alerted and firing every type of weapon they possessed, the Japs retreated, having lost over half their original strength. The whole action had lasted only about ten minutes, but the sergeant remarked when the medical officer was tying up his injured finger that "it was the busiest ten minutes he ever remembered."

We then kept on hammering away at Biak, with an occasional visit to Noemfoor, Manokwari, Babo, and Fakfak. With a final attack by nearly 150 heavy bombers and strafers dropping 317 tons of bombs, and which finished ten minutes before the troops went ashore, we made another westward jump of 200 miles from Wakde when we landed at Bosnek on the island of Biak on the 27th. The opposition at the landing beach was negligible and Major General Horace Fuller's 41st Division pushed west along the shore toward the three Jap airdromes about seven miles away.

However, when the landing forces were almost in sight of Mokmer strip, the nearest of the goals, the Jap force, which later turned out to be somewhat over 10,000 crack troops, rallied and halted our advance. The flat coastal shelf, which was anywhere from a mile to a hundred yards wide from the landing beach until it widened out to three or four miles in the airdrome area, was dominated by 500-foot-high limestone cliffs that rose almost vertically from the plain. These cliffs were honeycombed with caves in which the Japs had established themselves, with light artillery and machine guns commanding the route over which our troops had to pass. We reached Mokmer strip on June 7th, but it was not until the 20th that we occupied the two old Jap airdromes of Sorido and Borokoe, five miles to the west of Mokmer, and several weeks more before the last remnants of the original Jap force had been mopped up.

In the meantime the engineer troops that had landed right after the infantry in order to fix up the former Jap fields were out of a job. Just to the south of Biak Island about two miles was another island named Owi, which was uninhabited, fairly free of tree growth, flat, well drained, and solid coral, except for about a foot of black topsoil. We wasted no more time moving in, and by the 21st of June the first troop carriers landed with personnel and equipment to set up shop for what soon became one of our most important main bases.

It was at Owi that I suddenly woke up to something that I should have realized a long while before. When we mentioned the place to the natives around Biak, they told us that no one lived there as the place was "tabu." We had heard that word many times ever since coming to New Guinea and had smiled and paid no attention to it. Just another silly native custom. In about ten days the troops on Owi began to develop scrub typhus. We had run into this disease several times before but now it suddenly dawned on us that in every case we had been warned by that word, but had been too dumb to understand what the natives were trying to tell us.

When we had first gone into the Dobodura area back in November 1943 we had selected Embi Lake near there, with its beautiful scenery and clear water, as an ideal camp site. The natives shook their heads and said they wouldn't go near the place. It was tabu. We thought at first that meant there were crocodiles in the lake but after dynamiting the place and being unable to raise anything, we moved in. In about ten days scrub typhus broke out and we had a lot of trouble. Hundreds of our troops ran high fevers for several days and we lost about two per cent of them.

The same story was repeated in some localities around Nadzab and up the Markham Valley, but for some unknown reason our superiority complex just wouldn't let us believe that there could be any sense to what we laughed at as the superstitious ravings of an ignorant, uncivilized native.

Luckily the doctors had just discovered a solution with which we could impregnate our clothing and which would act as a repellent to the little, almost microscopic mites that lived in the grass and caused us all the trouble. Wearing the impregnated clothing and burning off the grass cured the problem and we had practically no scrub typhus after that, but from then on if the natives said tabu about a place we took the proper precautions. A little late, of course, but we had learned a lesson.

The seizure of Wakde on May 17th, followed only ten days later by our landing at Biak, evidently scared the Jap high command. Nimitz's carriers on the 19th raided Marcus Island, 1100 miles southeast of Tokio, on the 20th hit the Kuriles at Shimusku, about the same distance to the northeast of the Japanese capital, Marcus again on the 21st, and Wake Island on the 23rd. Our bombers were raiding Yap and Palau almost daily and our reccos were beginning to appear over Mindanao in the

Philippines.

Whatever his line of reasoning the Nip began to show signs of annoyance. He moved bombers and fighters into Sorong, on the western tip of New Guinea; and Babo, on MacCluer Gulf, which had been deserted after a week of pounding by our 380th Group bombers from Darwin, became active again. Almost daily from the 1st of June to the 5th, Jap fighters and bombers appeared over Biak, to the gratification of our fighters who had begun to complain at the lack of targets. The 49th Group, in particular, was thinking hard about that magnum of brandy. During that five days they picked up another twenty victories, while our bombers and strafers attacking Babo and Sorong destroyed another forty Jap planes on the ground and in the air and wrecked both airdromes.

On June 2nd our reconnaissance planes picked up a Jap convoy leaving Davao, the big port at the north end of the bay of the same name in southern Mindanao. It was headed southeast toward the Halmaheras. Two battleships, five cruisers, and ten destroyers were sighted, and the reports said the decks of most of the cruisers and destroyers were "black with troops."

Shortly after midnight on the 4th, ten B-24s of Lieutenant Colonel Ed Scott's 63rd Squadron, using the new radar bomb-sight, made contact east of the Halmaheras and went to work. Two Jap destroyers were sunk and two cruisers damaged. The two battleships and three of the cruisers reversed their course and headed back toward Davao. The remaining two cruisers and eight destroyers continued on their southeasterly course. Late that afternoon they were northwest of Waigeo Island, heading in the direction of Sorong. Weather stopped us from attacking them on the 5th and 6th, but late on the 6th a reconnaissance plane reported they were still at Sorong and that it had scored a direct hit on one of the destroyers. That night the radar bombers of the 63rd Squadron reported scoring hits on two more destroyers or light cruisers northwest of Manokwari. Weather again interfered with our operations on the 7th.

On the morning of the 8th the weather cleared and we assigned squadron search areas to the B-25s from Hollandia and sent them out to find the Nip convoy. Two large luggers went to the bottom off Noemfoor, one 3000-ton transport and a small destroyer were sunk and six other small freighters or transports set on fire and destroyed at Manokwari, but these were not the targets we were looking for.

About noon Major William G. Tennile, Jr., leading ten B-25s of the 17th Reconnaissance squadron and escorted by eight P-38s, picked up two Jap light cruisers and five destroyers covered by six Jap fighters, north of Manokwari, headed east at high speed. This was the target. The decks were black with troops and destined to reinforce the hard-pressed Nips on Biak Island.

Tennile gave the signal for the attack in pairs, he and his wingman, Second Lieutenant Howard C. Wood, taking on one of the cruisers, while the other four pairs swept in at mast height on four of the destroyers. The concentrated gunfire of the two Jap cruisers was turned on Tennile and his wingman, who were both shot down before they got within bombing position. Their sacrifice, however, enabled the others to bore in and strike effectively. Four destroyers were sunk and the fifth damaged. Another B-25, piloted by Captain Sumner G. Lind, was shot down during the attack on the destroyers. The P-38s took care of all six of the Jap fighters and then strafed the decks of the crippled destroyer as it struggled to catch up with the two cruisers which had turned around and headed northwest at full speed.

Tennile was awarded the Congressional Medal of Honor posthumously. MacArthur called his performance "one of the most magnificent exploits of the war."

Wakde had been put in shape quickly and, within a week after its capture, was crowded with aircraft based there or refueling on the way to or returning from distant attacks. Our main heavy-bomber strength was still back at Nadzab, as the Hollandia fields had not turned out as well as we expected and were only suitable for fighters and light bombers. The Maffin Bay area was too soft to be practicable for quick airdrome construction, so we were crowding Wakde to the limit and hurrying development at Owi as fast as we could. I knew we were inviting trouble by parking aircraft almost wingtip to wingtip on Wakde, but we had to have our fighters forward to maintain cover over the shipping constantly unloading at Biak, and besides the flow of airplanes from the United States to the Pacific was coining along nicely. I no longer had to rebuild my wrecks from the ground up as I had been forced to do a few months previously, so we stacked them in and relied on attacking the Nip airdromes to keep the Nips from attacking us.

The Nip, however, refused to play the way I wanted him to and for a week, beginning the 5th of June, until I got some night fighters operating at Wakde, he put over some mean night attacks. On the night of the 5th a Jap bomber blew up a bomb dump and had everyone on the island believing that an earthquake had hit the place. Between the Jap's bombs and our own we lost several hundred drums of gasoline and three P-47s. Two nights later I lost ten more airplanes on the ground, and, the night of the eleventh, the devils dug holes in the runway that took us all the next day to fill in and burned up another four airplanes. The Jap bombers came over at such frequent intervals all that night that the men spent nearly the whole time in their foxholes.

Whether someone told the Nips that we had installed night fighters at Wakde or not I don't know, but we had very little trouble after that.

About this time I spent several days at Nadzab with Whitehead and the staff, going over the plans for our operations to follow the clean-up of Biak. I had already discussed with General MacArthur the advisability of taking over Noemfoor in order to get room to utilize the aircraft of the Fifth Air Force and the RAAF, as well as those of the Thirteenth Air Force and the Royal New Zealand Air Force, which were scheduled to come under my command on June 15th. The General had agreed and set July 2nd as the date of the landing. About August 1st, another jump was to be made to Sansapor, on the north coast of the west end of New Guinea, about 200 miles west of Biak. Airdromes were to be prepared there to support our next move, which would be to capture the island of Morotai, the northernmost of the Halmaheras, on September 15, the same date that the Navy was scheduled to land at Palau.

I told Whitehead that on June 15 I was announcing the organization of the Far East Air Forces, which I would command, in addition to holding the job of commander of the Allied Air Forces, and that I was turning over the command of the Fifth Air Force to him. Streett would run the Thirteenth, with his headquarters at Los Negros, where the most of the Thirteenth's bomber units were now located.

I also told Whitey that I was putting his name in to General MacArthur for two stars. Whitehead thanked me, but then I said, "I'm pushing back to Brisbane tomorrow morning at six o'clock, so tell Sergeant Raymond to report to my ship at that time. I'm getting fed up with eating the cooking at Lennon's Hotel." Whitey immediately replied, "How about you keeping command of the Fifth and never mind about another star for me and letting Raymond stay here? I'd rather keep him than be a major general." I believe he really meant it.

Whitey, who used to have the worst mess in the SWPA, had been living high since I had let Raymond stay up in New Guinea and now he hated to think of losing the cook. No arguments were any good as far as I was concerned. I told Whitey that by this time I was sure that Raymond had trained a couple of cooks for him, but that, anyhow, I wanted to last through the war and if I missed his cooking much longer I would be a dyspeptic invalid and be sent home. Whitey groaned a few times and swore that his delicate system would collapse if Raymond didn't stay but sent word out to the kitchen for the lad to pack up and be on board my airplane the next morning. Whitey snorted when his aide came back and remarked that Raymond seemed pleased at the news. I felt so badly about the whole thing that I left Whitey two bottles of Scotch to console him.

After a series of carrier raids on Saipan, Tinian, Rota, Pagan, and Guam in the Marianas, during which approximately 140 aircraft and thirteen vessels were destroyed, the Central Pacific Forces under Admiral Spruance landed at Saipan on June 15, 1944. Fighting desperately, the Jap garrison resisted until wiped out. The island was declared secured on July 9, 1944.

On the 19th of June the Jap fleet came out of hiding and ventured east of the Philippines, close enough to Spruance's fleet to launch their carrier aircraft. They then turned and went back west through the channel between Formosa and the Philippines. They evidently intended their aircraft to strike Spruance's fleet and then continue on to their fields in the Marianas, from which they could conduct further operations or return to the Philippines by way of Yap, Palau, and Mindanao.

If the Nips had made one properly timed mass attack with the 400 planes that they launched, they might have had a chance of inflicting some real damage to Spruance's Task Force 58, but as usual they dribbled their efforts in piecemeal and in a series of uncoordinated attacks. They paid a real penalty for their dumbness as our carrier boys shot down practically every one of them. Spruance then moved west and launched his planes against the Jap fleet, which was now without air defense. The Nips got away, but the two-day operation cost them 428 airplanes, three aircraft carriers, and two tankers. In addition, three other carriers, a battleship, three cruisers, and three destroyers were damaged. Task Force 58 lost 122 aircraft and seventy-two crew members, and two carriers and a battleship suffered superficial damage.

Tokio Radio promptly announced that Premier Tojo had resigned and General Koiso had taken his place, but it didn't mean anything to the kids.

On June 15, 1944, I announced the formation of the Far East Air Forces and that Whitehead was succeeding me as Commander of the Fifth Air Force. His nomination to the rank of major general had already gone in and I had been assured that there would be no hitch in getting him his second star. Sergeant Raymond was back on the job and I was already contemplating a few alterations in my uniform to make it more comfortable in the general vicinity of the waistline.

The last two weeks of June saw just about the last of the Nip air force, as far as New Guinea was concerned, and the imposition of an air blockade of the few ports left to him that spelled isolation and short rations for his garrisons for the rest of the war. At Babo, Manokwari, and

Sorong, 10,000 tons of shipping went to the bottom in a series of three skip-bombing and strafing attacks, beginning the 15th. Sixty Jap aircraft were destroyed on the ground and in the air around Sorong on the 16th and 17th by our B-25s and P-38s. By the end of the month all that remained at all three airdrome areas was the remains of burned-out and destroyed buildings, heavily cratered runways, and a ring of wrecked aircraft around the strips and in the dispersal areas. Halmahera was now the nearest place where the kids could go to build up their scores.

On July 2nd the 158th Regiment of the 32nd Division landed at Noemfoor Island. Following our established pattern, we had beaten the place apart all around the landing area. Ten minutes before the troops went ashore, 300 instantaneously fuzed 1000-pound bombs crunched along the line of Jap beach defenses. When our troops went ashore there was not a shot fired. There were still a few Nips alive in their machine-gun positions, but they were so stunned by the blast effect of the heavy bombs that they sat by their machine guns staring straight ahead, numb with shock, while our infantry gathered them in. It was one of the few times in the war that we had a chance to take Jap infantrymen prisoners without trouble.

An immediate result, however, was a hurried call for help. One of the prisoners under interrogation said there were 5000 Japs on the island. We hadn't believed there were over 1500 and had landed only about 3000 men. The 503rd Parachute Regiment was hurriedly flown up to Hollandia and on the 4th of July the C-47s dropped 2424 paratroopers on the old Jap airdromes at Kornasoren and Kamiri, which the landing force had captured the day before.

It was a good job and nicely executed, but unnecessary, as things turned out. Whether the Jap lied deliberately or was "bomb happy," I never found out, but the actual count of Nips on Noemfoor was actually only about 1000 and these were never reinforced, although Manokwari was only about sixty miles away.

19. SANSAPOR AND MOROTAI
July—September, 1944

On July 4th one of the war correspondents just in from Nadzab came into the office and asked me if I knew that Colonel Charles Lindbergh was in New Guinea. I admitted that I didn't. He laughed and said that GHQ didn't know it, either. I immediately sent word to Whitehead that I thought it would be a good idea to have Lindbergh flown to Brisbane, as I'd like to see him.

He came into the office the next day. It took about two minutes of conversation to confirm what I had already suspected. He wanted to get some ideas about fighter design and was particularly interested in the P-38 and whether or not a two-engined fighter could really stand up in combat with a single-engined job. He had some affiliation at the time with one of the airplane companies and had gotten permission from the Navy Department to visit and make observations at first hand in the South Pacific. After spending a few weeks in that theater, flying with the Navy and Marine pilots in the Solomons, he thought that it would be a good idea to come over to the Southwest Pacific area, where the P-38s were actually indulging in combat every day. Without bothering to get permission or clearance from anyone, he had simply gotten a ride to New Guinea and checked in with Whitehead at Nadzab, where he had been staying since the first of July.

I could imagine how his unauthorized appearance in our theater might be looked upon by some of the GHQ crowd, so I said, "Let's go pay General MacArthur a call and get a legal status for you before we do anything else."

We went directly to the General's office where I introduced Lindbergh. MacArthur was the soul of cordiality and, after chatting for a while, asked Lindbergh if there was anything he could do for him. I butted in and said that I wanted to look after him as I had an important job that would keep him busy every minute that he could spare. If anyone could fly a little monoplane all the way from New York to Paris and have gas left over, he ought to be able to teach my P-38 pilots how to get more range out of their airplanes. If he could do that, it would mean that we could make longer jumps and get to the Philippines that much quicker, so I wanted to take him under my wing, issue the necessary orders, and put him to work. Lindbergh nodded with that charming kid grin of his that is one of his best assets. General MacArthur said, "All right, Colonel. I'll just turn you over to General Kenney, but I warn you. He's a slave-driver."

As soon as we got back to the office I got Lindbergh fixed up with enough pieces of paper to legalize his status and then we talked about the job. He was quite enthusiastic about its possibilities and thought that with a little training he could increase our operating radius of action nearly fifty per cent. We were now operating the P-38s to a distance of about 400 miles from our airdromes. If Lindbergh was right, we could stretch it to 600.

The first hitch came when I told him that I didn't want him to get into combat. He was actually a civilian, with a lot of headline possibilities. Those headlines would not be good, if he should get shot down, and they would be still worse if by any chance the Japs should capture him. He thought that it would be hard to check on how well the pilots were absorbing his teachings if he couldn't go along to watch them and, besides, he wanted to observe the P-38 in combat to get the answers in regard to the comparison of single-engined versus two-engined fighters. I said that the Nip aircraft were not flying over New Guinea any longer, except for an occasional night bomber, so I'd let him go along with the fighter escort to the bombers and strafers making attacks anywhere except on Halmahera. Up there, a fight would be a certainty. I reiterated I didn't want him to get into combat. My top priority, as far as he was concerned, was to show the P-38 pilots how to get more range out of their airplanes.

Lindbergh went back to New Guinea and went to work. For the next six weeks he did a superb job for me. Flying constantly with the P-38 squadrons, he preached and practiced his technique of economical operation of the engines. The 600-mile radius of action soon became a reality and the kids got more and more enthusiastic about Lindbergh and his ideas. He spent most of his time with Colonel Charles H. MacDonald's 475th Fighter Group and pretty soon that outfit was beginning to talk about the possibility of even 800 miles as an operating radius of action.

During the early part of August the group, then based at Biak, was covering a bomber raid on the Jap oil depot at Boela, on the island of Ceram. It was a very successful, although rather uneventful, raid except that a lone Jap airplane suddenly loomed up directly ahead of Observer, Lindbergh's P-38. I had told him that, of course, if it came to a matter of self defense, I could not expect him to refrain from shooting. How much elasticity his conscience suddenly acquired in regard to the business of self-defense I don't know and never asked him, but anyhow Lindbergh fired a burst and the Jap went down.

I knew about the story shortly after the group landed, but as long as no one put in a claim for official credit for the destruction of an enemy airplane I pretended that I hadn't heard of the occurrence.

A week or so later, to prove the long-range capabilities of the P-38, Lindbergh, accompanied by Colonel MacDonald, Lieutenant Colonel Meryl Smith, and Captain Danforth P. Miller, headed off for Palau, a little over 600 miles to the north of Biak, their departure point.

They arrived over the main island of Babelthuap at 15,000 feet, dove to the tree tops, strafed a patrol boat, and headed south over the main Jap airdrome. Several Nip fighters took off and Lindbergh gained the shocking information that a Jap airplane on his tail was something he could not get rid of. Luckily he had three experts at the art of handling such situations. They extricated him from his dilemma by the simple method of shooting down the Japs before the Japs shot down Lindbergh. MacDonald got two, Meryl Smith got one, and Miller probably destroyed another, which went down on fire but was not seen to crash by the quartet hurriedly extricating themselves from the aroused hornets' nest. The little episode over Ceram had made shooting down a Jap look easy. Lindbergh knew better now. He mentally filed the picture away somewhere in his brain. The next time he wouldn't get caught like that at low altitude by a Nip airplane that could out-maneuver him.

There wasn't to be any next time, however. I told him he couldn't go out on any more combat missions and I wished he would go home. I was, and still am, exceedingly fond of Charles Lindbergh, but I was getting worried. I owed him a debt of real gratitude for increasing the combat range of our fighter planes. It was going to pay heavy dividends for the rest of the war and I appreciated what he had done, but I was getting worried for fear he would get shot down. If that happened it would hurt the Air Force and it would certainly bring down a lot of criticism on MacArthur and on me for allowing him to go out on such missions.

Lindbergh agreed. He said he had taught the kids all that he could, had learned the answers about fighter tactics that he had come out to get, and was ready to return to the United States. I asked him not to tell anyone back home about being in combat as long as the war lasted and said that, if the story leaked out, I would tell the newspapermen that there was no record of his ever having flown on combat mission, let alone having shot down a Jap airplane. Lindbergh said he had no intention of telling the story as he too was anxious not to have any publicity in regard to his activities in the Pacific, particularly while the war was on. After thanking me for the opportunities I had allowed him and for legalizing his status, he left for home on August 21st.

During July we drove airdrome construction at top speed at Owi Island, Biak, and Noemfoor, in order to get our planes in position to take the Jap air strength out of the Halmaheras before we went into Sansapor. This operation was scheduled for July 30th and I didn't want to see it postponed. We would need one or two airdromes fixed up there to support the invasion of Morotai the 15th of September, and to keep an eye on the Jap air bases in the Netherlands East Indies as we moved into the Philippines.

By the middle of July we had 200 aircraft on Owi and 275 more at Mokmer. They were crowded in, but there was no help for it until we got more runways and more dispersal areas built at both places. The two fields on the north coast of Noemfoor at Kamiri and Komasoren were coming along fast. We already had two Aussie fighter squadrons there and I planned to have Streett move his headquarters and his two heavy-bombardment groups to Noemfoor as soon as the construction permitted. The two P-38 fighter groups and the B-25 strafer group I scheduled for Sansapor. In the meantime, about half my aircraft were so far away from the war that they could not reach any worthwhile targets. I encouraged leaves to Australia and for a few weeks Sydney, Melbourne, and Brisbane saw more of the United States Air Force than at any time before or since.

On July 17, I flew to Los Negros in the Admiralties to see how Streett's

Thirteenth Air Force bombers were coming along and to inspect the RAAF Wing commanded by Squadron Leader Davison. Everything was in good shape, in spite of the tough missions and the tougher living conditions. The lads told of constantly killing two or three varieties of poisonous snakes around the camp area, and the breed of mosquito that carries elephantiasis was causing a little worry as there were many cases among the natives on Los Negros.

At mess that evening Streett pulled a surprise on me when he brought my son, Second Lieutenant Bill Kenney, in to spend the evening with us. I knew that Bill was a member of the 868th B-24 radar-equipped, Bombardment Squadron of the Thirteenth Air Force, but I didn't know that he had come up to the Admiralties from the Solomons where I had last heard of him. The newspaper crowd, of course, came around to get photographs to doll up the "father meets son" story but had trouble getting any statement out of Bill except that he was "glad to see Dad." We were glad to see each other and sat up after everyone else had gone to bed, discussing every subject from how the war was progressing to when the Army was going to issue beer to the troops.

After all the trouble I had gone to to keep him out of my command, I was now his commanding general, after all. I warned him that he would probably have to stay a second lieutenant until everyone else had been promoted as I couldn't allow any story to start that "the general's son was getting the breaks." Bill said it was all right with him. He understood.

While air activity over New Guinea had ceased, except for an occasional night raider over Noemfoor or Biak, and Jap air strengths throughout the Netherlands East Indies had dwindled to perhaps 150 miscellaneous aircraft scattered over a dozen airdromes, the Jap was holding his strength of between 175 and 200 airplanes in the Halmaheras and had enormously increased his air force in the Philippines. On May 1st our figures showed only fifty operational planes in the whole of the Philippines, but by July 25th the number was up to 1400, of which 800 were listed as operational combat types and 600 as training aircraft. What the Nip kept so many aircraft in the Halmaheras for, I couldn't figure out. He certainly was not using them, although our packed fields at Owi, Biak, and Wakde should have tempted him. The only smart thing about his set-up there was that he kept his planes pretty well dispersed on six widely separated fields. He had evidently learned one lesson from his Hollandia experience.

On the 27th we made our first of a series of raids to eliminate Halmahera as an air base. Sixty-two B-24 bombers and forty-eight B-25 strafers escorted by sixty P-38s struck the blow. Fifty Jap planes were destroyed on the ground and the P-38s got fifteen of the twenty to twenty-five Jap fighters that intercepted. We lost two P-38s but recovered both pilots. An analysis of the photographs indicated that the Japs had already withdrawn some of their planes before our attack. Their fighters were unskilled and uneager. Our heavy bombers came in at 6000 feet, so that the Nips had plenty of warning. They might have been low on gasoline. I hoped so, for our attack burned up so much that they had a lot less than before we hit them.

We had expected trouble with the Halmaheras but the attack of the 27th left everyone contemptuous of the capabilities of the Nip air force. Some of the fighter pilots, who had hurried back from leave in Australia in order not to miss the show, said they might better have stayed in Sydney.

On July 30th the 6th U. S. Division landed at Sansapor. The landing was heavily supported by our aircraft, but that wasn't necessary. There were no Japs in the air or on the ground to interfere with the operation. In order to make it nice and comfortable for the troops when they went ashore, we sprayed the whole area to a depth of a mile inland with DDT, the day before the landing. The miracle insecticide had just recently arrived in the theater and we wanted to give it a real service test, for if it worked we had a requirement equal to that of any place in the world.

Brigadier General Earl (Diz) Barnes, the head of the Thirteenth Fighter Command, went ashore with the troops as the Air Force Task Commander to supervise the construction of the airdromes and take charge of air operations out of Sansapor when that base was established.

Colonel Robert R. Rowland, commanding the 348th Fighter Group, dropped in to see me on his way back to the United States for a thirty-day leave. He said things seemed to have slowed down a bit and he thought this was a good time to get a rest. Rowland had lost quite a lot of weight following an attack of malaria. The atabrine treatment that he was still following had given him the typical yellow-ochre complexion that a lot of the kids had from taking the drug.

I told him that I wanted to get him back when his leave was over but that if he had a recurrence of malaria while home and the doctors found out about it they wouldn't let him return to the Fifth Air Force. Rowland said, "General, I know that, but they are not going to catch me. I've got about two thousand atabrine pills with me and I'm going to take them religiously until I get back. I don't want to miss out on going back to the

Philippines with the gang."

I had told everyone in the theater that we would have our Thanksgiving dinner this year in the Philippines and the strange part of it was that the Air Force to a man believed it. The rate that we had been moving since the occupation of the Admiralties had made the story more plausible.

Major General Barney Giles, who was then Arnold's Chief of Staff, and General Hull of the War Department General Staff, and several other officers from Washington spent the week of August 6th with me. I took them up to see General MacArthur, who gave them his views in no uncertain terms when the suggestion was made that the Philippines might be bypassed. He still could not see the virtue in blockading and condemning to starvation millions of Filipinos and the thousands of American prisoners and internees, even if the idea were sound from a military standpoint, which he did not believe it was. I could not help but agree with him. He predicted that thirty days from landing at Lingayen he would be in Manila.

I flew the party all over the theater, from Port Moresby along our route of advance all the way to Noemfoor, landing there as well as at Owi, Hollandia, Nadzab, and Port Moresby.

Giles sent a wire to Arnold urging again that some B-29s be allotted to me but Arnold wasted no time refusing to consider the proposition. The Washington crowd were all impressed with the difficult conditions under which we were fighting and with the speed of our advance during the past ten months. I told them it was good to hear but what I would really like was a batch of replacements for about 20,000 jungle-weary Air Corps troops, who had been sweating it out now for two years. Then I also would like some extra crews for my airplanes so that I could afford to let some of them go to Australia on a week's leave about once every two or three months. The visitors were still polite but began talking of the manpower shortage. They left on the 13th and the next day I returned to Brisbane.

We continued working away at the Halmaheras and by the end of August we had just about crossed the airdromes there off our target list. From the 7th of August on, the Nips put no aircraft in the air during our attacks. We made heavy attacks on the 9th and 10th, destroying about thirty-five aircraft on the ground, and thereafter we kept watching the Jap airdromes to see if the enemy either repaired the runways or brought any more planes in. On August 26th only twenty-one aircraft were sighted, scattered on the six airdromes in the Halmaheras. It was probable that even these were out of commission as they had not been moved since last photographed a week before.

About this time Washington had decided that the European war would be over by November 1, 1944, and increasing attention was being paid to planning for a possible speeding up of the war in the Pacific. We were called on for another schedule and finally Major General Richard Marshall, MacArthur's SOS chief, was sent to Washington on August 27th with all the supply and shipping requirements for the following schedule of operations:

Morotai Island	15 September 1944
Talaud Islands	15 October 1944
Sarangani Bay, Mindanao	15 November 1944
Leyte	20 December 1944
Mindoro	15 January 1945
Aparri, Luzon	31 January 1945
Lingayen, Luzon	20 February 1945

I was informed that as soon as the European show was over, Far East Air Forces would be augmented. Six heavy-bombardment groups, one medium group, four fighter groups, and three troop-carrier groups were to be in the theater, ready to work, three months after VE Day. That would add about 1200 more aircraft to the 2000 now with Far East Air Forces and the 1200 planes of the Australian, New Zealand, and Dutch Air Forces, which were part of my Allied Air Force Command. As far as aircraft were concerned the shoestring days of 1942 and 1943 were over.

On September 1st I moved my Allied Air Force and Far East Air Forces Headquarters from Brisbane to Hollandia, a distance equal to that from Washington, D. C., to San Francisco, California. The previous morning the transport aircraft had started loading personnel, baggage, office furniture, filing cabinets with essential records, communications equipment, and even stoves and rations. We were operating at the new headquarters that evening by six o'clock and 1000 people who had eaten on the previous day in Australia sat down to dinner in Dutch New Guinea 2400 miles away.

Whitehead had already moved his Fifth Air Force Headquarters to

Owi Island on August 10th. With the runways there finally lengthened to permit fully loaded take-offs, on September 1st we made our first daylight strike on the Philippines since early 1942. Fifty-seven B-24s dropped 105 tons of bombs on the airdromes at Davao, destroying twenty-seven Jap fighters and eleven bombers on the ground. We lost two B-24s by antiaircraft gunfire. There was no interception.

The following day sixty-five B-24s took off escorted by fifty P-38s and put 150 tons of bombs on the docks, stores, warehouses, and barracks in the Davao area. The bombing was excellent. Only three Jap fighters took off to intercept. All three were shot down. We had no losses.

The fighter escort was provided by P-38s from the Fifth Air Force operating out of Biak and Owi. They, of course, had the benefit of Lindbergh's instruction and, as they could land back at Sansapor on the return from Davao, did not worry about the 1300-mile round-trip distance involved. In the meantime, Brigadier General Barnes' P-38s from the Thirteenth Air Force had started operating from Sansapor and, under Barnes' direction, the pilots had been practicing the Lindbergh technique of extending range. They asked the Fifth Air Force fighters if a few, say six, of the P-38s from Sansapor could go along to see how this long-flight stuff was done. Of course, if they found they were getting low on fuel at any time, they would turn around and go home. The Fifth Fighter lads politely said yes, they thought it would be good experience for the Sansapor amateurs.

Much to the annoyance of the Fifth's fighters, Barnes' P-38s were still with them as they swept in over Davao, but the air was rent with cries of anguish and wrath when the Sansapor kids shot down one of the three Jap planes that intercepted. It was worse than robbery. It was taking a dirty advantage of their good nature. To add insult to injury the P-38s of the Thirteenth Air Force then proceeded to strafe the waterfront around Davao and an airdrome to the north of the city. The Fifth's P-38s followed them through the strafing and returned with them to Sansapor, but to show their superiority as long-range operators, refused to land and continued on to Biak. They, of course, had an alternate landing place at Noemfoor, which would have saved them nearly a hundred miles, but they all had enough fuel to make their home base, although some planes' tanks were bone-dry by the time they had taxied to their dispersal areas. Another rivalry had started which still goes on when representatives of the two outfits meet at an Air Force reunion.

Beginning a couple of weeks before the formation of the Far East Headquarters on June 15th, I had started building up a WAC detachment and by the time we were ready to move to Hollandia I had nearly 200 women working in my headquarters, running the telephone and teletype sections, interpreting photographs, running the file systems, and, what was more important, releasing men to the combat squadrons and groups.

They must have been scheduled originally for Alaska, as they arrived clothed for cold weather in heavy woolen skirts, coats, and shirts. I had them turn their clothing in to the quartermaster and re-equipped them with cotton clothing. GI trousers were substituted for the skirts, and the tailors of Brisbane kept busy for two weeks making the gals look presentable. Each Wac was issued a pair of rubber boots.

Just before leaving for Hollandia I inspected the detachment. They were as fine a looking body of troops as you would find anywhere in the theater. Their equipment was in fine shape, everything was properly packed, and they were anxious to take off for New Guinea and show that they could do their part in the war.

I told them all about the climate in New Guinea, with its humidity, heavy rainfall, mud, its hundreds of species of crawling and flying bugs and a much lower standard of eating than they were accustomed to. I warned them that, as we moved forward, my headquarters would be subject to bombing, it would be the roughest of living conditions, they would probably stand in line in mud up to their ankles in the mess line, the atabrine line, and to get to the shower. I hoped they would now understand the issue of rubber boots. There were no beauty parlors and no department stores in New Guinea. All I could promise them was a lot of hard work and I really meant hard work, as I had already released two men for each Wac that was standing in front of me. If any of them didn't like the prospect ahead of them I wanted to know it then as airplane space was too valuable to waste on someone I could not depend on.

No one said a word. I turned them over to Captain Blanche Kline, their commanding officer, who marched them to the airplanes where they proceeded to load themselves and their equipment as though they had been doing this sort of thing all their lives.

As they had been warned that in New Guinea eggs were just something that you reminisced about, the Wacs decided that a simple way to solve the problem was to go into the egg business for themselves. They all chipped in and bought thirty hens from an Australian poultryman, who said they were the finest egg producers anywhere in the country and that, if they didn't give complete satisfaction, the Wacs could bring them back and he would gladly refund their money.

The thirty hens formed part of the WAC equipment flown to Hollandia and, as they wandered around scratching for worms or bugs in the enclosure on the edge of the WAC camp, they were at least as great an attraction as the girls themselves to the passing natives and GIs, who paused to stare at the two unaccustomed sights. The Wacs talked about the bacon and eggs, omelets, and souffles that they were going to eat. Some thought it might be better to save the eggs and raise chickens. The thought of fried chicken brought a lot of recruits to that side of the discussion. One thing, however, soon began to worry both parties. The hens hadn't laid an egg since arriving in New Guinea. Among the men were several "experts," who were called on for advice. Changes were made in the diet. Every article of the ration was, in turn, added to the feed which had been bought in Australia and which was supposed to make any hen concentrate on laying a minimum of one egg a day. Nothing worked. Finally one day someone made the observation that there should be some roosters in the flock. The Wacs grinned happily. Of course. Why hadn't anyone thought of it before? That was the obvious answer.

A delegation went over to the nearest village and, after a few simple barter transactions, involving such things as lipstick, face powder, and a few articles of clothing, three slim native roosters were acquired and put into the enclosure with the Australian hens. The hens displayed a fair amount of enthusiasm for the newcomers but reciprocity was decidedly lacking. The roosters, who back in their native village were fighting cocks, evidently got homesick; they refused to eat, no matter how much they were tempted by the assorted delicacies that the now-worried Wacs kept putting in front of them.

Finally, when it looked as though the roosters were about to die from their self-inflicted hunger strike, the Wacs held a meeting. The decision was quickly made. They would return the roosters and go out of the egg business. That night the Wacs had a dinner that became part of the history of the organization, something to look back upon whenever the rations were worse than usual, something to remind you of meals that you used to eat long ago and meals that you looked forward to eating again some day when the war would be over.

Our camp site at Hollandia was scenically the most beautiful of any of the locations we had anywhere in New Guinea. It was built part way up the slope of Cyclops, the 6000-foot mountain mass just to the north of Sentani Lake that separated it from the Pacific Ocean. The camp was about half a mile from the lake and looked down upon it from an elevation of about 500 feet. Across the blue waters of Sentani, which was a winding lake over twenty miles long and varying in width from half a mile to a hundred yards, the deep green hills of central New Guinea formed a backdrop of peaks, ravines, and jungle growth that was almost unreal. It was, at least, theatrical. Little cone-shaped green islands, with native houses on stilts in the water clinging to their shores, dotted the lake. To complete the picture, directly in back of the camp and perhaps two miles away, a 100-foot waterfall seemed to spring out of the center of Cyclops Mountain, dark and forbidding, with its crest perpetually covered with black rain clouds. After we had sprayed the area with DDT and gotten rid of the flies and mosquitoes, it wasn't too bad. Even though we were all anxious to get moving forward again as soon as possible, when we finally did pack up to go, nearly everyone said, "I hope the next place is as good as Hollandia."

General MacArthur arrived on September 11th to open up his advanced headquarters which was on the crest of a hill about a mile east of my place. The greater part of his staff had preceded him a few days.

I told him that Arnold had refused to give me the B-29s which I wanted to put on Balikpapan, which was producing the greater part of Japan's aviation fuel. With the heavy concentration of Jap aircraft in the Philippines, if I could destroy the Edelanu and Pandasari refineries at Balikpapan, the Nips might not be able to get their planes off the ground to oppose us when we went in. With the sudden increase in aircraft strength in the Philippines during the past two months I doubted that the Nips had been able to build up sufficient reserve of gasoline to keep going very long. I proposed to set up a series of big attacks as soon as the Sansapor airdromes were finished, which should be during the last week in September. I needed that place as a take-off point or a place to land at on the way back, as the round trip from Biak to Balikpapan and return was around 2500 miles. In the meantime, I would smash all the Nip airdromes in the Celebes to cut down on the Jap capabilities of interception. The distance was too great to have the fighters escort the bombers unless a field could be constructed at Morotai in time. We could not delay too long, as even after I destroyed the Balikpapan refineries and the storage tanks there, it might be a month or two before the pinch would be felt in the Philippines. The General agreed that Balikpapan was a primary target and said to go ahead with my plan. I talked with him at some length about omitting the Talaud operation and jumping directly from Morotai to Sarangani Bay, a month ahead of the present schedule. He said he was leaving on the 13th on a cruiser with Admiral Kinkaid to go on the Morotai landing, which was to be two days later, and as soon as he returned we would get the staff together and discuss means of hurrying the schedule along.

I ordered Major General Streett to move his headquarters and his two heavy-bombardment groups, the 5th and 307th, to Noemfoor at once. In order to give the Thirteenth Air Force some prestige, I put Streett in charge of the task of taking out Balikpapan and told Whitehead to attach to Streett whatever heavy-bomber strength he needed to do the job. Streett was to do the detailed planning, make the arrangements for refueling at Sansapor, and set the dates for the attack. The only provisos were that I wanted those two refineries destroyed and I wanted it done as soon as possible.

On the 13th I flew over to Owi and inspected Whitehead's show. It was the same old businesslike, dependable, well-organized outfit. That trio of Whitehead the commander, Wurtsmith looking after the fighters, and Crabb handling the bombers was hard to beat. The Japs could have given some excellent testimony along that line.

I told Whitey to put his bombers to work on the Jap airdromes in the northern Celebes, from Menado to Sabang, and get rid of their big field and air depot at Kendari.

Whitehead and I discussed the speeding up of our schedule by omitting the Talaud operation. I insisted that the Jap air force was no longer the threat that it had been and, if we could knock out Balikpapan and the Navy big carriers kept on destroying Jap aircraft and shooting down Jap aviators, we could go almost anywhere anytime, if we were willing to gamble a little to speed up the war and get it over with. If the Navy would give us air cover until we could build an airdrome and get our fighters in place, I would be willing to go to Leyte direct, for in the present state of the Jap air force I believed one of our fighter groups, with adequate radar warning, could lick everything the Nips could throw at us. I said to have his staff prepare a study omitting the Talaud operation and have it ready for me by the 16th.

Whitehead was inclined to agree with me but Cooper was quite worried about our necks being stuck out so far. If the Japs did manage to knock off some of our carriers or the carriers had to leave before we got our fighters in place, the Nips would have a field day taking our shipping to the bottom as it was trying to unload troops and supplies. He did not believe that the Jap air force was as ineffectual as I claimed it was.

Cooper was nervous and tired from the strain of the past six months. He was not sleeping and had lost a lot of weight. He needed a rest badly and I decided that it wouldn't be long before I'd have to send him home. In regard to our argument, he had a lot on his side, of course, but we were winning and I felt this was the time to keep moving, while we had the enemy on the run. I believed that the Jap would put up a real fight for the Philippines. He had to, to stay in the war. With our bombers operating from the Philippines, we could cut his whole line of supply to the Netherlands East Indies. Without that supply line he didn't have the vital resources necessary to maintain his forces at war. However, with the backbone of his air power broken, I did not see how he could prevent us from taking the Philippines away from him anytime we wanted to make the attempt.

That evening the Navy communique said that Halsey's carrier aircraft had destroyed another 200 Jap planes on the ground and in air combat over the Philippines.

On the 15th, the 31st Division and a regiment of the 32nd Division landed at Morotai. There was no opposition and the engineers went to work immediately on airdrome construction. We repeated the procedure started at Sansapor and sprayed the area with DDT while the landing operation was going on. Admiral Kinkaid's baby carriers took care of the air cover but it was not needed. There was no Jap air interference. The Halmahera airdromes were nothing but airplane graveyards by this time.

20. BALIKPAPAN
September—October, 1944

On September 15th, I went up to GHQ to talk with Sutherland about skipping Talaud and going directly to Sarangani Bay on October 15th. I found out that my idea was already old stuff.

During the carrier attacks of September 13th one of Halsey's pilots had been shot down and rescued later on in the day. The pilot had talked with some of the Filipinos, who said there was only a handful of Japs in the whole Leyte area and that they had very few airplanes.

Halsey radioed to Nimitz recommending that both our Talaud and Sarangani Bay operations and the Navy plan to capture Yap Island be scrapped and that the next objective be Leyte. The matter was put up to the Joint Chiefs of Staff, who were then attending the Second Quebec Conference. Nimitz offered to turn over the XXIV Army Corps, then loading in Hawaii for the Yap operation, to MacArthur if it would help him to get into Leyte at an early date. The Joint Chiefs wired MacArthur asking for his views.

MacArthur was on a cruiser off Morotai, with radio silence being preserved. There was no way of getting in touch with him, but a decision had to be made and an answer sent to the Joint Chiefs immediately. Quite naturally everyone was reluctant to make so important a decision in General MacArthur's name without his knowledge of what was going on, but it had to be done. I argued that whatever we had been ready to do on October 15 could now be switched to Leyte, as long as the Navy would take care of the air cover until we could get our land-based air in place. I didn't believe Halsey's information entirely as our intelligence showed the Japs had about 35,000 troops in Leyte, and I was sure that, in spite of the losses Halsey's carrier boys had inflicted on the Nips in the past couple of days, the Jap had plenty of aircraft in the Philippines. However, I did agree with him that the Jap air threat was not great enough to stop us from landing on Leyte. The fact that Halsey's carrier attacks had destroyed 400 airplanes on the ground and only fifty in the air during his last series of raids reinforced my belief that the Nip Air Force was about through as an effective fighting unit.

A radio went to the Joint Chiefs the morning of the 16th in MacArthur's name stating that if he could have the two infantry divisions of the XXIV Corps, he would make a landing on Leyte, October 20th. In a few hours the answer came back telling us to cancel Talaud and Sarangani Bay and go into Leyte on the date set by our radio. I sent a courier message to Whitehead informing him of the change in plans and told him to send some of his staff to Hollandia right away to work with my staff and GHQ. A lot of fast work had to be done to scrub all previous planning and start over again. Another thing that complicated the job was that we would not have as much shipping available on October 20th as we would have the middle of December, the original date for the landing in Leyte. This would mean still more adjusting and tightening up on the amount of stockage of supplies we would take ashore initially to keep us going until the ships could turn around and return with another load. It was going to be a tight squeeze.

Whitey sent word that his planners were on the way and that the attack on Kendari that morning by three heavy groups had been highly successful. Two-thirds of the depot buildings had been destroyed. Many fuel fires had blazed up and the place pretty well wrecked. Eight Jap planes were airborne but refused to fight. It was certainly not like the old days. The Jap was making plenty of airplanes to replace his losses but he hadn't found out how to replace the combat aviators we had shot down.

Whitehead wanted to cancel the Balikpapan strikes, but I refused. I told him it would be put on as I had directed, on October 1st or sooner, if the Sansapor field was ready prior to that date. Whitehead also was getting tired and needed a rest badly. I advised him, as long as there wouldn't be anything big taking place except the Balikpapan show for another couple of weeks, to go to Australia, rest up, and be back on the job about the 10th of October. He said he really was tired and would leave in a day or two.

In addition to the big refineries at Balikpapan the Japs were operating some small refining facilities at Lutong in North Borneo, where they also maintained a large stockage of oil. I told Streett to list that place as a secondary target for any missions in that area which didn't get to their primary target for any reason.

On the night of the 16th, Major Weston, the Operations Officer of 868th Bombardment Squadron of the Thirteenth Air Force, left his desk to do his share of the squadron's work and took off on a shipping search of the Makassar Straits, with his secondary target the Lutong oil facilities. Not finding anything on his shipping search, he arrived at Lutong at daybreak. Weston said everything seemed peaceful and quiet, so he took his B-24 down on the deck and flew between two rows of big 50,000-barrel oil tanks. The boys firing out the waist window on the right set fire to four tanks and the gunners firing out the waist windows on the left got seven tanks. Weston then did a 180-degree turn and came back between two rows of barracks and gave them the same treatment as the oil tanks, while the Japanese were firing out the windows with small-caliber rifles. There were no casualties in the crew but 164 bullet holes were counted in the B-24 when it landed.

That crew had accomplished in twenty minutes a job that I would have considered excellent for a whole group of bombers if I had sent them out for the special purpose of destroying the oil storage at Lutong. We had no way of knowing how full the tanks were at the time they were burned and destroyed, but the Nip had lost at least a half million barrels of storage capacity.

General MacArthur arrived back at his headquarters about noon on the 17th, full of enthusiasm about the Morotai operation, and asked what had been going on since he left on the 13th. Sutherland was worried about what the General would say about using his name and making so important a decision without consulting him, but his worries were wasted. MacArthur not only approved the whole scheme immediately but wanted us to explore the idea of another landing on the same date at Lingayen Gulf. He was way ahead of the most optimistic of us. However, when we explained that the forces available and the shipping available would only take care of a five-division effort, which would mean two at Leyte and only three for Lingayen where the Nips could oppose him with at least five, he said okay but that he wanted to be in Lingayen by Christmas. He said also to explore the possibilities of a prior operation in Mindoro and possibly Aparri. If Leyte proved to be an easy operation, he wanted to move fast.

General Arnold had been trying to get Bertrandais away from me to run his big depot at Wright Field, Ohio, for some time. He had not offered me a suitable replacement and besides I needed Bertrandais, so I had refused. On the 17th Hap offered me Major General Clements McMullen for Bertrandais and promised to make Bert a major general I wanted McMullen and had been trying to get him for the past two years. He was tops in the supply and maintenance field and a personal friend of twenty-five years' standing. Besides, I could not get another star for Bertrandais and I would not stand in the way of his promotion. The exchange was made and I asked that McMullen be given a top-priority flight to me right away.

Dick Bong returned from taking the aerial-gunnery course in the United States and reported to me for instructions shortly after I moved to Hollandia. He said he really had learned so much about shooting that he was ashamed of the way he had wasted ammunition in getting the twenty-eight victories he was credited with. He was all ready now to go out and prove it, if I would just give him an assignment to a fighter squadron. I told him that I wanted him to visit all the fighter outfits and teach them what he had learned back home. If he could teach a hundred kids to get one Nip apiece that would be far better than asking him to shoot down that many. He didn't look too happy about the job I'd given him but cheered up a little when I told him he could go along once in a while to see whether or not his pupils had absorbed their instructions. Even when I insisted that I didn't want him to do any shooting except in self-defense, he didn't seem too worried. I told him that while he had gone, the boys had learned to get a lot more range out of the P-38 and he should get that knowledge as soon as possible or he would not be able to accompany them all the way on some of the long trips. He went away happy. He had a job again.

Major Tommy McGuire of the 475th Group came in to see me a few days later to see if I would overrule the doctors and let him fly again. McGuire had arrived in the theater back in the spring of 1943 to find that Bong had eight victories. Every time he shot down a Jap he'd find that Bong had also gotten one. He'd always been eight behind. When Bong got up to twenty-eight and I sent him home, McGuire had just made his score 20. He said, "Now is my chance. Dick will be away for at least two or three months. By the time he gets back he'll be Number 2 on that scoreboard."

Japs got scarce about that time, however, and just as the group moved to Hollandia and it looked as though business was about to pick up again, Tommy had a siege of dengue and then an attack of malaria. He had been in the hospital ever since and had just managed to get released the day before, but the medicos had stipulated that he do no flying for another ten days.

"I'm still eight behind," mourned McGuire, "and by the time those fool docs let me fly again there is no telling how many that guy Bong will knock down. General, how about letting me go?"

I told him to take it easy until he was right again. There was no need for him to worry as I had put Bong on a gunnery instructing job and hadn't assigned him to a combat outfit. Anyhow, no one was going to get many Nips until we got to the Philippines. There simply were no fights in the theater within range of a P-38 any more. He was disappointed because I wouldn't overrule the doctors but seemed a little more cheerful as he said he guessed he'd go back to his squadron and see what had been going on since he went to the hospital.

On September 30, sixty-four B-24s from the 5th and 307th Bombardment Groups of Streett's Thirteenth Air Force and the 90th Bombardment Group of the Fifth Air Force hit Balikpapan with eighty-five tons of 1000-pound bombs. Direct hits were scored on the Pandasari refinery by the 5th Group and considerable damage caused. Huge fuel fires sent smoke up to 18,000 feet and a large tanker at the dock was set on fire. The group was not intercepted. The 307th, which was almost an hour later in getting to the target, found it obscured by clouds and there were from thirty to forty Nips waiting to intercept them. The Japs fought savagely, returning again and again to the attack. The 307th had four bombers shot down. Our rescue Catalinas picked up three complete crews and half of the fourth. The group had two other badly shot-up airplanes collapse when they landed at their home base. The 90th Group got in on the tail end of the fight. They had one bomber shot down but we recovered the crew. They also found the target obscured by clouds. The Nips lost nine fighters definitely and probably eight others.

Streett sent the 5th and 307th over again on October 3rd. Forty B-24s dropped fifty tons of bombs. Once again the 5th Group, flying a tight formation, scored several direct hits on the Pandasari refinery, the cracking plant, sulphuric acid plant, and the oil storage, putting the target out of business for some time. From twenty-five to thirty Japs intercepted. The 5th Group shot down three Nips with no losses to themselves.

The 307th ran into trouble again. Their formation was loose and they were again late getting to the target. From forty to fifty Japs attacked and again swarmed all over the formation. The 307th got nineteen definite and five probables but it cost us five more bombers. A submarine picked up one complete crew and four men of another plane.

That evening I got word that the Japs were flying eighty-five more fighters into Balikpapan from Singapore. These were Jap Navy pilots and the best they had. Another message to Macassar had called on the base there to help out, too. It looked as though the Japs were really serious about keeping Balikpapan operating.

I sent word to Streett that the next attack would be by both groups of the Thirteenth and three groups of the Fifth and to hold it for a few days until I could see if we could speed up the airdrome construction at Morotai and get some fighter cover.

On the 6th I flew to Noemfoor to inspect the Thirteenth Air Force heavy-bombardment groups and see how the morale was standing up after the losses of the two Balikpapan strikes. They were getting close to the breaking point. I got the kids together, discussed the importance of the mission and how it would deprive the Nips of gasoline and make our job easier in the future. I said I was going to send three groups from the Fifth Air Force along at the same time and some fighters to cover them over Balikpapan. They looked a lot happier when I left.

Back at Fifth Air Force Headquarters at Owi I told Wurtsmith to figure out what we could do on fighter cover from Morotai, 830 miles from Balikpapan, which I expected would be ready for fighters in about three days. I also wanted to know how far they could escort the bombers if the fighters operated out of Sansapor, which was 1000 miles from Balikpapan. Up to that time the longest fighter mission had been made by P-38s operating out of Sansapor, accompanying the heavy bombers to Davao and back to Biak, which had involved a total flight distance of about 1650 miles.

While I was over inspecting the bomber outfits, Squeeze talked the proposition over with the fighter pilots. When I saw him later on, he told me he had it all arranged. Fifty of his pilots said they would accompany the bombers from Sansapor to Balikpapan and cover them during the bombing. Of course, they wouldn't have enough gas to get any more than a third of the way home but if I would just send six or seven rescue Catalinas out to meet them, they would fly as long as they could and then simply parachute down and get picked up by the Cats. I asked Wurtsmith where he found fifty kids that crazy. He said, "If you will let them go, I'll get you two hundred in half an hour." I vetoed that scheme. I couldn't afford to lose the planes, anyhow, but if the water should happen to be too rough for the Catalinas to land, I'd lose the kids, too.

Later that evening Squeeze came back and said as soon as we could operate from Morotai the fighter support could be furnished. Not only could the P-38s make it, which I had been sure of, but the P-47 pilots were going along, too, if they could get the runway lengthened a little, by carrying two 300-gallon, droppable, auxiliary gas tanks hung on the wings instead of the standard 165-gallon tanks. I sent word to Fighter Hutchinson, who was my Air Task Force Commander at Morotai, to get another 500 feet of runway done by the afternoon of the 9th if humanly possible. I asked for too much, however. It was not until the 11th that I had the job done.

On October 10th we really went after Balikpapan and the Japs went after us. It turned out to be one of the toughest fights of the war. One hundred and six heavy bombers escorted by thirty-six P-38s made the third of our raids on the Nip refineries. The Pandasari refinery was badly damaged. The photo interpreters said it was out of action for months. The lubricating oil refinery was damaged but could be repaired in a month. Three of four large storage tanks were destroyed and ten out of thirty-eight smaller tanks wiped out.

The Japs put ninety fighters into the air against us. They lost sixty-one of them. We lost four bombers and one fighter. The complete crew of one bomber was recovered.

Two of the Jap planes shot down figured as numbers twenty-nine and thirty on the score of Major Richard I. Bong, who had flown up from Nadzab, where he had been instructing the newly arrived pilots in gunnery, to see how his pupils were doing when they came to shooting for keeps. Bong assured me that he had been forced to get both of them "in self-defense." I remarked that the gunnery instruction he had gotten back home had evidently done him some good. The kid grinned and said, "General, it really is a good course, but just to make sure, I pushed the gun barrels into the Nip's cockpit and pulled the trigger."

I wired General Arnold:

"During the strike on Balikpapan by five heavy-bomber groups on October tenth Major Richard Bong in a P-38 accompanied the escorting fighters to observe the results of the gunnery instruction he had been giving since his return from the United States. While conducting his observations, he was forced in self-defense to shoot down two Nip aircraft. While regrettable, this brings his official score to thirty enemy aircraft destroyed in aerial combat. Have cautioned Bong to be more careful in the future."

Hap replied:

"Congratulations to Major Bong on his continued mastery of the manly art of self-defense. Feel sure your warning will have desired effect."

On the 14th we just about finished Balikpapan off for the rest of the war. One hundred and one heavy bombers, escorted by sixty P-38s and P-47s, did the deed. The Edelanu refinery was completely destroyed and a lot more damage done to the Pandasari plant. We stopped worrying about whether or not the Nip would get any aviation fuel out of that place for many months to come.

The Nip only got fifty fighters in the air, but he fought with the same toughness that he had shown on all three of the previous raids. It cost him forty-three airplanes. We lost two bombers and five fighters but rescued four of the fighter pilots. It was the last time we were ever intercepted over Balikpapan. Those Jap pilots had been the best they had, but they were all dead.

Major Tommy McGuire finally got out of the clutches of the doctors in time to make this raid. Two Nips tried to argue the question of air superiority with Tommy. They lost. That brought McGuire's score up to twenty-two—still eight behind Dick Bong.

Every once in a while, when reading our daily intelligence summaries, I would wonder whether Americans were crazier than Aussies or vice versa. An Australian B-25 operating out of Darwin was sent on a combination reconnaissance and propaganda leaflet-dropping expedition to Batavia, Java. The pilot decided that at low altitude he could accomplish both missions better than any other way. The following extracts from his report certainly show that he carried out his original idea and that he thoroughly enjoyed the flight:

"Just after dawn, proceeded west at two hundred feet altitude along the north coast of Java, following the approximate line of the railway and dropping pamphlets at villages. Railway bridges were intact. Passengers at stations showed surprise but there was no sign of any air-raid alarm.

"Entered Batavia from the southeast suburbs at altitude of 50 feet. Apparently no alert had been given. People were walking normally in the streets and looked up in surprise. Trams were just beginning to run.

"The aircraft then turned south over Batavia again and the pilot visited the dog-racing track, one of his peace-time haunts.

"From Batavia the aircraft went south to Buitenzorg where a barracks was strafed. Japs were seen running out of the barracks. At Buitenzorg it was noticed that a large Jap flag was flying from the Governor-General's palace. Some 50-caliber fire was seen to enter the front portico. On leaving the palace the rear gunner gave it a burst. The palace and surroundings showed no change and there were still deer in the park."

What it was that made those kids always want to do even more than you called on them for, I don't know, but whatever it was, both the Australians and the Americans had it.

Whitehead came back from leave on the 11th looking fine. He swore that he felt so good he could last for two more wars before he would need any more rest. With the Leyte show due to break soon I was glad to see him in such good shape. I told him to send Cooper home before he broke down. Coop had looked as though he was on his last legs when I saw him a couple of days before.

On the 13th I got a real reinforcement. Major General Clements McMullen reported to me for duty. I had been practically running my own supply and depot maintenance organization for the past year, but now I could relax and let McMullen do it for he was just about the best in that game. I told him to take care of the needs of both the Fifth and Thirteenth Air Forces and to move his headquarters from Brisbane to Hollandia as soon as my headquarters moved to the Philippines. At that time I wanted everything in Australia cleared out. In the meantime I was not going to worry about supply any more.

My next job was to take care of Cooper. He was heartbroken at the idea of going home just at this time. I told him I was trying to save him for a lot of work later on but if I kept him on he would crack and I'd probably have to ship a corpse back home. I wanted him to do nothing for a month in the United States and then go to England, check in with General Spaatz who was running the big air show there, and make arrangements for the units that were to come to me from Europe as soon as Hitler folded. When that job was done, if he would let me know the date he was available, I would radio Arnold to send him back to the Far East Air Forces. Cooper felt a little better after our talk. He left for home on the 15th.

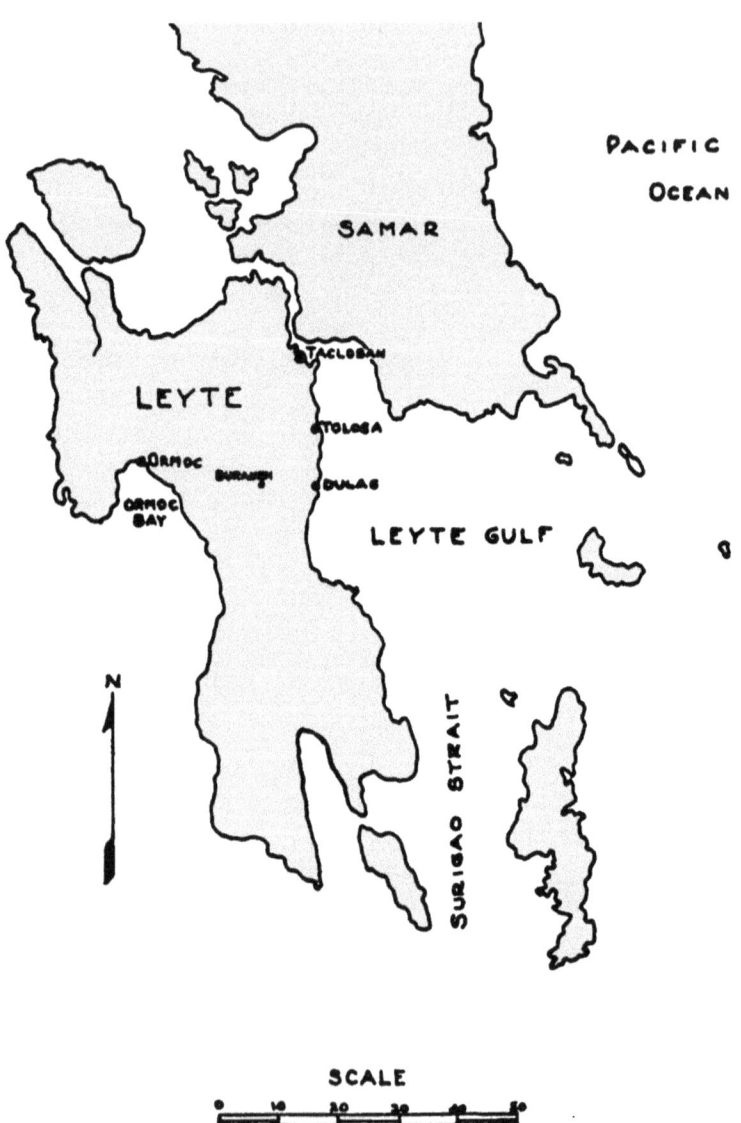

21. BATTLE FOR LEYTE GULF
October, 1944

ON THE 16th, with General MacArthur and General Sutherland, I left for the Leyte operation on board the cruiser Nashville. Halsey radioed that he would support us in the Leyte area with two fast carrier groups instead of the four originally planned. He had been hitting Formosa and Luzon every day since October 10th and had claimed the destruction of 807 aircraft and the sinking of twenty-six vessels. He had lost eighty-nine planes, but, in addition, the heavy cruiser Canberra had been torpedoed on the 13th, the light cruiser Houston on the 14th and again on the 15th. The carriers Franklin, Hancock, and Hornet and the light cruiser Reno had also been damaged.

That evening the reports from Philippine guerrillas stated there were 8000 Jap troops in the Tacloban-Dulag area where our landings were to be made. I wished that I could put about four heavy groups over the landing beaches to strike ten minutes ahead of the actual landing but the Morotai airdrome construction had not gone ahead far enough to permit anything like that number.

More guerrilla reports from Leyte during the 17th gave the total Jap strength on the island as 20,000, which included the Nip 16th Division and some marines. Tokio Radio broadcast during the evening that the

Nips had taken Halsey's fleet apart, sinking eleven carriers and damaging six, as well as shooting down 1000 of his planes and sinking twenty-three other Naval vessels. Halsey reported that the Canberra and the Houston were being towed out of the combat zone and that some other vessels had been damaged but not enough to take them out of action, while his aircraft losses were less than 100. His fast carriers were moving south and would be into position to support the Leyte show.

On the 18th the guerrilla reports upped the Jap strength to 23,000 Japs on Leyte. It began to look as though by the time we got ashore the number would probably be up to our original estimate of around 35,000. During the day the minesweepers in Leyte Gulf swept up sixty mines, with about forty per cent of the area covered.

At three o'clock in the afternoon we picked up the main convoy just east of Palau. The whole ocean looked full of ships as far as the horizon on all sides. It was a sight that I wouldn't have missed for anything, although just before dark, when we suddenly changed course to avoid hitting a floating mine, I wished I were on an airplane instead of a ship.

We arrived off the entrance to Leyte Gulf at eleven o'clock at night on the 19th. The battleships of Kinkaid's 7th Fleet during the day had entered the gulf and spent three and a half hours shelling the Jap shore positions. Dulag was reported in ruins. The minesweepers had cleared the place and everything was all set for us to go in the 20th on schedule. One of the minesweepers had been hit by shellfire from Jap shore batteries but not badly damaged, and one of our destroyers had bumped into two mines. It hadn't sunk but was in bad shape.

Halsey notified us that two of his four fast carrier groups would go to the Admiralties on the 21st for refueling and return to Leyte on about the 30th or 31st. I hoped that it wouldn't take long to get airdromes going at Tacloban and Dulag. The Nips still had a lot of planes in the Philippines and could bring plenty more in from Japan if necessary to attack our shipping in Leyte Gulf if we let our air cover get too thin. Kinkaid had eighteen baby carriers, but they, too, would have to withdraw to get fuel and bombs and rest their crews after a few days of operation. I still thought the Jap was licked and could not stop us but he could make our show a bit expensive for us if we didn't keep air control over the Leyte area.

At daybreak on the 20th we entered the gulf. A few floating mines were sighted and at eight o'clock a periscope was reported up ahead of us. Two destroyers swarmed over the spot dropping depth charges. I was glad to see them working but still didn't feel good about being on a ship. The surprising thing to me was the absence of Jap aircraft. One lone Nip appeared at nine-thirty, coming in low from the direction of Dulag. He turned toward a destroyer which promptly shot him down.

The battleships, cruisers, and destroyers of Kinkaid's fleet shelled the landing-beach areas for another three hours and at ten o'clock the troops started going ashore. The X Corps, consisting of the 1st Cavalry Division and the 24th Infantry Division, beached about six miles south of Tacloban, while the XXIV Corps, comprising the 7th and 96th Infantry Divisions, landed in the Dulag area. Opposition to the actual landing was negligible in both cases as the Japs had evacuated their shore positions.

General MacArthur, General Sutherland, President Osmena of the Philippines, and myself, with a few American and Filipino staff officers, went ashore about two-thirty in the afternoon in the area designated as Red Beach, where the 24th Division had landed and worked its way inland about 300 yards from the water's edge where they were being held up temporarily by a few Jap pillboxes. Four of the big landing craft, beached where we went in, had been hit by Jap mortar fire and one was burning nicely when we landed. One light landing craft had just been sunk. There seemed to be a lot of Nip snipers firing all around the place and the snap of the high-velocity small-caliber Jap rifles sounded as though some of them were not over a hundred yards away.

MacArthur calmly walked around wearing his old Philippine marshal's cap, his only weapon the big corncob pipe he was smoking. We passed a lot of troops lying behind the boles of the palm trees staring into the bushes and occasionally firing at something. We passed two of them on our way to find General Irvin, the division commander. I was walking along about ten feet behind MacArthur when one of the GIs looked up. He nudged the soldier next to him and said, "Hey, there's General MacArthur." The other lad didn't even look around as he drawled, "Oh, yeah? And I suppose he's got Eleanor Roosevelt along with him."

We found Irvin, discussed the situation with him, and then went back to the beach where a radio-broadcasting station had been set up to announce the landing to the people of the Philippines and the rest of the world as well. MacArthur made a brief announcement of the fact that we were back and then turned the microphone over to Osmena. We then returned to the Nashville.

Just after we left, the Japs tried to break through to the very part of the beach on which we had landed. They didn't quite make it. A few Nips got within a few yards of the shore but were mopped up before they could be reinforced. It was a good thing we left when we did.

On the way back, a Jap torpedo bomber flew low over our heads and smashed into the cruiser Honolulu. There wasn't much doubt that this was a deliberate suicide, or "kamikaze," attack. There had been several instances of ramming reported during the war, but most of these could have been unintentional. Of course, the Japs had talked a lot about this method of warfare for some time but this was the first case that had come to my attention that really looked like a planned and premeditated ramming attack. The cruiser immediately took a list and they started backing her toward the beach. In the meantime three more Nip airplanes strafed the place on shore we had just left, causing some damage to vehicles, boats, and supplies. Our fighter cover appeared too late and the antiaircraft fire was too wild, so the three Nips got away.

That evening about seven o'clock, four more Jap planes came over. Every vessel in Leyte Gulf seemed to be shooting at them but they dropped no bombs and none of them got hit. We had no aircraft in the air as they had gone back to their carriers to land before dark.

I had estimated that the Japs could put over three or four raids a day, using their old standard formation of twenty-seven bombers and thirty to fifty escorting fighters. They had enough aircraft in the Philippines to make that effort, but the reaction so far had been pitiful. With around 600 vessels in Leyte Gulf, the target should have tempted the Nip to make a much bigger effort. Kinkaid's baby carriers had a total of over 400 airplanes, but not over a third of these at best could be counted on to be in the air at any one time, and this number would decrease each day with operational losses and airplanes getting out of commission for normal mechanical troubles. I hope we would get airdromes ashore soon so that I could get my fighters up from Morotai, where I had told Whitehead to assemble the 49th Fighter Group as the first to come forward.

The 21st started off with a raid at six o'clock in the morning by three Jap planes which picked out the cruiser Australia as a target for a kamikaze attack. Two of the planes were shot down by antiaircraft gunfire, but the third crashed into the bridge, killing or wounding the captain, his executive officer, and about twelve others, besides doing considerable damage to the ship. That evening the Australia and the Honolulu left for the Admiralties to effect repairs.

At ten o'clock that morning, I went ashore with General MacArthur at White Beach, where the 1st Cavalry Division under Major General Mudge had landed. The Japs had evacuated some excellent concrete pillboxes and earth-covered log bunkers when the Naval gunfire and rockets began to drop among them. A few snipers had been left behind but our troops had cleaned them out and advanced up the coast, where they had captured the old Jap airdrome just east of the town of Tacloban. Mudge reported that his casualties so far had been five men killed and fifteen wounded.

We drove along the coastal road to the Tacloban airdrome, passing a few dead Jap snipers on the way. Mudge already had the bulldozers and graders at work filling up bomb craters and getting the ground leveled off. The place was nothing but a sandspit a little over a mile long and about 300 yards wide. We would have to cover it with steel mat and, before that was laid, we needed some coral or rock ballast on top of the existing soil. One of the engineers said they had located a coral deposit about two miles away and as soon as they got the trucks operating would start hauling it to the airdrome. The place looked about big enough for one group of seventy-five fighter aircraft, or maybe a few more, if we parked them wingtip to wingtip. I didn't want to risk that again after our experience at Wakde, but if we didn't get airdrome space on Leyte pretty fast I knew I would have to take the gamble.

Mudge took Tacloban that afternoon. His troops moved so fast that they picked up some excellent souvenirs. The Jap headquarters was evacuated in such a hurry that the Nip officers left their baggage, clothes, swords, pistols, and personal belongings behind them.

Jap planes, coming over three to six at a time, raided the Leyte Gulf area at intervals all day. They scored a few hits on shipping, but we had no actual sinkings and most of the Japs were shot down by Kinkaid's carrier boys or by ships' antiaircraft gunfire. One of the baby carriers, the Sangamon, was hit by a suicide plane carrying a 500-pound bomb. The ship was damaged considerably but remained on the job in spite of it. A curious thing about those kamikaze attacks was that they were all made with small bombs, generally around 250-pound size, with a few 500-pounders. You would have thought, if they were going to lose the pilot and the airplane anyhow, that they would have carried a 2000-pound load or at least a 1000-pounder and be assured of doing some real damage. It was just another bit of Jap psychology that we couldn't figure out.

The day before, the censors had let a story go out on the radio that MacArthur and his top staff officers had come back to the Philippines on board the United States cruiser Nashville. The following evening, the 21st, Tokio Radio informed us on their broadcast that the brave kamikaze boys would see to it that the Nashville would never leave Philippine waters. I thought more than ever that a sailor's life was not for me, particularly during wartime. I would cheerfully have traded my nice comfortable quarters and excellent mess on the Nashville for a tent under a palm tree ashore and an issue of canned rations. MacArthur laughed at me and said it was good for me to find out "how the other half of the world lives."

On the 22nd I went ashore at Dulag with the General. Our 7th Division had worked inland a couple of miles and was now engaged with Japs all over the place. Mortar and artillery fire could be heard all around the perimeter of the area occupied by our troops. We drove out to the Dulag airfield, just west of the town, halting a couple of times. Once to wait for some of our sharpshooters to knock a Jap sniper out of a tree about seventy-five yards off the road, and then while the anti-tank boys stopped and set fire to a Nip tank that was coming along the road in our direction. We passed the tank on the way to the Dulag strip. It was still burning. The whole crew had been killed. All along the road, troops were wandering around looking for snipers and every once in a while someone would shoot at something. We went out on the east end of the airdrome and looked it over. The west end was being used as a field of fire by the Nips on one side and our troops on the other, but this didn't seem to bother MacArthur who wandered around discussing the situation with the ground troops and asking me what I thought of the place for an airdrome. I was glad when he finally decided to go back to the beach. The Dulag area didn't look too promising for an airdrome location. It was big enough but not high enough above sea level when the rainy season came, and that period was due in another couple of weeks. If we could get about a foot of gravel or rock over the existing surface, it would probably be an all-weather field but the engineers would have to work fast. By and large I didn't like the looks of the prospective airdrome situation in Leyte. We had two other former Jap airdromes under consideration, west of Dulag a few miles, at San Pablo and Burauen, which I intended to look at as soon as our troops chased the Japs out, but the Filipinos I talked to at Dulag told me that these also would be quite muddy as soon as the rains started.

Jap aircraft again raided the shipping during the day but they were still dribbling in planes three or four at a time and getting them shot down without accomplishing any real results.

Shortly after midnight of the 22nd–23rd, two of our submarines contacted a Jap naval force, estimated at eleven heavy vessels, off the north coast of Borneo. They attacked and reported sinking two heavy cruisers and damaging another. Another sub reported a heavy cruiser, a light cruiser, and two destroyers off the entrance to Manila Bay.

Around noon on the 23rd we landed at the dock at Tacloban and had a ceremony at the Philippine Commonwealth Building. Osmena was installed as president and Tacloban was named as the provisional capital, pending our recapture of Manila. General MacArthur pinned a Distinguished Service Cross on Colonel Kangleon, the Filipino guerrilla leader on Leyte, whose bands were cooperating with us. Osmena then announced that he was appointing Kangleon the governor of Leyte Province. The Commonwealth Building was in excellent shape. There had been some looting, but it was all ready for occupancy as government headquarters as soon as a little furniture could be moved in and the telephone communication restored. The town itself showed little evidence of having been in a war. It looked strange to see concrete docks and concrete or macadam roads and substantial buildings again. It made you feel as though the end of the war was in sight.

After the ceremony we went over to the best-looking house in Tacloban, formerly owned by a man named Price who had been killed by the Japs during the occupation. It was a goodsized concrete house in excellent shape and had been used by the Nips as an officers' club. MacArthur decided to take it over as his temporary headquarters and living quarters for himself and some of his staff. He assigned me one of the bedrooms.

As we walked in the gate, he asked what a large mound of earth, just off the walk to the main entrance, was doing there. Someone explained that it was the best dugout they had ever seen, about twenty feet underground, with ventilating system, electric lights, easy chairs, tables, and so on. MacArthur said, "Level it off and fill the thing in. It spoils the looks of the lawn," and walked into the house. I didn't think it made much sense, but I didn't say anything. It wouldn't have done any good, anyhow. I had tried to get him into dugouts before without any success.

I drove out to the airdrome to see how construction was coming along and found Photo Hutchison, who was there as the Air Task Force commander, all upset. Twenty-four landing craft had beached on the edge of the strip and unloaded supplies, guns, ammunition, and troops right on the airdrome, stopping all work. Another twenty-eight ships were reported coming in during the night to do the same thing. I got hold of Krueger and got him to promise to stop such foolishness and get some steel mat ashore so that we could surface the airdrome and start operating. I warned him that Kinkaid's carriers could only stay a few more days and that if we didn't have some fighters in place when the carriers had to leave, the Jap planes would have a field day. To make sure that no more stuff would be moved in on the Tacloban strip, I saw

General MacArthur, who passed the word to both Krueger and Kinkaid that he wanted the place used as an airdrome not as an unloading point. The trouble was that the ship captains wanted to unload and get away from the beach before some Jap aviator bombed them while they were sitting ducks and unable to dodge. I couldn't blame them for feeling that way, but we needed an airdrome and we needed it quickly.

The 24th started off with a bang. At eight o'clock in the morning a Jap fleet was picked up off the southern tip of Mindoro, headed east. It consisted of five battleships, ten heavy cruisers, two light cruisers, and fifteen destroyers. This was an entirely different sighting from the one of two battleships, four cruisers, and several destroyers, sighted early on the 23rd off the north coast of Borneo.

Between 8:30 a.m. and 9:15 a.m., fifty Jap planes made attacks on the shipping in Leyte Gulf. The cruiser Shropshire and a destroyer were damaged by near misses. One landing craft was sunk and a Liberty boat badly damaged, with around 200 casualties, from a direct hit. MacArthur and I had a grandstand seat, watching three Nips start their dives from about 15,000 feet toward the Nashville. One was shot down by a Navy fighter plane, a second was torn apart by a direct hit by the antiaircraft guns, but the third kept on coming. The Nashville was maneuvering nicely and the Nip plane crashed into the water and blew itself up about 200 yards astern of us. Very probably in that long dive the plane got going so fast that the controls would not respond and, once committed to the straight line path of his dive, the pilot could not deviate from it. I asked MacArthur how soon he was moving ashore. When he laughed, I remarked that I was serious about the matter. He said he'd move as soon as the house was fixed up. I decided that as soon as I got ashore I'd hurry that job up myself.

At ten o'clock that morning I left for Tacloban strip in a light landing craft to see how the work was coming along. I had a much better seat this time than on the Nashville. Between twelve and fifteen more Nip planes came over while I was on the way to Tacloban. About 500 yards from us a medium-sized landing craft was sunk and a large one hit and badly damaged. Two of the five dive-bombers making the attack were shot down by the ship's guns. Then a couple more strafed from south to north all along Red Beach and set a few fires. The whole sky was covered with antiaircraft bursts but the two Nips calmly turned around and strafed the beach from north to south and then flew away to the west. Just as we were about a hundred yards off the Tacloban strip, another Jap plane coming in to strafe the place was shot down and crashed on the edge of the beach in front of me.

The Battle for Leyte Gulf   343

The airdrome was still in a mess, all cluttered up with troops and supplies. I got hold of Krueger's brigadier generals, who was looking after the unloading of supplies, and told him to hurry up and clear the airdrome. Photo Hutchison came over while I was talking to him. I told Hutchison that, beginning at daybreak on the 25th, he was to take his bulldozers and push back into the water anything still left on the place which interfered with getting an airdrome built. Some of the troops and equipment belonged to antiaircraft units. I told the brigadier he could save himself some trouble by leaving those outfits on the edge of the airdrome, where they could take up position to defend it. He agreed. We then mobilized every soldier and Filipino we could get our hands on and started clearing the strip. By nightfall we were beginning once again to make an airdrome and two bargeloads of steel mat had arrived to start the final surfacing.

When I got back to the Nashville, there was lots of news. All day Halsey's carrier aircraft had been attacking the big central Jap fleet coming east from Mindoro, and had reported one large battleship of the Yamato class on fire and down at the bow, a Kongo-class battleship badly damaged and on fire, two other battleships heavily hit, one light cruiser capsized, two heavy cruisers torpedoed, and another heavy cruiser hit by bombs.

In addition, a light cruiser had been hit off the entrance to Manila Bay and one of three destroyers off Panay had been sunk.

Halsey's force, which was about 130 miles northeast of Manila, had been attacked during this time and the carrier Princeton heavily hit and set on fire. The cruiser Birmingham, in close, trying to help put out the flames, was damaged and a lot of her personnel killed when a heavy explosion took place on the Princeton as the fire reached the fuel tanks. The Princeton was abandoned and sunk immediately after. The Japs had lost 150 planes.

Kinkaid's carrier aircraft in the meantime had picked up the southern Jap fleet, steaming east in the Sulu Sea, consisting of two battleships, one heavy cruiser, and four destroyers, and had scored hits on everything except two of the destroyers. Later on in the day a reconnaissance plane picked up another Jap force, in the Sulu Sea to the northwest, consisting of two more heavy cruisers, a light cruiser, and four more destroyers.

About 4:30 p.m., another complication was introduced into the picture when a reconnaissance plane reported a third Jap fleet, 130 miles east of Cape Engano on the northeast tip of Luzon, headed south. The force was reported to consist of two large carriers, one light carrier, one heavy cruiser, three light cruisers, and three or more destroyers. Another report came in later stating that in addition there was another group with them consisting of four battleships or heavy cruisers, five heavy or light cruisers, and six destroyers.

Relying on Halsey's 3rd Fleet to take care of the Jap central fleet, Kinkaid moved his whole force, except twelve baby carriers, south to engage the Jap southern fleet, which was heading east into the Mindanao Straits. The twelve small carriers were to stay off Leyte Gulf and furnish cover for the shipping in the gulf and support Krueger's ground operations on Leyte.

At 8:30 the evening of the 24th, Halsey radioed Kinkaid that he was going north with his three carrier groups to attack the Jap northern fleet. We knew that his other carriers under Admiral McCain had already left for Ulithi to refuel, get more bombs and ammunition, and rest the crews, which meant that there would be no aircraft to hit the Nip central fleet if they tried to come through the San Bernardino Straits. However, that fleet had no carriers and Halsey's battleship and heavy-cruiser force should be able to handle the situation, providing that the Nip land-based air didn't suddenly come to life and make an all-out effort against him. I didn't like the situation at all. MacArthur wasn't saying much, but he didn't act too happy, either. We kept pretty close to the radio most of the night.

Early the morning of the 25th Kinkaid's 7th Fleet contacted the Jap southern fleet in the Surigao Straits between Leyte and Mindanao. The Jap force consisted of two battleships, three heavy cruisers, one light cruiser, and eight destroyers. The Nips were practically wiped out in the action. Both battleships, one heavy cruiser, and three or four destroyers were definitely sunk. Practically all the rest of the Jap fleet suffered heavy damage. The light cruiser managed to escape from the battle area, but on the 26th was sunk by B-24s of the 13th Air Force, off the island of Negros. One of the damaged destroyers was also sunk later in the day by carrier aircraft from Kinkaid's 7th Fleet.

In the meantime, disaster suddenly stared us in the face. The Jap central fleet, now reported consisting of four battleships, seven heavy cruisers, and between twelve and fifteen destroyers, and possibly two light cruisers, emerged from the San Bernardino Straits, cut south down along the east coast of Samar, and shortly after daybreak engaged the twelve baby carriers that Kinkaid had left behind when his fleet went south to engage the Jap southern fleet. Kinkaid immediately radioed Halsey for help from the big ships of the 3rd Fleet, as the old battleships of the 7th Fleet were outclassed by the more modern battleships of the Jap central fleet, and in addition Kinkaid's ships were low on ammunition, after having participated in the bombardment of the shore prior to the original landings in Leyte Gulf as well as in a surface naval engagement earlier that day. For the first time we now learned that when Halsey had turned north with his carrier groups, he had taken the rest of the 3rd Fleet along with him. That was how the Jap central fleet had gotten through the San Bernardino Straits unmolested and was now firing broadsides at the baby carriers. Kinkaid kept calling for help and asked where the 3rd Fleet was. Halsey replied that he was ordering McCain to turn back and launch an air attack against the Jap central fleet as soon as possible. Kinkaid knew that McCain's aircraft could not possibly get into the battle, even at extreme range, before afternoon and he still wanted to know where the 3rd Fleet's fast battleships were. About 10:00 a.m. Admiral Nimitz in Hawaii got into the argument. Halsey finally turned his 3rd Fleet around and headed back south, leaving Admiral Mitscher with four large and four medium carriers to deal with the Jap northern fleet.

The Jap gunnery against the little carriers was unbelievably bad. While they made hits on a number of them, only one, the Gambier Bay, was sunk by gunfire. The destroyers Hoel and Johnson and the destroyer escort Roberts were sunk by the Jap warships when these light craft did the best they could to protect the carriers.

The engagement went on, with our units withdrawing south under fire until about 9:30 a.m., when suddenly the Jap fleet turned around and headed back toward the San Bernardino Straits.

About an hour later six suicide attacks were made on the crippled baby carriers. The Saint Lo, already damaged by gunfire, was sunk by a Jap plane which dove into the deck. Two more kamikaze planes hit the Kitkun Bay and badly damaged her. Two other attacked the Kalinin Bay, which was already in bad shape. One of the Nips missed and crashed into the water. The other was shot down, but, as it crashed, skidded along the deck of the carrier and over the bow into the water. The last of the Nips dove at the Fanshaw Bay, which was also in bad shape, but luckily the antiaircraft fire caught the Jap plane and blew it up before it could crash into its target.

Every baby carrier that was still operable kept sending its aircraft out all day, flying patrols over the 7th Fleet and the shipping in Leyte Gulf, and attacking the withdrawing ships of the Jap central and southern fleets. Over a hundred Navy aircraft, with no decks to land on, flew into Tacloban and Dulag when their gas got low. The Tacloban strip at that time was soft and under construction and in no shape for landings, so that twenty of the first sixty-five planes to land were wrecked. At Dulag forty planes landed and eight of them cracked up. At Tacloban we hurriedly rolled the surface, gassed up the Navy aircraft, and loaded them with bombs and ammunition. They then took off and joined the rest of the 7th Fleet aircraft in hunting down the withdrawing Jap ships. We had no gasoline or ammunition or bombs at Dulag, so arrangements were made to send some down in barges so that the aircraft could fly the next day.

These lads of Kinkaid's did a great job during that day of October 25th. In spite of all they had gone through and in spite of the fact that almost every Liberty boat and landing craft in the harbor shot at them coming and going whenever they got in range, they turned in an excellent score for the day. Three Jap heavy cruisers and a destroyer were sunk and two other cruisers heavily damaged.

In the meantime, Admiral Mitscher's carrier boys had contacted the Jap northern fleet which turned out to consist of: one large carrier, three medium-sized carriers, two battleships converted to carry aircraft, three light cruisers, and eight destroyers, or seventeen vessels in all.

Mitscher did pretty well, sinking all four of the carriers, a light cruiser, and two destroyers.

McCain's aircraft made one strike around one o'clock that afternoon and another about three o'clock. Some damage was claimed on withdrawing units of the Jap central fleet and one of the already badly damaged battleships in the Sibuyan Sea was sunk.

We breathed easy again. The Jap plan was a bold but suicidal attempt. As things had turned out, the Nip had lost a large part of his naval strength, but he certainly missed the boat when, after getting Halsey to go after the northern or diversionary force and leave the San Bernardino Straits open for the passage of the main effort of the Jap central force, that commander had abandoned his mission with victory in sight. If he had kept on going, he could have entered Leyte Gulf and shot up a couple of million tons of our shipping before anyone could have stopped him. He probably would have lost most of his force doing it, but the whole scheme was a suicide show, anyhow, as far as the Jap navy was concerned. On the other hand, while it would not have won the war for Japan, it would have resulted in the loss of vital shipping necessary to keep the war in the Pacific going, would have cost us a heavy casualty toll, and would have stranded around 150,000 American troops in Leyte without their food supplies and ammunition and opposed by a strong Jap force which was constantly capable of being reinforced from the neighboring islands. The Jap fleet would probably have lost four more battleships, two or three more cruisers, and maybe another ten destroyers, but the United

States would have suffered a setback so serious that it might have delayed the prosecution of the war for another five or six months. The situation in Europe at that time was demanding so much of our resources that there would have been no relief from that direction. The Battle for Leyte had cost the Jap three battleships, four aircraft carriers, six heavy cruisers, five light cruisers, ten destroyers, and probably another 250 aircraft. We had lost the carrier Princeton and two baby carriers, two destroyers and a destroyer escort, about 150 airplanes, and several cargo vessels and landing craft, besides major and minor damage to plenty more naval and merchant vessels; but that was cheap compared to what we would have suffered if the Jap had been a little smarter. It proved once more that he had no business starting a war with the United States in the first place. He just wasn't good enough to play in the big leagues.

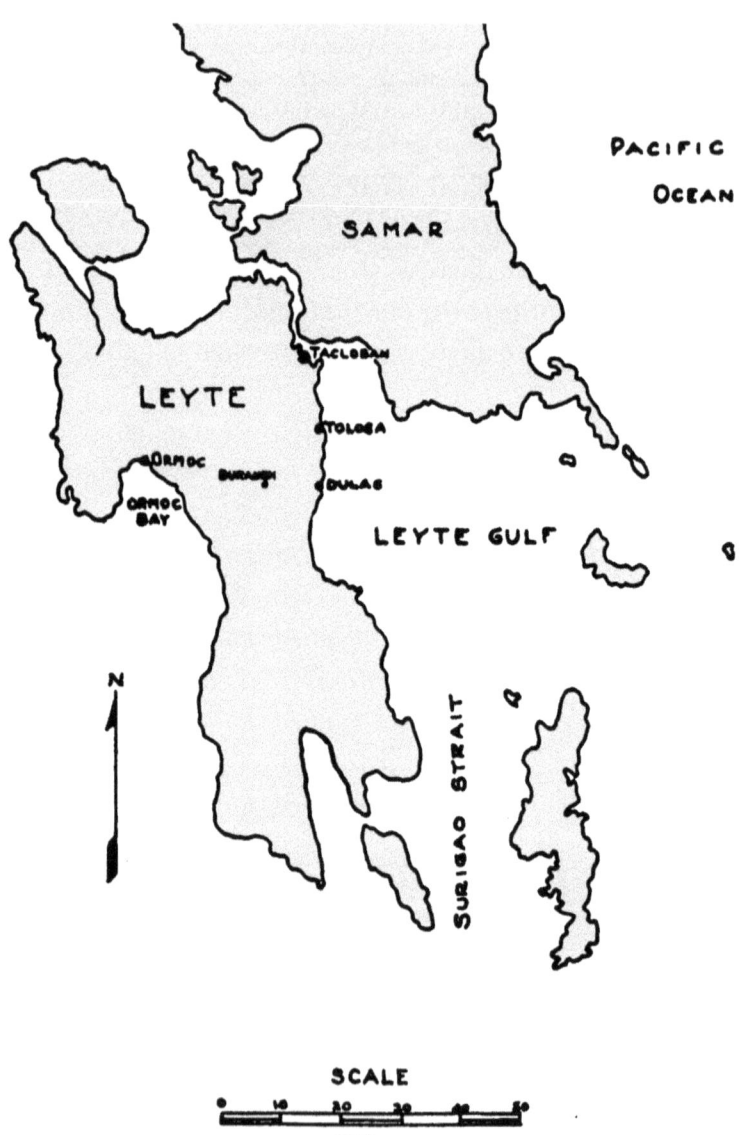

22. THE PHILIPPINES: I LEYTE
November—December, 1944

I HAD BEEN ashore all day the 25th. General MacArthur moved into the Price House in Tacloban to stay and I went out to the airdrome, where I spent most of the time getting in and out of slit trenches because of Jap air raids. Over a hundred planes came over the area during the day in flights of six to twelve airplanes, making eleven raids in all. Most of their attacks were directed at the shipping, but after dropping their bombs they came over to strafe the airdrome and drive us to shelter. One formation of four came in low on a strafing attack, but when the leader was shot down in flames right on the end of the runway, the other three turned over the harbor and hit a large landing craft and a patrol boat, both of which had to be beached to keep them from sinking. A Liberty boat unloading at the dock at Tacloban was hit and badly damaged, a large warehouse destroyed, and several holes made in the pier itself. Two bombs hit within a hundred yards of the Price House. It was a busy day that finally quieted down after the last episode at seven o'clock that evening. It was just beginning to get dark when some Navy torpedo bombers came in to land at Tacloban. We had been notified that there would be six. All of them let down their landing gears and turned on their running lights. I was standing there with Colonel Photo Hutchison watching them and, as the first plane landed, we suddenly realized that there were six still up there in the circle. A Nip bomber had tagged on the end of our formation in the dark and had let down his landing gear and turned on his lights like the rest. Just then, instead of landing, the Nip, who was the next plane in position to turn in for his final approach, headed out over the bay, slapped his bomb-load into a big landing craft loaded with drums of gasoline, and got away. We didn't need any landing lights for the next half hour. The landing craft burned to the water's edge and sank.

Tokio Radio said in their broadcast that evening: "General MacArthur and his staff and General Kenney have established their headquarters ashore in Tacloban in the Price House, right in the center of the town. Our brave aviators will soon take care of that situation."

Before turning in that night, I dropped into General MacArthur's room for a chat. He looked up as I came in and put down the book he had been reading. It was a story of the life of General Robert E. Lee.

"George," he said, "I've been reading about a remarkable coincidence. When Stonewall Jackson was dying, the last words he said were, 'Tell A. P. Hill to bring up his infantry.' Years later when Lee died, his last words were, 'Hill, bring up the infantry.'" He paused, lit his pipe, took a few puffs, and continued, "If I should die today, or tomorrow or any time, if you listen to my last words you'll hear me say, 'George, bring up the Fifth Air Force.'"

During the night the Japs sent the first of a series of convoys into Leyte with reinforcements, landing them at Ormoc on the west side of the island. From five to nine destroyers and five barges made up the convoy. It was estimated that 1000 troops were put ashore. We didn't have anything to stop them.

On the 26th the Japs made twelve more air raids, using a total of around 150 planes. The first one was at 5:15 a.m. One bomb hit a Filipino house on the edge of the back yard of the Price House, killing twelve Filipinos and wounding ten others. Mud, rocks, and debris were thrown all over our house. Another bomb hit the war correspondents' house, killing "Ace" Bush of the Associated Press and wounding four others.

Out on the airdrome the Nips kept us busy all day getting in and out of slit trenches again. Four times we had no warning at all. They came in over the tops of the trees strafing and dropping fragmentation bombs as they flew the whole length of the strip. We had laid about a thousand feet of steel mat by three o'clock in the afternoon, in spite of all the interruptions. I was out on the mat with Colonel Woods, the boss engineer, and a map of the area spread out on the top of my jeep explaining where I wanted the dispersal areas located, when I happened to look up. About 500 yards away, just lifting over the palm trees at the south end of the airdrome, were four Nip planes with the lights beginning to twinkle from the machine guns mounted in the wings. We dropped flat on the ground behind the jeep as the Nips swept along the runway strafing and dropping bombs. The jeep didn't get touched but a truck ten feet away had its tires punctured, the gas tank set on fire, and the driver wounded. Two men were killed in an antiaircraft machine-gun pit about fifty feet to one side of me. Two Navy planes were set on fire and destroyed and two other men killed and ten wounded. Over three-quarters of the bombs failed to explode or we would have taken a terrific beating. One of the duds landed in front of a big tent in which thirty or forty Navy flyers were gathered around having a drink of coffee.

The attacks on the shipping did practically no damage. The bombing was the worst I had ever seen the Jap do and there were no kamikazes. Just before dark two more bombs were dropped about a hundred yards from the Price House. I had just come back from the airdrome when the attack started. MacArthur made the remark that it proved I was the one the Nips were after. He claimed that since I had left after breakfast that morning, all attacks had been out my way and no one had bothered the Price House until I came back there.

President Osmena had come over to pay General MacArthur a visit. After the "All clear" was sounded, the three of us walked out onto the porch and chatted while we watched the people walking along the street in front of the house. Suddenly Osmena said, "See that man across the street picking up cigarette butts? He is a Jap. A Filipino wouldn't be doing that." We told one of the guards to pick the man up. Osmena was right. It was a Nip sergeant who had been told to put on civilian clothes and hang around Tacloban to get what information he could that might be of value to the Jap commander on Leyte. He knew enough of the Tagalog language to get by. There had been quite a few collaborators among the residents of both Tacloban and Dulag during the Jap occupation and we already had locked up nearly a hundred of them since coming ashore. That Nip spy would have had no difficulty in transmitting any message that he wanted to send to his own headquarters. After interrogation he joined the rest of the Nip prisoners in the little camp that we had established on the edge of town for that purpose.

By that evening we had 1500 feet of steel mat in place. I told Hutchison to work all night and get the length up to at least 3000 feet by noon of the next day. That would be a little close but a good P-38 pilot could land that short. I then radioed Whitehead that we had room for 34 P-38s at Tacloban and to have that number come up from Morotai landing at noon on October 27th. The last time I had seen Whitey, I had told him that I wanted the first fighter contingent in Leyte to be experts.

Admiral Kinkaid said his baby carriers could not do anything more until they had been refueled, and in many cases repaired. As I would have the only airplanes on the scene, beginning at noon on the 27th, MacArthur issued instructions that I was responsible from then on for all air operations over the Philippines. Halsey's carriers moved back to get reequipped and rest the crews, who had been worked pretty hard for the previous two weeks.

The Nip made single-plane attacks on the airdrome at intervals all night. The bombing was not very good but one man was killed and three wounded and the work was so interfered with that by noon we had only 2500 feet of steel-matted runway. During the final hour we rolled coral into the sand for another thousand feet at the end of the strip so that if the P-38s overran the mat they would not get wrecked.

About ten minutes before noon, MacArthur and I were at the table having lunch at the Price House. Suddenly we heard a low drone of engines. The General said, "Hullo, what's that?" I said, "That's my P-38s from the 49th Fighter Group," and got up from the table. MacArthur was already ahead of me and called for me to get in his car. We drove out to the field. All over the area you could hear soldiers and Filipinos yelling and cheering at the sight of those P-38s. They looked good to me, too, as the whole thirty-four of them came in over our heads flying a formation that would have been a credit to the Air Force at an inaugural parade back home.

The kids landed, taxied into position, and as they got out I introduced them to General MacArthur, who shook hands all around and patted them on the back. They got a great kick out of it and laughed when he said, "You don't know how glad I am to see you." The last plane to land taxied out to the end of the line opposite where I was standing with MacArthur talking to the gang. A chubby, towheaded lad eased himself out of the cockpit and started back behind the tail of the plane.

I yelled, "Bong, come here." The kid came over and saluted looking at me solemnly and then grinned shyly as he saw MacArthur standing there smiling at him.

"Who told you to come up here?" I demanded.

"Oh, I had permission from General Wurtsmith and General Whitehead," he answered.

"Did they tell you that you could fly combat after you got here?" I asked.

"No," replied the cherub, "but can I?" Everybody laughed now, including me.

I told him that we were in such a fix that anyone could fly combat that knew anything about a P-38 and had one to fly. Bong saluted happily and went back to his airplane to get it gassed up and ready to go. As soon as twelve airplanes were ready we put them in the air to cover the refueling of the others. Our radar had been installed that morning and, while it wasn't working very well yet, we could get about ten minutes' warning of an attack.

We worked feverishly laying steel mat all the afternoon and by four o'clock we had laid another thousand feet and were out of trouble. For some reason no Jap airplanes bothered us that whole time. It was the longest period without seeing a Jap airplane since we had landed a week previously.

At five o'clock five Nips came over on a raid. Lieutenant Colonel Walker, the group commander, Lieutenant Colonel Bob Morrisey, who had shot down the first Jap plane for the 49th Fighter Group back in March 1942 and who was now on Wurtsmith's staff, Lieutenant Colonel Jerry Johnson, the group executive officer, and Bong took off to intercept. Walker had engine trouble right after the take-off and came back mad as a wet hen. The other three did their stuff. Johnson got two of the Nips, Morrisey and Bong racked up one apiece, and the remaining Jap disappeared in the clouds and got away.

About half an hour later four more Nips ran into eight of our P-38s who were patrolling the area. This time none of them went back home. The kids were jubilant. They hoped that the Japs had lots more to send over. I told them not to worry. There were probably at least 500 Jap planes in the Philippines and, if thirty-four P-38s expected to gain control of the air over that number, they had their work cut out for them. They didn't seem much impressed at the odds. As a matter of fact, neither did I. I really believed that the 49th Fighter Group could lick the whole Jap Air Force from what I had seen during the past week. The Jap flyer of October 1944 was an exceedingly inferior successor to the aggressive, skillful, reckless fighter that we had taken on in 1942 and 1943.

Just before dark two Japs dressed in Filipino clothes were picked up just outside the airdrome. We had over 1500 Filipinos working on the strip so I told Hutchison to put armed guards on every P-38, day and night. They were probably all right, but I couldn't take a chance. We didn't have very many of those airplanes and couldn't afford to lose them, especially by sabotage. Besides lengthening the runway we had enlarged the parking area, so I radioed Whitehead to send up thirty more P-38s and my airplane, as I wanted to go back to my headquarters in Hollandia for a couple of days to make arrangements for moving it to Leyte as soon as possible.

The next morning, Hutchison told Bong to scout around over the area around Tacloban and see if he could find any suitable airdrome sites. Tacloban's capacity was definitely limited and construction at Dulag and San Pablo did not look too promising. Bong had hardly taken off before the radar warned him of two Jap fighters in the vicinity and gave him their location and course. Bong intercepted, fired a couple of bursts, and chalked up two more victories. The whole combat was seen from the airdrome so there was no question about their being officially credited to him. I wired Arnold:

"In accordance with my instructions Major Richard Bong is trying to be careful, but the Nips won't do their part. On the twenty-seventh, five hours after arriving at Tacloban, Bong was again forced to defend himself and number thirty-one resulted. On the twenty-eighth, while looking for suitable localities for airdromes in the vicinity of Tacloban, he was assaulted by two more Nips, who became number thirty-two and thirty-three. Unless he was bothered again today this is his latest official score.

Kenney."

Arnold answered a day or so later:

"Major Bong's excuses in matter of shooting down three more Nips noted with happy skepticism by this headquarters. Subject officer judged incorrigible. In Judge Advocate's opinion, he is liable under Articles of War 122 [willful or negligent damage to enemy equipment or personnel].

Arnold."

I posted copies of both radios on the group bulletin board.

I returned to Hollandia that afternoon and told Whitehead, who met me there, to go to Leyte, pick out a place for his headquarters, and move up as soon as possible. I said to keep Tacloban supplied with fighters as fast as parking space became available.

I told Colonel Hewitt, my engineer, to go to Leyte, build a headquarters camp for me at Tolosa, on the gulf about twelve miles south of Tacloban, and have it ready for me to move into by December 1st. The staff started packing up right away. They all wanted to go to the Philippines, in spite of the wonderful location at Hollandia.

That evening the second Jap convoy to Ormoc, consisting of eight destroyers or destroyer escorts and twenty-eight barges, carrying an estimated 2400 troops, landed without being attacked. All our P-38s were out covering the Leyte Gulf area and before we could get them back, gassed up, and loaded with bombs, it was dark. The boys had been busy all day covering the shipping and driving off the few scattered attackers that came over. Besides the two shot down by Bong, three other Nip aircraft were destroyed in air combat over Leyte, while the Thirteenth Air Force B-24s from Morotai destroyed twenty-three aircraft and damaged fifteen others on the ground at Puerta Princesa on the island of Palawan.

It was quite evident that the Japs realized the decisive nature of the struggle for the Philippines. The islands sat astride the routes from Japan to the Netherlands East Indies and if our aircraft once started flying over the China Sea and cutting off the supplies of oil, rubber, tin, and other essential raw materials to the factories in the homeland, the Rising Sun would soon start setting.

The Nip air force in the Philippines was reinforced and replacements flown in, in spite of constant heavy losses. Some of the replacements came from the Netherlands East Indies, which were stripped to the point that air combat south of Morotai practically ceased after we had landed in Leyte. With no air opposition we roamed the area from Morotai to Java and from New Guinea to Borneo practically at will, destroying the inter-island barge and lugger traffic, sinking the shipping, blasting supply establishments, and generally making things difficult for the Jap garrisons. Each day the number of aircraft out hunting over this area would run to between 200 and 400, but while they would report a lot of damage to shipping and shore targets, enemy aircraft were no longer mentioned.

The Nips kept on raiding the Tacloban field each night. Most of the bombs landed in the water on either side of the sandspit, but when one did hit the field it caused a lot of trouble. We were packed in too tight, but it couldn't be helped.

The light Philippine roads were breaking down under the hammering of our heavy vehicles and every engineer put on roads meant one less on airdrome construction. There weren't enough engineers and in a few weeks the rainy season would slow all construction work down to a walk.

On the morning of the 30th, Photo Hutch sent out a cry for help. He had been raided during the night and now had only twenty P-38s left. We told him to push his wrecks into the water and we'd send him some replacements.

Twenty P-38s flew up from Morotai that afternoon. About twenty miles south of Tacloban they were warned of the approach of ten to twelve Jap airplanes. The P-38s arrived just in time, shooting down six Nips and chasing the rest away.

Slim little Major Tommy McGuire got one of the victories—his twenty-third. He landed, casually told his crew chief to paint another little Jap flag on the side of his plane, and remarked, "This is the kind of a place I like, where you have to shoot 'em down so you can land on your own airdrome. Say, how many has Bong got now?" McGuire hadn't heard about the last pair that Bong had shot down. Much to his annoyance he found that his regular "eight behind" had suddenly become ten behind. Tommy rectified the situation early the next morning. Nine P-38s on a patrol spotted six Jap planes about fifteen minutes after take-off. Four of the Nips didn't get away. Two of these were credited to Major Thomas McGuire of the 475th Fighter Group. His score was now twenty-five. "Well, at least I'm holding my own," said Tommy.

During the Jap raid on Tacloban on the 30th, a hot fragment from a Jap incendiary bomb wounded Pappy Gunn in the arm and sent him to the hospital for the duration of the war. Pappy had come up on Photo Hutchison's Air Task Force staff and had been of inestimable help, handling operations and maintenance at the Tacloban strip, organizing Filipinos to help with the construction, and worrying our engineers into working faster. I hated to lose Pappy, for his priceless stories of his personal exploits were a constant source of entertainment. No one ever questioned their authenticity any more, as we had long since learned that as soon as we disputed Pappy's "facts," he shut up. We didn't want to miss the stories, so Pappy had free rein. By this time we had all decided that Paul Bunyan and Baron Munchausen were both second raters.

On November 1st the Japs sent their third convoy of reinforcements into Ormoc on the west coast of Leyte. Seven transport vessels, four destroyers, and two destroyer escorts unloaded around 22,500 troops between the 1st and 3rd. Photo Hutchison loaded bombs on the P-38s at Tacloban and sent them out to attack the Nip shipping. On the 1st they sent a 7000-tonner to the bottom and badly damaged another, besides shooting down five Jap escorting fighters. In the meantime Whitehead and Streett had sent fifty-eight P-38s, fifty-four B-24s, and twenty-six B-25s up from Morotai to clean up the Jap airdromes in the central

The Philippines: 1 Leyte 357

Philippines at Bacolod, Alicante, Carolina, and Cebu. We destroyed an even 100 Jap planes on the ground and shot down a total of thirty-six Jap fighters and eight bombers during the day. It cost us four B-24s and three P-38s.

Three of the Jap fighters that were shot down fell to the guns of Lieutenant Colonel Robert B. Westbrook, the Commander of the 347th Fighter Group of the Thirteenth Air Force, bringing his score up to twenty enemy planes destroyed in air combat.

On the 2nd the B-24s from Morotai sank another 7000-ton Jap vessel off Ormoc and scored hits on a medium-sized cargo vessel. Sixty Jap fighters were shot down and ten other enemy aircraft destroyed on the ground in a series of combats that lasted throughout the day. We lost five P-38s and a P-47. During the night the antiaircraft guns got another five Jap raiding aircraft.

The P-38s from Tacloban had a busy day on the 3rd. The weather stopped them from getting across the Leyte mountains until afternoon but about two o'clock they got away for an attack on the shipping at Ormoc. Direct hits were scored on a large freighter, which was left in a sinking condition. Forty Jap escorting fighters intercepted but the P-38s shot down twenty-four of them without loss to themselves. In the meantime seven more Jap planes were shot down in combat and ten destroyed on the ground by the B-25s from Morotai in another raid on Cebu. We lost a B-25.

That night twelve Nip bombers raided our shipping in Leyte Gulf, blowing up a Liberty boat. It evidently encouraged them to make another daylight raid on Tacloban. Shortly after daybreak forty Jap bombers and fighters were reported on the way. Twenty-seven P-38s shot down thirty-two Nips and chased the rest away. Once again we had no losses.

The 49th Group had been so busy for the past four days they had forgotten to count. At lunch someone said, "What is the score? We must be up to 500 by this time." The count was made. It was 535. No one knew who had gotten number 500 or just when.

"Aw, let's open that bottle at 600," said Colonel Jerry Walker, the group commander. Everyone agreed, but this time they intended to watch that score a little closer.

With the small numbers of aircraft we could find airdrome space for in Leyte, I couldn't do much more than cover the Tacloban and Leyte Gulf area, so I asked General MacArthur to request Halsey to hit the Jap airdromes on Luzon where the Nip continued to maintain his aircraft strength. MacArthur made the request and we got a message back saying that Halsey would hit Luzon with all his strength on November 5th and 6th.

From the 28th of October to the 4th of November we had shot down 350 Jap aircraft in combat and destroyed 120 more on the ground, but the Luzon fields were still heavily populated.

Halsey carried out his raids on Luzon on the 5th and 6th as he had promised and ran up a nice score doing it. In a series of attacks throughout both days the carrier boys shot down 113 Nip aircraft in combat and destroyed 327 more on the ground. In the meantime we had destroyed fifty-eight Jap fighters and a bomber in combats over Leyte and Cebu.

The night of the 9th the Japs sent their fourth convoy into Ormoc Bay. Three or four large merchant vessels with fifteen assorted destroyers and destroyer escorts, covered by fighters, comprised the force. At daybreak eight Jap planes hit Tacloban strip, putting several craters in the runway and destroying three airplanes that were being repaired. Photo Hutchison had the runway repaired in less than an hour and threw everything he had into an attack on the convoy. Every P-38 that could fly carried bombs to Ormoc. Four B-25s carrying couriers and mail up from Hollandia were pressed into service and thirty B-25s of the 38th Group came up from Morotai, getting into the combat at 11:30 a.m. To them went most of the credit for sinking Jap shipping that day, when the Nips lost three of their four large merchant vessels and between five and seven of the naval escort vessels. Seventeen Jap planes were destroyed in combat, dozens of barges sunk, and Jap troops and supplies being unloaded were strafed and bombed with heavy casualties to the enemy.

We had seven B-25s shot down. Eighteen of their twenty-five crewmen were later rescued. In addition, we lost four P-38s. During the air combat over Ormoc, Bong got number thirty-four "defending" himself against a Jap fighter. Tommy McGuire, who said as he took off that morning that he was only going along to protect his interests, also shot down a Nip—his twenty-sixth—still eight behind.

I flew up from Hollandia that morning and blundered into the Tacloban area just in time. Hutchison sent four P-38s up to escort me in and just before I landed I watched them knock down two Nip fighters that came over to take a look at my B-17. We had just completed landing at the strip when my P-38 bodyguard intercepted and shot down three more Jap planes, which fell in flames within a hundred yards of the Tacloban airdrome.

When I checked in at the Price House, MacArthur laughed when he saw me and said, "I knew you must have come back. Everything has been quiet around here ever since you left for Hollandia. I'm sure you are the one the Japs are after now."

That afternoon we got word of another Jap convoy out of Manila on the way to Leyte. Halsey was on his way south from his Luzon attacks and would be in position to hit the Nip vessels before they got to Ormoc. In addition the weather prophets said that we would be fogbound until midmorning around Tacloban and the weather would prevent anything from Morotai operating against the Jap convoy all day. I asked MacArthur to request Halsey to handle the new threat—Ormoc convoy number five. Halsey did the job the next morning. All four merchant vessels and four of the six naval escorts were sent to the bottom by the carrier lads. It was a good thing Halsey handled the show as my weathermen were right. The fog at Tacloban prevented flying until ten o'clock and nothing came out of Morotai to the north all day.

About noon our fighters started operating. They had another good day. Bong got two more Nips, making it thirty-six. McGuire shot down two to hold his place at twenty-eight. Major Jerry Johnson also let it be known that he was in the race, getting two more to bring his total to twenty. Altogether we brought down twenty-three definites and five probables during the day. We lost two P-38s but recovered one of the pilots.

On the night of the 10th Tokio Radio announced that General Tomoyuki Yamishita had been appointed commander of the Japanese Army Forces in the Philippines and quoted him as saying that he was going to annihilate the Americans. Yamashita was the general who had licked the British at Singapore and had taken over command in the Philippines in the last stages of the 1942 campaign against Wainwright, after MacArthur had been ordered to Australia by President Roosevelt. This news, coupled with information from documents captured a few days before in Leyte, indicating a big Jap attack in the near future, looked as though the Nip really intended to fight for the Philippines. Tokio Radio indicated how seriously the Japs considered the campaign by repeating almost daily that the battle for the Philippines was the battle for Japan itself.

The Jap strength in Leyte by this time had been built up to around 35,000 men with units from their 1st, 26th, 40th, and 102nd Divisions and the remnants of the 16th, which contested the landing with us originally. Those first three divisions were supposed to have come from Manchuria and had the reputation of being hard fighting troops, trained and toughened in the war in China. I was getting more and more worried about the airdrome situation. Tacloban could not handle many more than the 100-odd aircraft already on it. Dulag could take another seventy-five fighters. San Pablo was coming along slowly and the two airdromes to the west at Burauen and Bayug had limited possibilities, but while they might each have the capacity for a group of bombers or a group of strafers, they were too close to the Jap front. Whitehead had established his advanced headquarters in an old school building in Burauen only a mile and a half from the Nips and sent Squeeze Wurtsmith up to run it. Squeeze didn't seem to worry about it, but they were still killing snipers all around his area every day. I told him to put some more guards around his headquarters and take no chances walking around unarmed, particularly after dark. We had already picked up over 200 Japs dressed in Filipino clothes in Leyte since we had landed about three weeks before.

The morning of the 12th I began to wonder if MacArthur were not right about the Jap aviators having designated me as a target. I woke up at five o'clock with lumps of mud and gravel coming through my bedroom window, tearing down the shutters, and making a horrible mess of the place. A bomb had landed in the back yard. Luckily the Japs believed in the use of delay fuzes so the thing had penetrated about ten feet into the ground before going off. The only damage was to my only suit of clean clothes hanging in the bedroom and to the feelings of the Filipino boys at the Price House, who worked all day cleaning the place up.

About this time the rainy season descended upon us. Airdrome construction practically stopped. The bottom fell out of the roads and our engineers were frantically dumping rock into the holes to keep the traffic going. The streets in Tacloban itself were knee-deep in mud and the twelve-mile trip from Tacloban to Dulag was a matter of three to four hours in a jeep or a truck.

With the continued stubborn resistance of the Japs on the west side of Leyte, it became evident that in spite of General MacArthur's impatience to get going we would have to postpone the pickup of Mindoro and the big landing at Lingayen Gulf on the main island of Luzon. December 15th was finally decided upon for the Mindoro operation and, although General MacArthur refused to admit it for a while, it looked like January 9, 1945 for Luzon. I told the General that we didn't have enough airdromes in Leyte to utilize the Far East Air Forces and that Mindoro would have to be fixed up enough to handle the bombers, strafers, and fighters necessary to support him in Luzon until we could move our strength up there. The GHQ staff was willing to entrust everything to Kinkaid's baby carriers, but while they could give us the air support for the first few days they did not have the staying power to keep it up very long, and if the Japs did very much kamikaze attacking and the carriers were sunk or had to pull out we would be in a bad way. MacArthur agreed with me and I got some promise of shipping to move my equipment for two heavy-bombardment groups, a strafing group, and a couple of fighter groups into Mindoro, with enough engineers to get airdromes built in a hurry as soon as we got ashore. Luckily that island would be having a dry season for the next four or five months, so that runways could be constructed quickly.

On the 14th, under cover of extremely bad weather, our ground observers reported the arrival at Ormoc Bay of the sixth Jap reinforcement or supply convoy, consisting of six merchant vessels, two small freighters, and six destroyers. They evidently unloaded in a hurry and got away during the night as our reconnaissance planes on the afternoon of the 15th could not locate them.

On the 21st, again taking advantage of bad weather, one large Jap merchant vessel and six small freighters arrived at Ormoc. We managed to get a few P-38s over the top of the bad weather for an attack. Three of the freighters were sunk and the rest badly damaged.

On the 24th I had gone over to Palau to make arrangements there with Major General Moore, Marine Corps, to let my 22nd Heavy Bombardment Group use his field for operations against the Philippines. That group, which was back at Owi, was practically out of the war and I wanted to put them to work. I completed satisfactory arrangements and flew back to Tacloban and landed just as the Nip put on a forty-plane raid on the shipping and the airdrome. It was the second raid of about that size during the day. They did very little damage to the vessels in the harbor. A few bombs landed on the airdrome, destroying six airplanes and damaging five others. During the day we shot down thirty-five Nips and the antiaircraft guns got eight more. We lost three P-38s in combat and picked up two of the pilots.

That afternoon the eightth Ormoc convoy of seven merchant vessels and several small destroyers or patrol craft tried to land at Ormoc Bay. P-38s, P-47s, and P-40s, everything we had at Tacloban, went to work. By the time they had finished the Nip had just about lost everything. Five large transports had been definitely sunk and the other two were so badly damaged that they probably sank during the night. One escort vessel had been sunk and another heavily damaged. All vessels had been strafed throughout a series of attacks which had lasted all afternoon right up to sundown. The decks were covered with troops, and very few of them and no supplies had gotten ashore. MacArthur was jubilant. Krueger was so relieved that he didn't even protest when he was told to move his headquarters out of Tanauen and rebuild it down along the beach at Tolosa, near where mine had been opened that morning. Soon after the landing on October 22nd Krueger had pitched his headquarters in a pleasant palm-tree grove near the Tacloban strip. Every time the Nip bombers overshot the airdrome the bombs landed in Krueger's camp. After a couple of weeks he had gotten fed up with it and moved about four miles down the coast to a beautiful dry palm grove at the village of Tanauen. This now looked like the only spot suitable for quick airdrome construction, so he was asked to move again. His headquarters staff were not so keen about being ousted, but Krueger knew he needed all the air support he could get. The next day the engineers went to work as the 6th Army staff moved out.

It was on the 24th of November that the first B-29 raid on Japan was made from the newly established airdrome on the island of Saipan in the Marianas. It was a small raid but it was the beginning of a lot of trouble for the Nips right on their home grounds.

The next morning the P-47s from Tacloban blew up a 5000-tonner that had been beached on an island about fifty miles west of Ormoc during the attacks of the 24th. Four small transports packed with troops on their way to Ormoc from Cebu also went down before the P-47s, who had been taking skip-bombing lessons from the B-25 boys.

On the evening of the 26th, just after dark, the Nips tried a new stunt on us. Three transports, loaded with fifteen paratroops each, were picked up flying north along the coast south of Dulag. One was promptly shot down in flames by our antiaircraft guns. Another was forced to land in the water off Dulag and all the occupants killed or captured. The third crash-landed inland from the Dulag airdrome and the personnel escaped into the hills.

What those forty-five men were supposed to do we didn't know, but on the supposition that they were supposed to destroy our airplanes and supplies, we set up more antiaircraft machine-gun defenses and increased the number of guards around our airdromes. In addition, all members of our squadrons were cautioned to wear arms or have them close by, ready for action in an emergency.

MacArthur's headquarters issued instructions the next morning for everyone to carry arms and wear steel helmets. The order didn't apply to me, so while I kept on carrying a pistol in a shoulder holster as I had since coming to the Philippines, I omitted the steel helmet. Some of the GHQ generals did look funny wearing them. I noticed that MacArthur continued with his standard headgear, the cap denoting his rank of Marshal of the Philippine Army, and that his most dangerous weapon was still his corncob pipe.

The Jap attempt to destroy our aircraft by suicide raiders on the 26th was probably linked to the events of the next two or three days. On the 27th another epidemic of kamikaze attacks broke out, this time directed against Kinkaid's fleet in Leyte Gulf. Rain squalls in the immediate vicinity of the Tacloban strip interfered considerably with getting our air cover off on schedule all morning. Most of the Nip raids, which were by units of five to eight planes, were broken up, but one of them got a break in the weather at a time when it was tying us up and drove home an attack. Kamikaze planes crashed into a battleship, a light cruiser, and a destroyer, inflicting considerable damage but not putting the ships out of action. The fool Japs were still using small bombs for the same unknown reason. I was glad they were so stupid but still could not understand why they didn't choose to make their exits from this world with the kind of explosion that would come from a 2000-pound bomb instead of a 250-pounder.

On the afternoon of the 28th our reconnaissance planes picked up the ninth convoy to Ormoc, consisting of twelve merchant vessels and three small escorting naval vessels. The convoy was in two sections. The first, about fifty miles ahead of the rear section, consisted of a 7000-ton and a 5000-ton merchant vessel with one escort. We got in one attack before dark on the first section, sinking one of the escorts and getting a hit on a 5000-ton merchant vessel. I ordered the squadron of P-61 (Black Widow) night fighters, that we had just gotten into Tacloban, to heckle the convoy all night and see if we could keep them from unloading.

The heckling worked. Both merchant vessels were still offshore with the decks piled high with boxes and crowded with troops when our attack hit them just after daybreak. Both vessels were sunk and seven out of fifteen Jap airplanes flying as cover were added to the score of our 49th Fighter Group.

The second section of the convoy scattered. Part of the ships headed for Cebu and the remainder hung around Masbate Island, but during the 29th and 30th our B-25s, P-47s, and P-40s got them all. Jap troop losses must have been heavy, but the loss of supplies and equipment for the enemy forces on Leyte could not help but make his situation there still more critical.

Although the advanced echelon of my headquarters had been operating at Tolosa since the 24th, I continued living in the Price House with General MacArthur until my camp construction gang could get around to setting up a place for me to live and for Sergeant Raymond to run my headquarters mess.

On the 29th Raymond arrived and I notified General MacArthur that he could rent my room to someone else. He said he hated to see me leave but he would be down to visit me frequently as soon as Raymond got started serving meals.

That afternoon the troop carriers brought in the rest of my headquarters organization from Hollandia and we started operating again at full scale. Included in the new arrivals was my WAC detachment. All the stories I had told them about the mud were tame compared with what they ran into. It had rained hard all day and was still raining when they arrived. The trucks bringing them down the coastal road from Tacloban had taken three hours to make the six-mile distance over a road churned into a gray slime that lapped over the running boards. The road could be followed by the lines of wrecked and abandoned vehicles on either side, where drivers had slipped off the narrow rock base trying to pass other cars.

The camp site itself was on slightly higher ground, but it was still muddy and lumber for tent floors was scarce. As soon as the tents were up, everyone got busy digging slit trenches for shelter from the Nip bombers who still came over at intervals nearly every night. Going through the WAC camp that evening I noticed that most of the trenches were only about two feet deep. About seven o'clock that night a Nip bomber flew overhead on the way to the Tacloban airdrome.

The antiaircraft searchlights and guns went into action and the Wacs went into their trenches. The next morning I noticed that during the night all of them had become deepened to around four feet. Over in the enlisted men's area the same thing had happened, but after that first air raid, they had dug their trenches at least a foot deeper than the Wacs.

On November 30th Arthur Sulzberger of the New York *Times* arrived and spent three days with us at the Price House. Another cot was set up in my room and Sulzberger bunked in with me. It was interesting to talk to him and get the news from the United States and discuss the progress of the war but I doubt if either of us got over three hours' sleep a night for those three days. There seemed to be so much for both of us to talk about, as we lay there in the dark, that it didn't seem to occur to either of us that we were tired. The third night I did fall asleep while Sulzberger was talking to me, but I was really sorry to see him leave the next morning on his way back home.

Beginning just about dusk on December 6th we had quite a bit of excitement for the next five or six days. Escorted by eight fighters and accompanied by four bombers which laid a smoke screen effectively hiding the operation, about thirty-five Jap troop carriers parachuted between 250 and 300 paratroopers into the San Pablo-Burauen area. At the same time two more Nip troop carriers appeared near Dulag and another pair actually came in to land at the Tacloban strip.

The two at Dulag, each with eleven men aboard, were shot down and crashed south of the strip, killing all members of each airplane.

One of the two trying to get into the Tacloban strip was shot down in flames a mile south of the airdrome. The other, riddled with bullets as he came in for a landing, crashed on the airdrome itself among our parked airplanes, killing the eight paratroopers on board and the crew of three. In the crash and the fire which followed, six of our aircraft were destroyed and five more damaged.

Papers found on the bodies indicated that a total of thirty-nine transports, carrying a total of 463 men, was to be involved in the operation, which was to be timed with a drive eastward by the Jap ground forces in the hills just west of Burauen and was to destroy our aircraft on the Leyte airdromes so that they could not interfere with the landing of another convoy due at Ormoc on December 7th.

The instructions were carefully drawn up and the accompanying maps showed the exact locations of our airplanes, bomb and gasoline storage dumps, headquarters, and camps. The raiders, who were loaded down with incendiary grenades and small arms, were to land on the airdromes, rush to their appointed stations, destroy the aircraft, set fire to our gasoline and bomb dumps, kill the personnel on the fields, and attack the camps. Those landing at Dulag and Tacloban were then to resist in place until killed but were supposed to deny us the use of both fields throughout December 7th.

The paratroopers landing in the San Pablo-Burauen area were to destroy airplanes and fuel dumps also, but were then to set up light machine-gun defenses and hold their position until the Nip ground troops from the west broke through to join them. The plan also called for an aerial reinforcement, seven hours after the first jump the evening of December 6th, and a third wave of paratroopers to come over twelve hours after the initial show. These two reinforcements, however, never did materialize.

The paratroop force did pretty well for a while. They burned up a dozen light planes and a few hundred drums of gasoline and took possession of the San Pablo and Bayug strips and the strip just west of Whitehead's Fifth Air Force Headquarters at Burauen. The Nips held San Pablo until noon of the 8th when Sixth Army troops killed about half the force and drove the rest to the west where they joined the gang that still occupied Bayug and the Burauen field. Here they were reinforced by about 200 Jap ground troops that had broken through our lines just west of Burauen. It was not until December 12 that the last of the Japs was liquidated and order restored in the area.

Whitehead was actually cut off from me for five days, with the war going on all around him. On December 10th the Jap ground forces west of Burauen made another attempt to drive through to the rest of the Nips pocket just east and north of Whitehead's headquarters. They were repulsed but part of the fighting was in Whitehead's own camp. His men killed twenty-three Japs and suffered losses of two killed and seven wounded in the engagements. The Nips burned down several tents and looted others of personal luggage.

We managed to keep telephone and radio communication open during the whole period, although, at one stage of the shooting, Whitey had torn up the floor in his headquarters and moved his telephone into a dugout under the house. His caustic remarks about the way the ground forces had abandoned him and his praise of the fighting ability of his hastily organized private "army" of communications troops, medical troops, orderlies, and clerks were highlights of the week. In the meantime, with Jap paratroops roaming around only a few miles from my WAC camp, I set up a guard system of my own at Tolosa and borrowed a detachment of Rangers from Krueger for a week until the trouble was over. The gals got quite a kick out of it but went on with their work as though such things were a part of their jobs.

Early on the 7th we picked up the tenth of the Jap Ormoc convoys, off the northwest tip of Leyte. It was a remarkable coincidence, for our 77th Infantry Division was scheduled to land in Ormoc Bay at almost the same time that the Jap show was due to disembark. Kinkaid was accompanying our amphibious expedition with his destroyers, and every fighter airplane I had available was over the Ormoc area in support, beginning at daybreak. The P-38s of the 49th and 475th Groups did the fighter-cover work and we loaded bombs on sixteen P-40s, sixteen P-47s, and sixteen Marine Corsairs, temporarily attached to the Fifth Air Force at Tacloban. These forty-eight planes were our striking force for the day.

The Japs escorted their convoy and at intervals dive-bombed ours, but, although they used a total of over a hundred airplanes during the day, the largest number they had over the Ormoc area at any one time was sixteen and sometimes it was as low as four. On the other hand, our fighter cover started out with fifty P-38s on the first attack and the number never dropped below twenty all day. In addition, the fighters operating as bombers, once they had dropped their bombs, took part in combat, covering our shipping and shooting down Jap bombers that tried to attack.

The Jap convoy consisted of four merchant vessels between 3000 and 7000 tons, two small freighters, four destroyers, and three destroyer escorts.

The first attack hit them at 9:30 a.m. The Marines sank a destroyer, the P-47s hit a big merchant vessel and set her on fire from stem to stern, and the P-40s hit another merchant vessel which grounded as the Japs were trying to beach her. This vessel was burning briskly as the boys left for Tacloban to get another load of bombs.

The next attack began at 1:30 p.m. The P-40s promptly sank a destroyer, the Marines blew up one of the freighters, sank a second, and set fire to a third. The P-47s hit all remaining destroyers and escorts, and then everyone joined in expending their remaining bombs and ammunition on any Jap vessel still afloat.

At 4:30 p.m. a Navy reconnaissance plane from Kinkaid's outfit reported that all the merchant vessels were sunk or destroyed, one destroyer was on fire all over, one destroyer was on fire and settling fast at the stern, and the three destroyer escorts were all on fire and sinking.

At 5:30 p.m. the P-40s and the P-47s made the final attack of the day. They blew up what was left of the vessel that had run aground during the first attack, sank five barges loaded with Japs, and finally expended their remaining bombs and ammunition in a tree-top attack on Jap troops, vehicles, artillery positions, and camps to the north of Ormoc.

The Nips sank two of our destroyers by kamikaze attacks but none of our amphibious craft were hit. Two Marine Corsairs and one of our P-38s were shot down. We rescued all three pilots.

The Japs lost fifty-six airplanes during the day. Bong got two of them, making his score thirty-eight, and McGuire got two, holding his place with thirty.

Little, soft-voiced, black-haired Lieutenant Colonel Jerry Johnson, the best shot in the 49th Fighter Group or any other group in the world, according to most of the Fifth Air Force, left his desk at Group Operations and joined one of the afternoon flights as Bob Morrisey's wingman. Three Jap fighters suddenly appeared on Jerry's side and below.

"One, two, three. Count 'em," yelled Johnson over the radio and tipped his P-38 over. Before Morrisey could get in a shot, Jerry fired three short bursts and sent all three Japs spinning down in flames. The P-38 kids came back that afternoon saying that it was the best exhibition of aerial gunnery they had seen in their lives. That made Johnson's score twenty-three.

Lieutenant Colonel Charles MacDonald, the commander of the 475th Group, got three more, bringing his score to twenty.

Major Dunham, leading ten P-47s of his 460th Fighter Squadron, finished bombing a big Jap merchant vessel in the morning and found sixteen Jap fighters moving in to the attack. Fourteen of the Japs were shot down. Dunham got four of them and let his nine teammates split up the other ten.

December 7th was the day of the last important air engagement over Leyte.

That evening the 49th Group suddenly remembered that it was time to check on that score again. They totaled it up. Six hundred and twenty-three. Colonel Walker asked what they wanted to do about it. "Let's wait until we get a thousand," said Jerry Johnson. The gang nodded and put the magnum away. Now they wouldn't have to worry about keeping score for a long time. It was too much of a strain, anyhow.

On the 12th the Japs sent their eleventh and last convoy load of troops and supplies out to relieve their troops on Leyte. Six merchant vessels with five naval escorts made another disastrous attempt. Whether the Jap aviators had become discouraged or the weather had interfered with their escort work was hard to tell, but only twelve Nip fighters showed up during the day. Only four of them got away.

Five of the Jap merchant vessels and four of the escorts were sunk or destroyed and the remaining merchant vessel and escort damaged. It was the end of the Jap fight for Leyte.

Shortly after Major Dick Bong got his thirty-sixth officially confirmed destruction of a Nip airplane in combat I recommended him for the Congressional Medal of Honor. General MacArthur approved it and sent

The Philippines: 1 Leyte

the recommendation to Washington. We received word of the award on December 8th and General MacArthur agreed to present it to Bong on the 12th at a public ceremony at the Tacloban airdrome.

We lined up a half dozen P-38s in a half circle, with the crews standing in front of them. Out in front of a guard of honor, consisting of twelve fighter pilots, all of whom had a dozen or more victories to their credit, stood Bong, speechless with stage fright and shaking like a leaf. In a sky full of Jap airplanes all shooting in his direction Dick would be as cool as a cucumber, but there in front of everybody, with MacArthur ready to decorate him, Bong was terrified.

The two advanced toward each other, exchanged salutes, and halted a few feet apart. General MacArthur stepped forward, put his hands on Dick's shoulders, and said:

"Of all military attributes, the one that arouses the greatest admiration is courage. The Congress of the United States has reserved to itself the honor of decorating those amongst all who stand out as the bravest of the brave. It is this high and noble category, Major Bong, that you now enter as I pin upon your breast the Medal of Honor. Wear it as the symbol of the invincible courage you have displayed in mortal combat. My dear boy, may a merciful God continue to protect you—that is the prayer of your Commander-in-Chief."

Two thousand spectators—GIs, airmen, war correspondents, Filipinos—said, "What a guy!"

The ceremony over, vastly relieved, Bong ducked into the crowd and slipped over to Colonel Walker's to see what the 49th Group was going to have for lunch. He had been so worried that morning that he had missed breakfast.

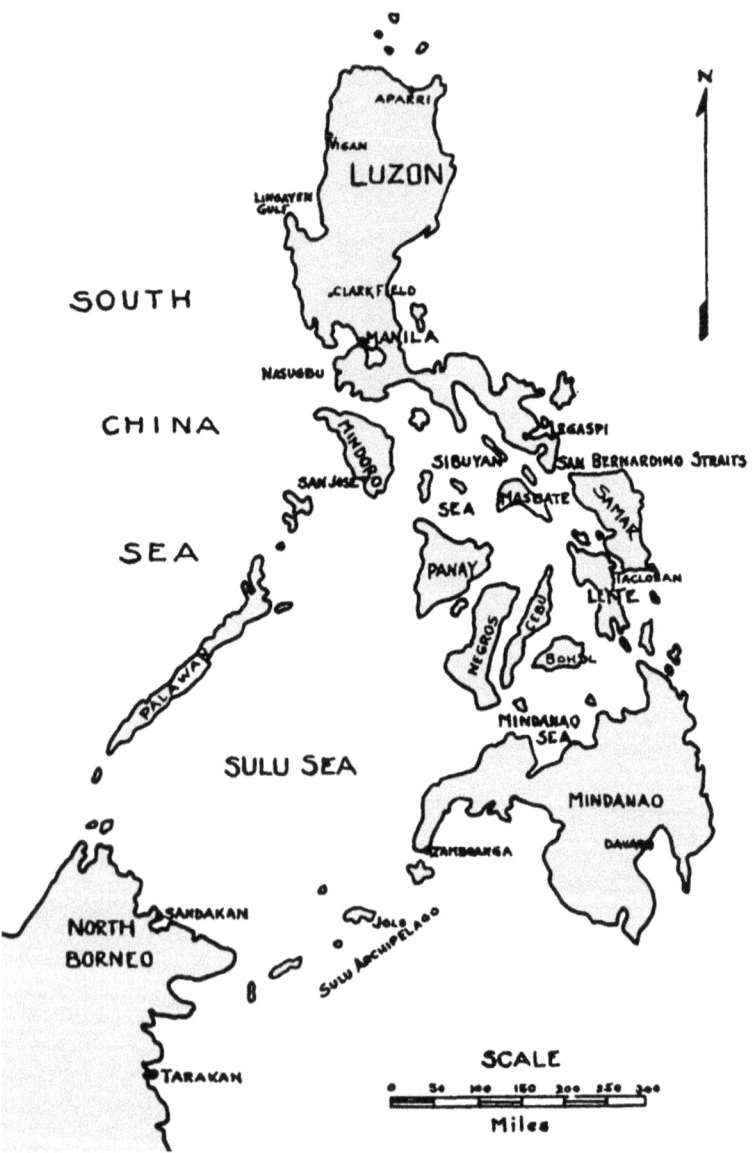

23. THE PHILIPPINES: II MINDORO
December, 1944

THAT MORNING the convoy carrying a regimental combat team of the 24th Division, the 503rd Parachute Regiment, and a lot of engineers and air troops left Leyte Gulf for the landing at San Jose on the west coast of Mindoro. Kinkaid assigned six baby carriers and several cruisers and destroyers to cover the convoy. The Fifth Air Force was to give it fighter cover down the east coast of Leyte, through the Mindanao Sea, and north through the Sulu Sea until past the island of Negros. In addition, heavy strikes were set up against all Jap airdromes in Mindanao and the central Philippines by the Far East Air Forces during the 13th, 14th, and 15th, while Halsey's carrier aircraft were to blast the Luzon fields during the 14th, 15th, and 16th.

On the 13th at 2:00 p.m., while passing through the Mindanao Sea, a lone kamikaze smashed into the Nashville, the cruiser MacArthur and I had come to the Philippines in and which was now being used by Admiral Struble as the flagship of the Mindoro expedition. One hundred and twenty-five men and officers were killed and 150 wounded. Colonel Jack Murtha was fatally wounded, Colonel Bob Morrisey seriously injured, and Brigadier General Dunkel, the ground commander, also received slight injuries. Admiral Struble transferred his flag to a destroyer and sent the Nashville back to Leyte Gulf.

Just prior to the attack our escorting fighters had intercepted twelve Jap bombers escorted by ten fighters and had shot down all twelve of the bombers and one fighter, with no loss to themselves.

During the day our airdrome attacks netted 121 Nip aircraft destroyed in air combat and on the ground.

The convoy was not bothered during the 14th and our fighters and bombers destroyed another 100 Jap planes in the attacks on the fields in the central Philippines.

We landed shortly after daybreak on the morning of the 15th at San Jose, Mindoro. No opposition was encountered and no casualties reported by Dunkel during the day. The ground was reported hard and dry and excellent for quick airdrome construction. The engineers were at work on two strips before dark.

The weather began closing down at Tacloban during the afternoon and we had to call our fighter cover back at three o'clock. About five, after the ships had landed their cargoes, the Nips sent over some more of their kamikaze boys, who smashed themselves into a destroyer and four landing craft. Two of the landing craft were sunk and the other vessels put out of action.

That afternoon I went down to Dulag to see how the airdrome was coming along. The engineers were trying to enlarge it, to give more dispersal area, as the planes were lined up wingtip to wingtip on each side of the runway, which was only a hundred feet wide to begin with. I stopped to chat with some of the 475th pilots, who were looking over their airplanes preparatory to going out on a patrol. One of them told me that Bong and McGuire had gone out on a Nip hunting expedition together that morning and had each shot down a Jap plane. I got in my jeep and drove over to the hut where McGuire was living with two or three other members of his squadron. I opened the door without knocking and walked in. Bong and McGuire, naked as the day they were born, were standing in a pair of tin washtubs, scrubbing each other's backs. They turned around, grinned rather sheepishly, and reached for towels. I sat down and, as they dressed, asked them what they had been doing that morning. McGuire took on the task of spokesman, while Bong kept nodding in confirmation as the story unfolded.

"You see, General," said Tommy, "that gang up at Tacloban didn't want Dick going along with them as he was stealing too many Nips from them, so he came down here to see if we would let him fly with the 475th.

We figured we were good enough so that we could take care of our own interests along that line so we said it would be okay. This morning he saw me getting ready to take off for a look at the Jap fields over on Mindanao and suggested that he go along. I had a hunch I shouldn't have let him come with me, but I had to be polite so I gave in. We picked up a wingman apiece and took off.

"We cruised all over the island, looking for something to shoot at, but the bombers and strafers have about cleaned the place out. We had just decided to call it off and go home when we spotted a couple of Oscars. They were on my side and I figured maybe Dick hadn't seen them so I barely whispered over the radio to my wingman to follow me and I dive to take one of the Nips. One nice burst and down he goes. I turn to knock off the other Oscar but this eavesdropping Bong had heard me talking to my wingman and had located the Nip. Before I could get in position I saw him blow up and Bong pulls up alongside of me waggling his wings and grinning at me like the highway robber he is. I'm still eight behind. I'll bet when this war is over, they'll call me Eight Behind McGuire."

Colonel MacDonald, the 475th Group commander, whose score was now up to twenty-three, and a couple of other pilots had come in while McGuire was talking. We all laughed at Tommy and his pretense of being sore at Bong. No one could help liking Bong, even his closest rival, Tommy McGuire, and no one could help liking McGuire, least of all, Dick Bong.

McGuire said they had a cold turkey that was only half gone and if I'd stay for lunch he'd see that I'd get my share of it if he had to tie Bong's hands. "Bing" Bong, as the kids had started to nickname him, was pretty good with a knife and fork.

I stayed for lunch, talked airplanes, combat, and how the war looked to be going, with the two top scorers and most of the rest of the squadron who kept dropping in, for the next couple of hours and then told them that I had work to do and couldn't hang around wasting my time gossiping with them any longer. I had enjoyed myself thoroughly but had made up my mind that Bong was going home as soon as he made it an even forty.

On the 17th McGuire allowed Bong to persuade him to go on another raid together. They each got an Oscar. Bong had his forty, McGuire his thirty-second.

I sent for Dick and told him—no more combat. He said he would like to keep on teaching the kids gunnery and there was one squadron of the 49th about to go into Mindoro that he thought he could do a lot of good with. I said all right, he could go over there as soon as the field at San

Jose was ready to land in, but when I said no more combat, this time I meant it. Furthermore, that as soon as he had finished checking up on that squadron I was sending him home. He nodded but didn't look too happy about it when he left.

The airdrome construction on Mindoro moved along even faster than we had hoped. On the 20th we moved the 8th Fighter Group over there just in time for them to get in a fight with air-raiding Japs, who helped swell the group's score by thirteen. We lost one plane but recovered the pilot. The kids added eleven more victories on the 22nd, nine on the 23rd, and twenty on the 24th. I told Whitehead to follow the 8th with the 49th and 58th Fighter Groups, the 3rd and 417th strafers, and either the 43rd or 90th Heavy Group. I told him also to move his Fifth Air Force Headquarters over there, too, as soon as his communications could be set up.

By the 22nd there were no longer any worthwhile airdrome targets left south of Luzon. That afternoon, with over 200 bombers, strafers, and fighters, we opened the campaign on the big Jap air base about sixty miles north of Manila, at Clark Field. Nearly 100 planes were destroyed on the ground and eight of the nine Nips that intercepted were shot down. We had no losses.

Grace Park airdrome near Manila was the target on the 23rd. Twenty-five more Jap planes were caught on the ground and destroyed. There was no interception and we had no losses.

Sixty Jap fighters intercepted our attack on Clark Field on the 24th. It cost the enemy thirty-three airplanes in combat and another fifty-eight on the ground. We lost a P-38.

On Christmas Day we plastered the Jap airdrome at Mabalacat, about ten miles northeast of Clark Field. Seventy Nip fighters rose to intercept. It cost them thirty-nine. We lost five P-38s but recovered three of the pilots. Tommy McGuire, leading his squadron of P-38s, shot down two Jap fighters to bring his score to thirty-four. For the first time, the "eight behind" jinx was out of the way.

On the 26th we raided Clark Field again. Only twenty Jap fighters contested this time. Our escorting fighters shot down thirteen of them. Four of the victims fell to Tommy McGuire, back in there still leading his squadron. His score now stood at thirty-eight, only two behind Bong's record forty.

The next morning I sent for McGuire and radioed Whitehead to find Bong and send him back to me, too. Bong had gone to Mindoro to finish up the course in gunnery instruction he was giving one of the squadrons.

I told McGuire that I was taking him off flying as he looked tired to me. Tommy protested, "General, I never felt better in my life. Besides I'm only two behind and—"

"That's just it," I said. "You are tired and you won't be rested enough to fly again until I hear that Bong has arrived back in the United States and been greeted as the top scoring ace of the war. As soon as I get that news, you can go back to work. If I let you go out today you are liable to knock off another three Nips and spoil Dick's whole party."

Tommy laughed but he didn't want to spoil anything for Dick, so it was okay. He'd relax, take it easy, get a lot of sleep for a few days, and then—he hoped the Japs wouldn't run out on us for a while yet. By the way, when was Bong going home?

"As soon as I can get him over here and loaded on a plane bound for the United States," I replied.

Bong came back from Mindoro the 29th. The weather had held him up the day before. I had a seat on a plane bound for San Francisco that night assigned to him. He packed up his stuff, I gave him a letter to General Arnold which gave him a courier's status with high priority for air travel, and said goodbye. He left at midnight.

Just before dark on December 26th a Navy reconnaissance plane sighted a Jap naval force of one heavy cruiser, one light cruiser, and six destroyers about eighty-five miles northwest of Mindoro headed toward San Jose. We had available on our two strips there twelve B-25s from the 71st Reconnaissance Squadron, the 58th Fighter Group (P-47s), the 8th Fighter Group (P-38s), and the 110th Tactical Reconnaissance Squadron, equipped with P-40s. Every airplane that could fly took off on the attack, which continued until after midnight. The Japs kept on coming and the planes kept shuttling back and forth, emptying their bomb racks and ammunition belts and returning for more. In addition to the difficulty of locating and attacking the Nip vessels in the dark, the enemy made the job still harder by bombing our airdromes at intervals through the night. In order to see what they were bombing and strafing, some of our pilots actually turned their landing lights on the Jap naval vessels. With neither time nor information for briefing during the operation, it was every man for himself and probably the wildest scramble the Nip or ourselves had ever been in.

At 11:00 p.m. the enemy fleet started shelling our fields and kept it up for an hour. Fires broke out in our gasoline dumps, airplanes were hit, the runways pitted, but the kids still kept up the attack. The P-47s couldn't get at their bomb dump because of the fire, so they simply loaded up with ammunition and strafed the decks of every ship in the Jap force. They said it was "like flying over a blast furnace, with all those guns firing at us."

Shortly after midnight the Jap fleet turned around and headed north. They had been hurt. A destroyer had been sunk and a cruiser and two destroyers heavily damaged.

The attack had saved our shipping at San Jose from destruction, which was probably what the Jap raid was intended for originally, but it had cost us something, too. The next morning's report showed twenty-five fighter pilots and B-25 crew members missing. We had lost two B-25s and twenty-nine fighter aircraft. During the next few days we picked up sixteen of the kids, who were still floating around the China Sea in their emergency life rafts.

I got General MacArthur to approve a citation for each of the units that took part in the show. To this day I haven't been able to figure out how they got away with it without losing much more than they did. A combination of sheer audacity, superb flying, and good luck had done the trick.

On the 30th Lieutenant Colonel Howard S. Ellmore, a likable, happy-go-lucky, little blond boy from Shreveport, Louisiana, leading the 417th Attack Group, the "Sky Lancers," caught a Jap convoy in Lingayen Gulf, off Vigan on the west coast of Luzon. In a whirlwind low-level attack, a destroyer, a destroyer escort, two large freighters of between 8000 and 9000 tons, and one 5500-ton freighter were sunk and a destroyer and two cargo vessels sunk.

It was a fitting climax to 1944, which had seen an advance from Finschaven to Mindoro, a distance of 2400 miles, equal to that from Washington to San Francisco. During that time my kids had sunk a half million tons of Jap shipping and destroyed 3000 Jap aircraft. Our losses of aircraft in combat during the year were 818.

1945

While our B-24s from Palau opened 1945 by smashing Clark Field, Ellmore's 417th Attack Group A-20s, heavily escorted by fighters, swept over Palaniz Bay, on the northwest coast of Luzon, where thirty small Jap freighters had been spotted the afternoon before by our reconnaissance planes. Every ship in the harbor received direct hits. Nineteen went to the bottom and eleven were heavily damaged. There was no Jap fighter opposition to either raid.

On the 2nd the Palau-based B-24s repeated their attack on Clark Field. There was still no Jap fighter opposition. On that same date our convoy, consisting of Kinkaid's whole 7th Fleet and an armada of Liberty boats and landing craft carrying four infantry divisions of Krueger's Sixth Army, left Leyte Gulf on the way to Lingayen. Over three million tons of shipping were involved, about one third of which were Naval supporting vessels. It is interesting to note that our intelligence people at this time estimated that the total merchant shipping available in the whole Japanese Empire was only 1,872,000 tons. Eighteen baby carriers accompanied the expedition to cover the convoy after they passed Mindoro and to support the troops going ashore in Lingayen Gulf until an airdrome could be built there for our fighters. Halsey's fast carriers of the 3rd Fleet were to raid the Jap fields in Formosa and the Ryukus on January 3rd and 4th, coordinate his attacks with mine on the Luzon airdromes on the 6th and 7th, and remain in support of the Lingayen effort to assist Kinkaid if he got in trouble. We hoped to have a landing field ready for our fighters at Lingayen by the 15th.

In the meantime, to make the Japs think we were going to land on the south coast of Luzon, instead of at Lingayen Gulf, we began pounding away at Legaspi, blew up bridges along the railroad from Legaspi to Manila, and strafed everything that looked like an airdrome south of Manila. Parachutes were dropped to simulate paratroop operations and fake expeditions, which would withdraw after the escorting patrol craft shelled the shore for a while, were sent against the Batangas coast.

On the afternoon of the 2nd, Lieutenant Colonel Ellmore, leading his 417th Attack Group for the third straight day, struck at the Jap shipping at San Fernando on Lingayen Gulf. Thirty-six thousand tons of shipping went to the bottom. Ellmore himself sank a Jap freighter and a destroyer escort, but as he pulled up from his last attack, he received a direct hit from the antiaircraft guns of a Jap destroyer. The airplane lost a wing and plunged into the water. We lost a great low-level attack leader—one of the best. The Sky Lancers had done a great job, but there was no celebration of the victory in their camp that evening.

On January 3rd, to help out the deception, landing parties from the San Jose-Mindoro area went ashore at places on both the east and west coasts of Mindoro. A Sixth Army detachment and a Filipino guerrilla force, with heavy support from the Fifth Air Force, occupied Marinduque Island, just northeast of Mindoro. Our fighters were told to pretend they didn't see them if any Jap reconnaissance planes flew over the show.

That night Tokio Radio admitted that we had "captured the unimportant island of Marinduque" and then gloated over the announcement that our attempted landings on the south coast of Luzon had been "everywhere repulsed with heavy loss."

The 3rd also saw the beginning of a determined series of kamikaze attacks against our expedition to Lingayen Gulf that lasted for ten days. The attacks were much better coordinated and executed than the ones we had experienced up to that time, and for a while they had Kinkaid pretty worried. Luckily the Nip was still using bombs too small for the job or we would have taken quite a beating. As it was, during the first two weeks in January, the Lingayen convoy lost five ships sunk, twenty-five badly damaged, and forty-two slightly damaged, with a casualty list of over 1600 men killed, wounded, or missing. The Japs lost between 225 and 250 aircraft.

Halsey's carriers raided Formosa and the Ryukus on the 3rd and 4th and whittled another slice off the Nip air strength. Halsey reported destroying 111 aircraft and sinking twenty-seven vessels against very feeble Jap air opposition.

We were getting no air opposition over Luzon either. Bombers, strafers, and fighters roamed the island almost at will, destroying airplanes on the ground, bombing the Jap shore defenses of Lingayen Gulf, shooting up troops and vehicles on the roads, and destroying rolling stock on the railroads. From the 4th to the 9th we got rid of over a hundred Jap airplanes on the ground, but shot down in air combat only three. The air combat days were over. Fighters, which had always done some bombing, became fighter-bombers.

On the afternoon of the 6th I got word that Bong had arrived back home and the newspapers had acclaimed him as the top scoring ace of the war. I sent for Tommy McGuire to have dinner with me that evening. After the meal, I told him that I believed he was now sufficiently rested and could go back to work shooting down Nips. I begged him, however, to take it easy and not try to get all the Jap airplanes left, in his first combat. If he would just play a hit-and-run game and get them one at a time he would live through the war, but if he started pressing or got careless his luck might run out. I didn't want to have to write a letter back to his parents.

Tommy assured me that he would be careful. He said that the next morning he and Major Rittmayer, a visiting P-38 pilot from the

Thirteenth Air Force, who had four to his credit, were planning to take along a couple of youngsters, who had just arrived in the squadron, for a sweep over the Jap airdromes on Cebu and Negros to see if they could stir up something. We said goodnight.

On the morning of the 7th McGuire and Rittmayer and the two new lads took off on the mission as scheduled. At 2000-feet altitude over Negros they sighted a lone Jap fighter plane, flying at about 200 feet off the ground. McGuire led his flight to the attack. The Nip turned sharply to the left and quickly maneuvered into position on Rittmayer's tail. Rittmayer called for help as the Jap fired a quick burst into him. McGuire pulled around in a frantic effort to get his guns on the Nip and save Rittmayer. He pulled his turn too tight in the attempt, and the airplane stalled and crashed to the ground. The Nip poured another burst into Rittmayer's already crippled P-38. Rittmayer went down in flames. The Nip ducked behind a hill and got away. The two youngsters, who had gone on the flight to get experience, in trying to stay with their two leaders had gotten out of position and couldn't catch him before he had disappeared from sight.

Once again an accident had deprived me of a great aviator and leader. No Jap had shot him down. I don't believe there was a Jap in the world could have shot Tommy McGuire down, but his loss was one of the worst blows I took in the whole war. I wrote to his father that night. It was not an easy letter to write.

We named a field in Mindoro after Tommy and later he was awarded the Congressional Medal of Honor posthumously.

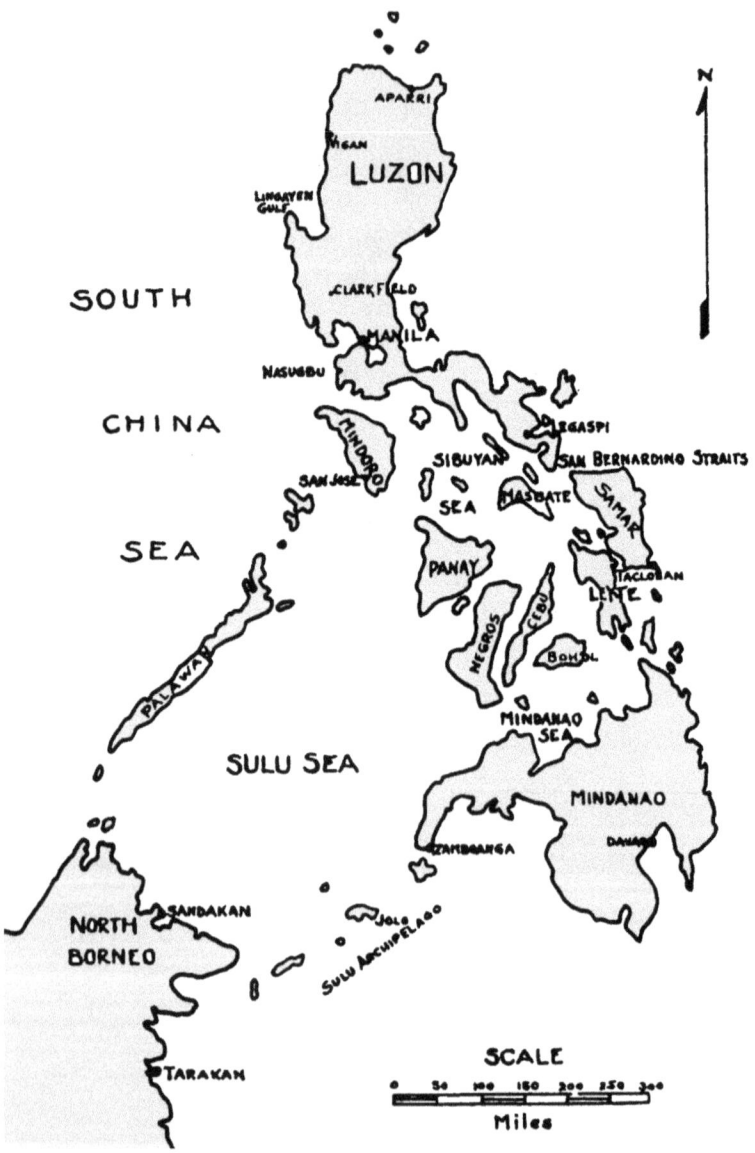

24. THE PHILIPPINES: III LINGAYEN TO MANILA
January—February, 1945

ON JANUARY 9th, four divisions swarmed ashore in Lingayen Gulf. On the east side of the landing area the I Corps, commanded by General Swift, landed the 6th and 43rd Divisions and on the west the XIV Corps, commanded by General Griswold, was composed of the 37th and 40th Divisions. The 25th Division remained afloat in the gulf as reserve. Opposition was negligible and by nightfall our troops had advanced southward distances of two to three miles. MacArthur went ashore with the troops and the next day established his advanced headquarters at Dagupan.

Arnold wired me on the 10th that he had to have Major General St. Clair Streett, who had been running the Thirteenth Air Force for me. I needed Streett but he had lost a lot of weight, his stomach was giving him trouble, and he was pretty worn down. I checked in with General MacArthur, who agreed with me that Streett should go home. I then wired Arnold that I would let him have Streett but as I would give the Thirteenth Air Force to Squeeze Wurtsmith, I would have to have someone to handle Squeeze's job of running the Fifth Fighter Command, so I wanted him to give me back Brigadier General Freddy Smith. Hap came through and said he would have Freddy sent out to me within the next thirty days.

That afternoon I received a radio from Whitehead's headquarters at Mindoro reading as follows: "New type plane, believed Jack, shot down this area. Pilot wearing Black Dragon gown, manacled by ankles to rudder controls. Pilot and plane beyond repairs."

"Jack" was the name we had given to the latest type Jap fighter plane. The reference to the Black Dragon gown indicated that the pilot was wearing his funeral clothes rather than that he belonged to the secret society of that name.

This, however, was another indication that the kamikaze business was not necessarily a purely voluntary affair on the part of the participants. We had just received the report of an interrogation of a Jap pilot captured a few weeks before that gave still further confirmation along this line.

The prisoner said that after finishing his flying training he had ferried an airplane from Japan to Formosa, where immediately on arrival he was greeted enthusiastically by the local commander and congratulated on his patriotism for "volunteering" for the kamikaze job. He was told that the next morning he would be sent on a mission to sink an American battleship in Leyte Gulf. The papers had already been made out promoting him two grades, his mother's pension would be increased accordingly, and she would be honored above all other women in his home town for having contributed her son to the cause of victory for the Emperor and Greater East Asia. The pilot was given special rations, an extra issue of rice wine, and the best living quarters at the airdrome. He said that he was not too enthusiastic about this method of volunteering and during the night had slipped out to his airplane and hidden his parachute under the seat, hoping that he would get a chance to jump out over territory held by his own troops after putting his plane in a dive headed in the general direction of one of our ships in Leyte Gulf.

The next morning he was given a map showing the location of the vessel he was to crash into and told that two Jap fighters would escort him to insure that he got to his target unmolested. Two bombs were wired onto his wings so that he could not release them and the fuzes were set so that they were armed soon after the take-off. That meant that if he tried to land, the jar would set them off and he would be blown up anyhow. He said he also understood that while the escorting fighters were to protect him on the way to his target, they were also there for the purpose of shooting him down if he did not do his stuff.

The Philippines: III Lingayen to Manila

On arrival over Tacloban, our P-38s intercepted the trio and promptly shot down the two Jap escorting fighters. The kamikaze lad said he then tried to head for Samar where there were still some Jap troops, but a P-38 set his plane on fire and forced him to jump where he was. He landed in the water just off the Tacloban airdrome where we picked him up and interrogated him.

While it was certain that the Japs did have bona-fide volunteers for suicide missions and large numbers of others would blindly obey orders to carry out such attacks, it was comforting to find that the Japs themselves didn't trust their "volunteers" too far. We found that in addition to sending fighters along to shoot the boys down in case they showed an indication of trying to dodge their assigned responsibilities and manacling the pilot's feet to the rudder pedals, they were taking an additional precaution. In all cases that we could check on during the January attacks on the Lingayen convoy, the canopy over the pilot's cockpit had been locked on the outside so that even if he did manage to hide a parachute in his plane, he would not be able to get out to use it.

On January 11th a couple of youngsters from the 82nd Tactical Reconnaissance Squadron took off from Mindoro in their P-51 (Mustang) fighters to look over the Jap airdromes in the northern part of Luzon and see whether or not they were occupied.

The leader was the squadron commander, Captain William A. Shomo. His wingman was Second Lieutenant Paul M. Lipscomb. Flying at 200-feet altitude, just southwest of Baguio, they suddenly saw about 2000 feet above them a twin-engined Jap bomber, escorted by twelve of the latest-type Jap fighters. They told me afterward they figured it must be some very important general or admiral being evacuated back to Japan—it might even be Yamashita himself. Neither of them had ever been in a combat in their lives, but they figured you had to start sometime and here was a wonderful opportunity.

Shomo, with Lipscomb hugging his wing, climbed to the attack and opened fire. Either the Nips didn't see them before the shooting started or they mistook the P-5 is for some of their own aircraft. It was probably the first time they had seen the Mustang, as it had arrived in the theater only a week or so before.

Shomo promptly shot down the bomber while Lipscomb destroyed a fighter. The fight was now on. The Nips had broken formation and now tried to get re-formed and do something about the two hornets that seemed to be swarming all over them, but they just weren't good enough. In addition to the bomber, Shomo got six fighters, while Lipscomb shot down four fighters. The remaining two Japs left at high speed for the north and a quiet place to land in Formosa. The kids flew around taking pictures of the eleven wrecked and smoking Jap planes on the ground and then headed back home.

I asked them, when they landed, why they let the other two Nips get away.

"To tell the truth, General," said the cocky, blond Shomo, "we ran out of bullets." Tall, lanky, drawling Lipscomb grinned and nodded confirmation.

I made Shomo a major and put in a recommendation to MacArthur for a Congressional Medal of Honor. Lipscomb I recommended for a Distinguished Service Cross and promoted to the grade of first lieutenant. Their awards came through a few days later. The record score in a single air combat for all time had been established. Seven victories in one combat, and particularly in the first combat, is still an astounding score.

An interesting angle to the story came that evening when I was chatting with the two youngsters. I asked them what they did for a living before they got in the Air Force. Lipscomb was a Texas cowboy. Shomo—believe it or not—was a licensed embalmer. Poor Nips.

That evening I got word that Lieutenant Colonel Robert B. Westbrook, the commander of the 347th Fighter Group of the Thirteenth Air Force and credited with the destruction of 22 Jap airplanes in combat, had been shot down while attacking a Japanese gunboat near Kendari. Westbrook was the top scorer of the Thirteenth when that Air Force joined me in June 1944. He had fifteen Nips to his credit then and had scored another seven victories in combats over the Netherlands East Indies and the Philippines.

With the practical abdication of the Jap from air combat, the fighter pilots were turning themselves into strafers and bombers to have something interesting to do. Westbrook had led a flight of eight P-38s to Kendari to strafe and bomb the shipping there. It was mostly small stuff, around two- or three-hundred-ton size, but a gunboat with a lot of antiaircraft guns on it was doing its best to keep us from molesting the cargo vessels. Westbrook let the rest of the flight work on the cargo ships while he and a wingman took care of the gunboat. Both of them were shot down in the attack. The wingman was picked up by a Catalina rescue plane later on, but Westbrook was lost. While his heroic act allowed the rest of his flight to destroy the cargo vessels, I didn't think it was worth it. I wouldn't have exchanged Westbrook for every Jap ship in the harbor, including the gunboat that shot him down.

His loss left Lieutenant Colonel Bill Harris, the commander of the 18th Fighter Group the top scorer of the Thirteenth Air Force, with sixteen victories.

As our troops moved south down the central Luzon plain toward Manila, Yamashita tried to rush his forces into position to stem the tide. With a complete absence of enemy aircraft over the Philippines, fighters, bombers, and strafers filled the air from dawn until dusk, shooting and bombing every Jap that showed his head and every vehicle on the roads. Bridges were taken out, railroad yards and rolling stock reduced to tangled scrap iron, locomotives destroyed, and Jap camps and garrisons strafed and burned.

Yamashita was tied to the ground. His one armored division was bombed and knocked out of action and his troops north of Manila driven off the roads and into the mountains. The futility of trying to conduct ground operations in the face of hostile and dominating air power was never more clearly demonstrated.

By the 15th of January, when our fighters first landed at the hastily constructed strip at Lingayen, the air attacks had destroyed seventy-nine locomotives, which was half of the prewar total in the Philippines, 466 railroad cars, 468 motor trucks, and sixty-seven staff cars.

On January 22nd thirty B-24s, escorted by fifty-three P-38s based at the new field at Lingayen, made our first attack on the Jap airdrome at Heito in Formosa. The fighters hadn't seen a Jap plane since they had moved to Luzon and were all smiles as they took off. There was sure to be a big fight over Formosa.

When they landed, however, they were disgusted. One lone Jap had taken off and gotten himself promptly shot down. The rest sat on the ground while the bombs and machine guns tore their airplanes apart, wrecked their buildings, and burned up their gasoline. The dream of getting that score up to a thousand began to fade. The Nip was quitting on them.

On January 27th our troops occupied Clark Field and we started cleaning up the wreckage of what had once been a powerful Jap air force and repairing the runways which had been badly cratered by our bombing of the past month. The six airdromes in the Clark Field area had all been treated alike. Wrecked and burned-out airplanes were all over the place. Rusty piles of burned gasoline drums dotted the edges of the field, and hangars, shops, and operations buildings were tangled piles of steel and concrete. Quite a number of undamaged aircraft were cleverly hidden away in the woods or covered with camouflage nets so that they had escaped detection. Everything was so dispersed, however, that it was simply impossible for the Nip to have assembled them on the airdrome for an operation unless he had several hours of warning. We counted over 600 aircraft in the Clark Field area. Most of these were wrecked and shot up so that they were only fit for the junk pile, but at least fifty of them were undamaged and another fifty could be repaired without much trouble. We found a few hundred drums of gasoline, buried in lots of ten to fifteen drums two or three feet underground. Boxes of instruments, radio sets, propellers, bombs, ammunition, and aircraft engines were also buried in the ground and hidden under the Filipino houses in nearby villages. The Nip had dispersed his stuff, all right, but how he ever expected to find it and use it to keep his air force going is still a mystery to me. His method of refueling his airplanes was evidently by hand. The only refueling gasoline tanker truck that I saw on a former Jap field in the Philippines was one at Clark Field which still had stenciled on its side the words "U S Army Air Force." I was beginning to feel annoyed that it was taking us so long to defeat so ill equipped and stupid a nation.

I went over to Clark Field, a couple of days after our troops had chased the Nips away from the quarters and barracks area of Camp Stotzenberg next to the airdrome, and found a couple of our enlisted men and a half dozen guerrillas bossing about 300 Filipino men and women working on the strip. They were hauling off the old Jap wrecked planes, sweeping the rubble off the concrete strip, and filling up bomb craters.

While I was driving around the place in my jeep with Colonel Hewitt, my chief engineer, a sniper in the high grass across the field opened fire. We got out of range in a hurry. The Filipinos scattered in every direction. One of the Filipino guerrillas, who had laid down his rifle and was helping in the dragging of debris off the runway, stopped his work, picked up his rifle, coolly went into the grass, and disappeared. A minute or two later we heard two shots. Neither of them sounded like the snap of the high-velocity Jap rifle. In a moment the Filipino appeared with a broad grin on his face, holding up the Nip's rifle and his cap. He then put down his own weapon and his souvenirs and went back to work. I drove over and told him what I thought of the exhibition. The rest of the gang who gathered around while I was talking seemed to be in complete agreement with me, and the guerrilla boy—he didn't look over eighteen—didn't seem to mind a bit being looked upon as a hero, particularly by some of the younger and better-looking of the women.

By the 10th of February the field was good enough for the troop carriers to start operations and I told Whitehead to begin moving an advanced echelon in to repair the buildings and get ready for movement of the Fifth Air Force headquarters into Clark Field as soon as practicable.

With his forces moving steadily south along the central Luzon plain toward Manila, MacArthur ordered Eichelberger, whose 8th Army forces had taken over the mopping-up job on Leyte Island from Krueger's 6th Army, to effect a landing at Subic Bay, drive east, and keep Yamashita's forces from withdrawing into Bataan. Following this operation, Eichelberger was to land at Nasugbu on the southwest coast of Luzon and drive north and east on Manila itself.

The entrance to Subic Bay was covered by Grande Island which the Japs had heavily fortified. Beginning the 19th and for ten straight days we blasted away at the island fortress and, on the 29th when the 38th Division landed at San Antonio, just north of Subic Bay, no shots were fired at them as they passed Grande Island and no opposition encountered on the landing beach. The next day Grande Island was occupied. It was a shambles that the last of the Jap garrison had evacuated the day before.

General Bob Eichelberger went ashore at Nasugbu with the 11th Airborne Division on January 31st and headed for Manila. To insure that the Japs would not block him off by defending the Taygaytay Ridge, about halfway to Manila, we dropped a regiment of paratroopers, the 511th, ahead of him. The drop was successful, Taygaytay Ridge was secured, and the advancing 11th Division linked up with the paratroopers on February 4th.

As Eichelberger was landing at Nasugbu, we got word that the Japs had quit feeding about 3700 internees held at Santo Tomas University in Manila and that in their already weakened condition, after three years on little more than starvation rations, it was doubtful if they could survive much longer.

General MacArthur sent for Major General Mudge, commanding the 1st Cavalry Division, which was on the east flank of Krueger's advancing forces, and told him to drive for Manila at top speed and rescue the people at Santo Tomas. We promised to cover him to the limit with aircraft all along the route. The 37th Infantry Division, which was the next nearest to Manila, was to follow Mudge into the city as soon as possible. The 1st Cavalry Division moved out shortly after midnight on February 1st, headed for Manila, 100 miles away.

Fighting off small bodies of Jap troops that tried to delay him, Mudge smashed his way forward and at 8:30 p.m. on the 3rd his armored cars clattered into the courtyard of Santo Tomas. The internees were ready and expectant. One of our planes just at dusk had flown over the university grounds and dropped a message attached to a pair of pilot's goggles. It read: "Roll out the barrel. Santy Claus is coming Sunday or Monday." This was Sunday. The university buildings were all secured immediately except one. The Jap commander, Lieutenant Colonel Hayashi and his detachment of sixty-three guards, holding 276 internees as hostages in the Education Building, refused to surrender. He stated that, if he were not given a safe escort so that he could join the rest of the Jap forces in Manila, he would not be responsible for the lives of the hostages. He also demanded that his men be allowed to take with them all of their individual weapons.

It was finally agreed that at daybreak on the 5th the Japs would be escorted to a point approximately a mile from Santo Tomas and released.

The Nips spent most of the night washing their clothes and polishing their equipment. Finally just before daybreak, shaved and clean, they formed up in the courtyard and, marching as though they were leading a Victory Parade, left to join their countrymen.

Among those rescued when Mudge's First Cavalry Division drove into Manila, were Mrs. Pappy Gunn and her three children, who had been interned ever since January 1942 when the Japs captured the city. All of them had lost weight but otherwise were in pretty fair shape. We loaded the whole family on a transport and flew them to Brisbane, to rejoin Pappy, who was undergoing treatment in our base hospital there for the injury to his arm from the Jap bomb fragment that had hit him on Tacloban strip on October 30th, 1944.

That afternoon the 37th Division arrived in Manila and the clean-up began. It had been hoped that the Japs would evacuate the city without destroying it, but the Nips decided that, if they couldn't have the place, they would cause as much destruction as possible before they left. Practically every building in the city was dynamited or burned during their retreat west across the Pasig River to the old walled city, where part of them made a last stand, and southeast through the suburban district to escape to the mountains. Bridges, churches, government buildings, office buildings were destroyed or damaged beyond hope of repair. Water lines, sewer lines, and power lines were blown up with land mines, and a city of nearly a million people left—without water, electricity or food—

to stare at their once beautiful capital city.

The wanton destruction of Manila was bad enough, but the Japs earned the undying hatred of the Filipinos for all time by their senseless orgy of pillage, murder, and rape of the civilian population as they evacuated the city and suburbs. People in the houses were called out and shot in the streets, the houses searched for liquor and loot and then set on fire. Crazed with alcohol, Japanese officers and men raged through the city in an orgy of lust and destruction that brought back memories of their conduct at the capture of Nanking several years before, when their actions had horrified the civilized world.

Eichelberger's driving 11th Airborne troops reached Nichols Field about three miles west of Manila on the 8th, where they ran into stubborn resistance. The Japs were slowly compressed into the old walled city, where they fought to the end. The last resistance collapsed on February 23rd. The campaign for Luzon now became simply a matter of hunting down and mopping up isolated Japanese forces, cut off from each other but well equipped with arms and ammunition and fanatically determined to fight to the last—to kill until killed. The worst of it was there were still a hundred thousand of them. Their position was hopeless but their code refused to admit surrender.

On February 15th one of our P-51 fighter pilots from the 3rd Air Commandos shot down one of our own DC-3 transports and earned a decoration for the job. He was returning from a mission against Formosa when he saw the transport circling with the wheels down for a landing at an emergency field on one of the Batan Islands, north of Luzon and about halfway to Formosa. The islands at that time were occupied by the Japs.

The P-51 pilot had to decide whether it was one of our own planes that was lost or a Jap-built DC-3 with American insignia. He flew up alongside and satisfied himself that the pilot was not a Jap. He then dived in front of the transport to keep it from landing on the Jap-held strip. The pilot of the transport circled again and again started to glide in for a landing. The P-51 pilot then decided on a desperate measure. Lining his sights on the left engine of the DC-3, poured a burst of machine-gun fire into it and knocked the engine out of commission. The transport pilot promptly ditched the plane in the ocean and the occupants got into their rubber boats.

The P-51 pilot then called Lingayen for a Catalina rescue plane and settled down to see that no one came along to bother the rubber boats. Luckily a Catalina was already on the way to that general area to pick up another fighter pilot who had been forced down in the water a few hours before.

In a short time the Catalina arrived and picked up the whole crew and passenger list of the DC-3. Among them were two nurses and two Red Cross girls on their way to Lingayen. The pilot had run into bad weather and gotten lost. He was getting low on gasoline and, thinking that he was in friendly territory, had decided to land. They were all quite put out at the action of the P-51 lad, until the situation was explained to them, but from then on the kid was the greatest hero of the war as far as they were concerned.

The P-51 lad already had painted on the nose of his airplane seven Nazi swastikas and one Italian insignia, which he had earned in the European theater, as well as a Jap flag for a victory in the Pacific. He added an American flag in memory of his latest exploit.

I awarded him an Air Medal for the job and told him I hoped he wouldn't feel called on to repeat that performance.

When General MacArthur had started his drive for Manila, I talked with him about Corregidor, the tadpole-shaped rocky fortress at the entrance to Manila Bay. Known for years as the Gibraltar of the East, it was garrisoned by an estimated 6000 Jap troops, provisioned for a long siege, bristling with artillery, and well-stocked with ammunition. As long as it held out we could not use Manila Harbor to bring in the tremendous tonnage of supplies needed to keep the war going. Landing supplies in the six- to ten-foot surf at Lingayen Gulf was impracticable. In fact, we were already running short of almost everything, as the stocks landed at the time of the invasion were used up. I was already beginning to fly gasoline up from Mindoro to the fields in the Lingayen area to keep the aircraft there going. We were about to bring the bombers and strafers in to Clark Field as soon as the runways were fixed up. There was no question about it, we were going to have to take Corregidor soon.

I proposed to slug the place to death with heavy bombs and then let the paratroopers take it. The General said go ahead. I sent word to Whitehead and to Squeeze Wurtsmith, who had taken over the Thirteenth Air Force, to "Gloucesterize" Corregidor. For ten straight days every airplane that we could spare bombed the Rock. Fifth Air Force and Thirteenth Air Force bombers, strafers, and fighters, and fighters and dive-bombers from the Marine Wing attached to me at Mangaldan, a strip ten miles east of Lingayen, shuttled back and forth all day. Over

4000 tons of bombs, hundreds of thousands of rounds of ammunition, and huge quantities of our new weapon, Napalm, or liquid fire, slowly crumbled the defenses of Corregidor.

On February 16th, 2065 troops of the 503rd Parachute Regiment landed on the rock on the old parade ground. Simultaneously a battalion of troops, totaling around 1000 men, crossed the channel from Bataan and landed on the small beach on the north side of the island. The opposition was negligible. A few scattered shots from snipers were the principal source of trouble. While mopping up of small groups of Japs who had taken refuge in the tunnels and caves which honeycombed the island continued for another two weeks, to all intents and purposes Corregidor was ours when the first paratroops landed. Every artillery position, every antiaircraft gun, even every machine-gun position had been smashed by heavy bombs. Manila Bay was open again to American shipping.

By the time the mopping-up operation was completed, over 6000 Japanese dead had been counted, while many more remained sealed in the caves that had been blown in by our bombs or dynamited by our ground troops after they landed. Our total death toll at the end of the operation was 210.

Corregidor had taken the most concentrated bombing of the war. It worked out to about 3000 tons of bombs per square mile. For comparison with some of our previous concentrated attacks, Wakde Island was the second on the list with 1000 tons per square mile, and Gloucester, which had given us the name signifying all-out bombing, was well down on the list with only 555 tons of bombs per square mile.

General MacArthur visited the Rock shortly after its capture and after inspecting the havoc caused by the continued air attack said, "Corregidor is a living proof that the day of the fixed fortress is over."

On February 20th I flew back to my headquarters in Leyte from Clark Field for a conference with Whitehead, Wurtsmith, Air Vice-Marshal Bostock, commanding the Australian air units, and Air Vice-Marshal Isitt of the New Zealand Air Force.

I assigned complete responsibility for all air operations south of the Philippines to Bostock, who would maintain his field headquarters at Morotai, and gave him operational control of the New Zealand units to assist him in this work. In case Bostock needed help, particularly at the time of any future land operations in the Netherlands East Indies, I told Wurtsmith to support him. Bostock would designate the targets and coordinate the timing of the American and Australian missions.

Wurtsmith's Thirteenth Air Force, with its headquarters to be established at Tolosa as soon as I moved to Manila, was to handle all air operations in the Philippines, except for those on Luzon. His primary job was to work with General Eichelberger's Eighth Army, to which MacArthur had assigned the mission of recapturing the Philippines south of Luzon. Landings had already been scheduled to capture Puerta Princesa on the island of Palawan on February 28th and Zamboanga on the southwestern tip of Mindanao on March 10th. In addition to his primary job, he would, of course, be subject on my call to support Whitehead's Fifth Air Force in its operations in Luzon and to the north.

I told Whitehead to move his headquarters to Clark Field as soon as the place was ready and to get his units into Luzon as fast as airdrome construction permitted. His jobs were to assist Krueger's Sixth Army in mopping up Luzon, smash the Jap air force and industrial targets in Formosa, and maintain an air blockade of the China Sea, all the way to the China coast, to stop all ship movements from the Netherlands East Indies toward Japan. Both Bostock and Wurtsmith, of course, had the responsibility for stopping Jap ship movements in their respective areas. In case Wurtsmith needed help at any time, I would call on Whitehead to place certain air units under the control of the Thirteenth Air Force as required by the situation.

I told my staff to prepare plans to move my headquarters to Fort McKinley, just southwest of Manila. The move would probably be about April 1st, as the place was terribly shot up and would need a lot of reconstruction.

Among those at the conference was Colonel Jock Henebry, who at that time was running my advanced combat training show at Nadzab, where the newly arrived replacement crews got a lot of gunnery and bombing training against the bypassed Jap holdings at Rabaul and around Wewak. Their operations also assisted the Australian ground forces, who were investing the Jap positions and mopping up isolated enemy garrisons all over New Britain and New Guinea. Jock had taken over the job on his return from leave in the United States in December 1944, from Colonel Carl Brandt whom I had given to Wurtsmith to head the Thirteenth Bomber Command then at Morotai.

Jock contributed two stories of his recent activities. The Australian patrols around Rabaul had been having a lot of trouble with Jap police dogs, who had been trained to give the alarm and even attack and try to hold intruders at bay until the Nips could get to the scene and dispose of or capture them. The Aussies had definitely located the dog kennels and had asked Jock to bomb them. Jock, who had spent over 600 combat hours in the Southwest Pacific making low-level attacks on Jap airdromes, troops, and shipping, said he had carried out the mission successfully but he hoped there would be no publicity. He didn't want anyone to know that he had been reduced to bombing dog kennels. The only reason he was mentioning it was that the news was bound to leak out sooner or later and he wanted us to know the true story and the reason for it.

A few days before, while visiting one of our old strips in the interior up around Mount Hagen, Jock had been approached by a delegation of native warriors. The chief, who boasted that he had an army of 500 spearmen and bowmen, said he wanted some air support in his projected campaign against a neighboring tribe that a few months previously had made an unprovoked raid on his tribe and had gotten away with a lot of women and pigs. He said he was pro-American so the supposition was that the enemy tribe must be pro-Jap. As an ally he wanted cooperation and, as an added inducement, he was perfectly willing to split on a fifty-fifty basis with Jock all the women and pigs of the enemy tribe that he captured.

Jock exhibited some high-power diplomacy in wiggling out of the proposed assignment. The presentation of gifts of costume jewelry, gold-lipped shells, and tobacco, followed by a pig roast and native dance, put everyone in good humor and the chief decided to postpone his campaign indefinitely.

On February 23rd a daring and spectacular raid rescued 2146 internees and prisoners from the Jap camp at Los Banos, 25 miles inside enemy-held territory, on the southwestern shore of Laguna de Bay, the big lake south of Manila which is the source of the Pasig River.

A detachment of paratroopers from the 11th Airborne Division jumped directly into the Jap camp, taking the Jap guards completely by surprise. The paratroopers killed the guards in the immediate vicinity, rescued the prisoners and internees, and fought off the Jap garrison until the arrival of guerrillas, who had been quietly infiltrating the area for the previous two nights, and a battalion of the 186th Infantry, which had moved the night before in boats up the Pasig River and across the lake.

The Jap commander, his staff, and the entire garrison of 243 were killed. American casualties were two killed and two wounded. The internees were taken to the waiting boats and evacuated to Manila.

Jack Burnell, a tobacco-firm executive, one of those rescued, said, "It's been a long time we've waited for just such Hollywood-American stuff."

On the 25th I returned to Manila, landing this time at Nichols Field, where we had a detachment of service troops getting the place ready to set up an air depot. Hundreds of Filipinos were working with our troops, clearing away the debris caused by the bombing and the fighting when the Japs had temporarily held up the advance of Eichelberger's troops. Guerrillas were patrolling the whole area, flushing Japs out of all kinds of hiding places. While I was there, two Nips were captured who had been living under a small bridge across a creek on the edge of the field. Three others were found hiding inside old wrecked Jap airplanes and killed. I drove to Neilson, the next field, a couple of miles to the east, with Colonel Hewitt, my chief engineer, and had to wait on the road until the guerrillas got rid of two more stray Japs, who had hidden in a clump of bushes and were firing at vehicles on the road to Fort McKinley.

We then drove to Fort McKinley. The place had really been smashed. A few of the old quarters could be repaired and the foundations of most of the old post buildings were good. We could construct on top of those. A large hospital had not been touched, the water system was only slightly damaged, and the roads could be quickly repaired. The guerrillas had taken over and were still clearing Japs out of numerous caves and tunnels on the post. They were vitally interested in liquidating Nips but were doing nothing about the indiscriminate looting that was going on. I got hold of the guerrilla leader, who promised to put a stop to it and issued instructions to his lieutenants to carry out the orders, even if they had to shoot to do it. Our conversation was interrupted by a guerrilla sergeant, who came up to us and asked politely if we would move off about fifty yards to one side. They had cornered about fifteen Japs near the old officers'-club swimming pool and wanted to start shooting, but we were in the line of fire. I decided I had some important business to attend to back at General MacArthur's field headquarters at San Miguel, about twenty-five miles east of Clark Field.

On the 27th I attended the installation of President Osmena at the Malacanan Palace in Manila. General MacArthur and most of his staff, General Krueger and his commanders, Filipino leaders, and newspapermen filled the big reception room. MacArthur made a brief speech, the chief point of which was the complete turning over of civil government to the Philippine Commonwealth. Osmena followed MacArthur at the microphone, praising the Americans for liberating the Philippines and asking for complete independence on August 13, 1945, which was the anniversary of the American capture of Manila from the Spaniards in 1898.

The ceremony was a most colorful one in a wonderful setting. The Malacanan Palace, the old residence of the Spanish governors, is one of the most beautiful buildings of its kind anywhere. It had scarcely been touched by the war and its carved woodwork, crystal chandeliers, paintings, furniture, rugs, and hangings were all intact. It was a real oasis in the midst of desolation. Madame Osmena and a number of the wives of top Filipino officials were there in their charming native costumes to add color to an impressive ceremony. The Filipinos had endured a lot from the Nip since December 1941, had fought loyally and desperately on our side ever since, and they were grateful to us for liberating them, but it was quite evident that they were a proud people who passionately wanted independence. MacArthur said he intended to endorse Osmena's request favorably to Washington.

On March 1st we got word that General Miff Harmon and his whole crew, including several members of his staff, were missing on a flight from Guam to Hawaii. Miff had taken over the job of running the Central Pacific land-based air units under Nimitz and was also deputy commander for General Arnold of the B-29 show in the Marianas. This latter command, called the 20th Air Force, the Strategic Command, had been placed directly under Arnold by the Joint Chiefs of Staff, who decided on the targets to be attacked and directed Arnold to operate the show.

For over a week, search aircraft hunted all over the area where Miff's plane had sent its last message but no trace was ever found of the airplane or any clue as to what happened. Whatever it was, it must have been sudden as otherwise we would have heard from the radio operator who had sent regular messages every half hour from the time the plane took off on the trip.

Since the fall of Corregidor we had raided Formosa almost daily with from fifty to a hundred airplanes. The Jap air force there was going the way of all the others, except that this time they were taking it lying down. Every mission reported airplanes destroyed on the ground, rail and motor transportation destroyed, warehouses burned down, and railroad yards plowed up, with no interception by the enemy fighters. On March 2nd we put over a simultaneous attack with 125 bombers, fighters, and strafers on six Formosan airdromes, destroying another forty Jap planes on the ground. The operation was worth mentioning because a lone Jap airplane had attempted interception of the raid. He was shot down so quickly, with so many people shooting at him, that the kids involved finally had to draw straws to see who should be given credit for the victory. All our planes returned.

Ever since the 21st of February we had been sweeping the China Sea and the China coast, from Saigon to Shanghai, looking for Jap shipping. By March 1st the air blockade had cut the volume of that shipping to one third of what it had been a month before. The kids were complaining that the hunting was getting so poor that they hardly felt like calling it combat time any more.

The main effort of the Far East Air Forces was in support of the ground operations. Whitehead was giving Krueger's troops an average of over 200 planes a day, bombing ahead of the infantry and smashing Jap resistance centers as fast as they were located. Wurtsmith was putting out another hundred or more missions a day to keep Eichelberger going in his mopping-up operations south of Luzon.

We received a report that Samah airdrome, on the island of Hainan off the south China coast, was loaded with airplanes. I didn't have to ask for volunteers on that mission. The job was to keep everyone from going. On March 6th ninety B-25s and sixty P-38s made the strike. They didn't find the target as lucrative as they had hoped, but they took care of what there was. The photographs showed they got just about all there was to get. Twelve Jap fighters intercepted. The score the kids sent in was eight definites and four probables. Twelve other assorted aircraft were destroyed on the ground, several gasoline fires set, and every building on the field burned or wrecked. We had no losses.

That day a wire came in from Arnold asking that I be sent to Washington for a conference about March 15th. Arnold had had a bad heart attack during February and I had heard from some visitor from the States that he was expected back in his office on March 15th. I had a lot of things to take up with him, so I suggested to General MacArthur that this was a good time for me to get away for a couple of weeks while only mopping-up operations were going on. Nimitz was planning to land in Okinawa about the 1st of April and MacArthur's staff was busy working out the schedule for some operations in Borneo, but these would be still later. I said I would like to leave the 12th of March and expected to be back in Manila by the 26th at the latest. The General said to go ahead and that when I got to the United States I was to buy myself some four-star insignia, as he had sent in a recommendation that I be promoted to the rank of full general.

I left on March 12th.

25. THE THIRD TRIP TO WASHINGTON
March, 1945

I ARRIVED IN Washington on March 14th. Arnold was still down in Florida recovering from his heart attack, so I checked in with Major General Barney Giles, his Chief of Staff. I had written to Arnold asking to let me change over my B-24 groups to B-29 outfits, so that I could drop ten tons of bombs per airplane instead of four. Airdrome space was beginning to be a problem to me. Also, with the 3000-mile range of the B-29, I could handle a much greater dispersion of targets than with the B-24, which had about one half the range of the B-29. Giles told me that the decision was against me. I asked about the B-32, a Consolidated Aircraft bomber that had been built as an ace in the hole in case the B-29 had not turned out successfully. Giles said they were building about 200 of the B-32s but the assignment would be up to Hap Arnold. I made an appointment to fly to Florida and see Arnold on the 17th. Barney told me that the President had sent my name to the Senate, recommending I be made a four-star general. Joe McNarney, Spaatz, Omar Bradley, and Walter Krueger were also on the list.

I had a long talk with General Marshall in regard to the landing in Japan proper, on the westernmost island of Kyushu, which had been tentatively set up for October or November 1945 with the code designation of Olympic. I said that we could land there any time we could get the ships to take in the troops. That we had enough troops, navy, and air power in the Pacific to do the job, as Japan was through. She had lost her air power, her navy, and her merchant marine. I didn't believe it was necessary to wait for Hitler to fold nor did we need any help from the Russians to beat Japan.

General Marshall did not agree with me. He said the Japs had a lot of fight left in them and a big army that we would have to defeat. He suggested that we might have to land in China first. I told him that the same effort that would be necessary to effect a landing in China would put us in Kyushu. It was common knowledge that the Japs had been putting out peace feelers for some time and I believed that there was a good chance that they would quit by the 1st of July or by September 1st at the latest.

General Marshall called in several members of his staff, who discussed the general situation in the Pacific. The main question to be settled was whether MacArthur or Nimitz would be in charge of the actual invasion of Japan. That decision had not yet been made by the Joint Chiefs, but I gathered from the conversation that MacArthur would probably be the one.

The question of the influence of typhoons on the date of the operation seemed to be worrying everyone. I told them that typhoons occurred anytime during the year off the south coast of Japan and that while it was true that there normally were more of them in July, August, and September, our weather people could forecast them ahead of time. Moreover, we had a real typhoon hit us in the early stages of the Leyte campaign and it had merely delayed our unloading for a day. I suggested that the timing be left to MacArthur and Nimitz, who were on the spot and could be continuously studying the various factors involved. Everyone listened but they would not buy my thesis that the Nip was as badly off as I claimed. The occupation of Kyushu was tentatively set for December 1945 and the landing in Tokio for March 1946.

Washington analysis of Japan's air strength was based on the number of aircraft. Their figures showed around 2500 airplanes in Japan proper and a production rate of around 2200 per month. I agreed that the Nips had a lot of planes but claimed that they were out of really qualified combat pilots and, furthermore, had a shortage of gasoline that kept their aircraft on the ground where we were burning them up as fast as we got within range. I didn't agree with the production figures. I thought that the Japs were making only about 1000 planes a month on account of a shortage of aluminum, which had been aggravated by our cutting the sea lanes to Malaya where the bauxite came from and by our destruction of the plant at Takao in Formosa which had been producing twenty percent of Japan's aluminum.

I saw General Arnold in Miami on the 17th. He looked rested and in good shape and declared he was ready to go back to work but the doctors wouldn't let him. I noticed that they reminded him, as we sat down to lunch, he had to take a nap in two hours.

I told him about how things were going in the Pacific and broached the subject of assigning me enough B-32s to equip one of my heavy groups. If he would give me the plane I would give it a real test so that he could make a decision whether to go on with production or abandon it. He finally promised to send them out to me, beginning in June, when about twenty would have been delivered from the factory. He ordered one flown to Bolling Field the next day so that I could look it over. If after inspecting it, I still wanted it, he would turn the production over to me until I had two B-32 groups.

The story I had heard before coming to Washington, that the B-25s and A-20s were to be discontinued in favor of the new attack bomber, the A-26, turned out to be true. I told Hap that the equipment we had was good enough to win the war and I did not want to start testing and experimenting with the new A-26 this late in the game. Furthermore, we had had four of them out in the Pacific already and they had not proved to be as good for our work as the older models we already had. Hap agreed to keep me going on B-25s and A-20s but asked me to equip one group with the A-26. I said I would.

I returned to Washington and the next day inspected and flew in the B-32. It was a nice job, about twenty miles an hour slower than the B-29, but could carry ten tons of bombs from Clark Field to Kyushu. I told Giles to line them up for me and Arnold's proposition was set up as an approved project.

On the 20th I called on President Roosevelt and spent an hour and a half with him. He looked tired. His complexion was almost gray and his hands shook as he held some pictures of bombed-out Corregidor that I showed him. He told me he had lost about twenty-five pounds and had no appetite. I couldn't help noticing the difference from the year before when he had looked the picture of health.

I started to thank him for sending my name in for four stars, but he said, "I sent it in because you have more than earned it. Now, sit down and tell me all about how things are going out there."

We talked about the war in the Pacific and he showed the same keen interest in every detail of each operation, and again his knowledge of the geography was so good that we seldom had to refer to the maps. He asked me when we could invade Japan. I told him anytime we wanted to and gave him the same reasons I had given General Marshall and the staff. He asked if General MacArthur felt the same way. I said I was sure he did but would prefer that the General give the answer himself. The President asked me to convey his best to MacArthur and tell him that he would appreciate a letter on the subject as well as his views on giving the Philippines their independence on August 13, 1945, as Osmena wished. He said he expected to make the trip to the Philippines and take part in the actual ceremony when the date was decided upon.

As I shook hands with him to leave, he thanked me for coming in, congratulated me on my job in the Pacific, and then said, "I suppose you would like to know whether MacArthur or Nimitz is going to run the campaign when the landing is made in Japan." I admitted that I was a bit curious. He laughed and said, "You might tell Douglas that I expect he will have a lot of work to do well to the north of the Philippines before very long."

I settled a lot of problems with the Personnel Section that afternoon and then flew to Dayton for a conference with Lieutenant General William Knudsen on modifications for aircraft coming my way and on the continuation of the P-38 in production. There was another drive at that time to stop building any more of them and to substitute P-51 Mustangs. I told Knudsen that the reasons I had given him in September 1943 for wanting the P-38, still held. We still had a lot of water to fly over and I wanted a fighter plane that could bring the kids back if one engine quit. Knudsen promised me he would not let anyone shut off P-38 production below the number required to keep me going.

I left Hamilton Field, California, on the evening of March 22nd and arrived in Manila on the 26th. I went immediately to see General MacArthur.

After giving him the story on my Washington trip, I said, "By the way, I heard a rumor that you are going to command the show when we go into Japan."

"I don't believe it," quickly replied the General. "My information is that Nimitz will be in charge and that I am to clean up the Philippines and then move south into the Dutch East Indies. Who gave you that rumor, anyhow?"

"A man named Franklin Delano Roosevelt," I replied.

MacArthur tried to keep the same expression but it was no use. He was as pleased as I was. He would have taken the decision the other way, like the soldier he is, but, of course, he wanted to lead the final drive on Japan—the biggest drive of the war. I wanted him to lead it, too. I didn't have anything against Nimitz but I thought MacArthur was the better man for the job.

He wanted to know when the decision would be announced. I told him that the "rumor" had said, "Soon."

On the 31st a radio from Washington announced that I had been confirmed by the Senate as a full-fledged four-star general.

On April 1st, General Krueger wired that the support he was getting from Whitehead on Luzon was superb and Eichelberger was loud in his praise of Wurtsmith's Thirteenth Air Force in his campaigns in the islands to the south of Luzon. The ground forces were following the bombs and tickled to death. For the period from January 28th to March 10th Whitehead's support missions had dropped over 12,000 tons of bombs and fired eight million rounds of ammunition. In addition to this work, during that same period he had bombed Formosa heavily almost daily and, in carrying out the blockade off the China coast, had sunk over 150,000 tons of Jap shipping.

Wurtsmith's operations had resulted in expending 5000 tons of bombs and firing over two million rounds of ammunition to help Eichelberger's forces in their mop-up campaigns. The hunting along the Indo-China coast and around Hainan had not been too lucrative, but Squeeze's planes had sent over 100,000 tons of Nip shipping to the bottom during that time.

The only complaint that I got was from the fighters. They wanted to know how soon we could go somewhere else where there would be some Jap planes in the air once in a while to keep them in practice.

We finally finished counting all the wrecked and damaged airplanes the Nips left on the Luzon airdromes when we chased them out. It reached a total of over 1500. Included were aircraft of every type the Jap factories had turned out during the last two years of the war.

The reason for so large a number being caught on the ground and destroyed or knocked out of commission by our attacks was partially given by a translation of the diary of a Jap pilot picked up by the guerrillas a month previously.

The writer had finished his training in a small, light, fixed-landing-gear biplane on December 26, 1944. During the next fifteen days he was checked off on a bomber, got in six flights by himself of about an hour each, and three days later, on January 13, 1945, joined a group of twenty-two planes which had been ordered to fly from Japan to Clark Field for operations there. One plane was cracked up and lost en route. According to his story, the remaining twenty-one sat on the ground until the 27th of January, unable to fly on account of our incessant attacks, which destroyed fourteen of the group's planes and killed a lot of the combat personnel.

On that date the survivors were ordered to evacuate Clark Field and go to Formosa. The seven remaining planes took off but the writer had motor trouble and had to land in the water on the west side of Lingayen Gulf. He swam ashore and was captured by the guerrillas.

Even if the others managed to get to Formosa, it still made the score for that group—no missions flown, six planes left out of the original twenty-two.

If we had fought the war like that back in 1942 and 1943, we would have been run out of the Pacific.

On April 6th word came from Washington that a new command structure was set up for the Pacific. MacArthur was the commander of all Army forces, and all Navy forces were to come under Nimitz. The B-29s which formed the 20th Air Force were to continue under the control of Arnold, who was to operate them under the directives issued by the Joint Chiefs of Staff.

As a preliminary to the capture of Okinawa in the Ryukus, the Central Pacific forces under Admiral Nimitz landed at Iwo Jima, about 700 miles south of Tokio, on February 20th and, after a bitter fight that lasted until the 16th of March, completed the capture of the island. Our losses were over 4500 killed and missing and around 15,000 wounded. The Japs lost 20,000. Airdromes were hastily constructed and three groups of fighters from the Seventh Air Force were stationed there to cover the big B-29 raids from Guam, Saipan, and Tinian in the Marianas.

On April 1st the 10th Army, commanded by General Simon Bolivar Buckner, landed at Okinawa, where for eighty-two days the Japs fought desperately to stave off our occupation of this vital airdrome area only 400 miles southwest of the island of Kyushu. They realized that the next jump would probably be to Japan proper. Okinawa, with plenty of room for staging troops and with airdrome sites which we could develop extensively enough to handle up to 5000 airplanes, was so important strategically that the Nip garrison died almost to a man before our occupation was completed.

On April 7th the radio announced that Russia had denounced her old neutrality pact with Japan. Under the terms of that agreement each had agreed that they would give the other three months' notice if they decided to abrogate the pact. Three months would bring the date up to July 7th, when the Russians could declare war on the Nips and still have the act clothed with some semblance of legality under international law. It looked as though the Russians had decided that Hitler couldn't last much longer and were getting the decks cleared so that they could hop on the Nips without being accused of being treaty-breakers.

The transfer of all Army units under Nimitz's control to MacArthur's jurisdiction required a lot of adjustments, so after going over the plans for effecting these transfers with General MacArthur and his staff for two days, on the 13th I flew to Guam with Sutherland, Chamberlin, and a few staff officers from GHQ for a conference with Admiral Nimitz and his staff.

Admiral McMorris, Nimitz's Chief of Staff, and Admiral Sherman, the Deputy Chief of Staff, met us as we landed and told us that President Roosevelt was dead. Although he had looked like a very sick man when I had seen him the previous month in Washington and I had remarked to General MacArthur on my return that I didn't believe he had long to live, the news still came as a distinct shock.

Our discussions at Guam lasted until the evening of the 15th. The jurisdictional problems involved on many of the islands in the Pacific, where there were both Army and Navy units and installations, were so complicated that the dates and methods of transferring them from the jurisdiction of one theater commander to the other could not be settled at that conference. A few agreements were reached, however. Among others, the Navy agreed to push airdrome construction at top speed in Okinawa with their own construction battalions and the Army promised to send in all the engineers they could spare to help out.

I wanted to move the whole Fifth Air Force to Okinawa, as well as the Seventh Air Force, which had been operating under Nimitz but was to pass to my control when the capture of Okinawa had been completed. The Seventh was scattered all over the Pacific and I proposed getting it together as a fighting force. This was agreed to in principle, but due to the shortage of shipping it looked as though it would be about the end of July before the various units could be united at Okinawa, even if airdrome space could be ready there at that time.

The Seventh had three fighter groups at Iwo Jima, equipped with P-47s whose primary mission was to escort the B-29s of the 20th Air Force during their daylight operations over Japan. They received orders from General Chaney, an Air Force officer who was the island commander and who operated under directives from Nimitz. One B-24 bombardment group was in Hawaii, another on Saipan, and the third about to move to Saipan from Palau, where it had been working for me. Two medium groups were also in Hawaii, one equipped with B-25s and the other awaiting a shipment of the new A-26 strafers with which it was to be equipped. If I could get this dispersed show together on Okinawa, it would add considerably to the punch I wanted to put on Kyushu before MacArthur's troops hit the beaches.

We set up the procedure for further meetings to iron out the problems we hadn't agreed on and returned to Manila on the 16th. I wished again that some way could be found to organize all our military forces into a single department of national defense.

On the 17th I had a party at my headquarters at Tolosa for a visiting delegation of newspapermen who were touring the theater. Pat Robinson of INS, who had gone back to the United States in February 1943, was there with Gill Robb Wilson of the *Herald Tribune,* an old friend of mine, dating back to World War 1. There were also William Shipper, North American Newspaper Alliance; Frank Sturdy, Chicago *Tribune;* Gilbert Cant, *Times Magazine;* James Leary, Chicago *Daily News;* Earnest Barcell of the UP; Sherman Montrose, Newspaper Enterprise Association; Dick Pearce, San Francisco *Examiner;* Herman Edwards, Portland *Oregonian;* Matthew Weinstock, Los Angeles *News;* Harold Street, Associated Press; and Nelson Pringle of the Columbia Broadcasting System.

After the party got well under way I said something about newspapermen who sank so low that they started writing poetry and then handed Gill Robb Wilson a book of his own, *Leaves from an Old Log,* and asked him to read us a couple of poems. He actually blushed, but he did oblige and then stood for a lot of kidding from the gang for the rest of the evening. Gill's surprise when I handed him the book that he had autographed for me years before was something to see.

The Third Trip to Washington

On the 20th General MacArthur announced that the entire central Philippines, the Visayas, had been cleared of the Japanese forces. Bob Eichelberger had really done a fast job and had already landed at several points on the big southern island of Mindanao. He was loud in his praise of the help he had gotten from Squeeze Wurtsmith's Thirteenth Air Force and the Marine Wing I had attached to it for operations. Bob sent word to me that the Air Force was still making good our boast that when we got through with the Nip beach defenses, the troops could go ashore with their rifles on their backs.

MacArthur radioed me: "Please accept for yourself and extend to all officers and men involved, my heartiest commendation for their brilliant execution of the Visayan campaign. It is a model of what a light but aggressive command can accomplish in rapid exploitation. MacArthur."

On April 26th I flew to Morotai for a conference with General Blamey and General Moreshead of the Australian Army, Air Vice-Marshal Bostock, and Brigadier General Carl Brandt, on the air support for the landing at Tarakan on the northeast coast of Borneo, scheduled for May 1st. The place was not heavily defended and the Aussies were using only one brigade, the 26th, for the job. The main reason we wanted the place was to build an airdrome there from which we could furnish support for the next two operations, one to capture the Brunei Bay area on the northwest coast of Borneo and the other to capture Balikpapan on the east coast. Wurtsmith was up to his ears in the support of Eichelberger's 8th Army forces in Mindanao, so I took Brigadier General Tommy White, his Chief of Staff, along with me.

The two Australian generals greeted me as warmly as ever and in a few minutes the business part of the conference was over. The Aussies were quite easy for me to deal with. In the first place, they knew what they wanted and they had confidence in my furnishing everything I could to help them get ashore and take the objective with minimum losses. They never argued about time schedules or the number of airplanes I would put into the show. They told me the targets they wanted taken out, how much opposition they expected, and asked me to just tell them whether or not I could help them out and what time the bombing and strafing attack would be over so that they could go ashore.

Moreshead had some bad news. General George Vasey, the brilliant commander of the Australian 7th Division, had been killed some days before in the crash of a transport in Australia. I hated to hear it. Vasey was one of my favorite generals of anyone's army and one of the ablest leaders I have ever come in contact with.

I told Tommy White to help Bostock's Australians out with both heavy-bombardment groups and the B-25 strafing group of the Thirteenth Air Force, in a continuous series of attacks on the Tarakan defenses, and to wind up with an all-out attack ten minutes before the Aussies went ashore. Bostock was given directions to go in with the ground troops and see that the old Jap airdrome was reconstructed in a hurry so that we could move in fighters and strafers to cover the Brunei Bay and Balikpapan operations from Tarakan.

Blamey and Moreshead were quite appreciative and promised that they would give Bostock all the help in their power in the airdrome construction.

That evening Brandt called in his staff and the group and squadron commanders for a talkfest after dinner. It was fun to sit around discussing tactics, strategy, the state of the war, and flying in general with those alert, intelligent young combat leaders.

One of them, Major B. E. Harris, the commander of the 868th Bombardment Squadron, the radar bombing unit, was one of the keenest kids I had run into in a long while. He had piled up an astounding personal total of 1200 combat hours in Europe and the Southwest Pacific theaters and still looked as fresh as a daisy. He told about an attack he had planned and led against the dock area at Soerabaya, Java, the night before. It showed imagination and skill of a high order.

Seven B-24s had made the 2400-mile round-trip flight from Morotai to Soerabaya. One of them came in at 15,000-feet altitude to attract the Jap searchlights and antiaircraft guns and two others, flying about two miles apart, synchronized their arrival over the target at the 6000-foot level with that of the first B-24 to further confuse the Nips. About two minutes behind these three decoys, flying just over the water, came the other four bombers, which skip-bombed the docks with fifty-gallon drums of Napalm, the jellied gasoline incendiary that we had used so successfully in Leyte and Luzon.

One large merchant vessel, two small cargo vessels, and the whole dock and warehouse area in the path of the attack were instantly a mass of flames. Harris said they could see the fires still burning when they were seventy-five miles away on the return trip.

On the 28th of April I moved my whole headquarters of 250 officers, 500 enlisted men, and 450 Wacs to Fort McKinley, Manila, by air. The engineers had not rebuilt all the buildings damaged in the recent fighting so that the majority of the personnel were put in tents. The plans called for building me a shack of my own, but in the meantime my headquarters commandant had found a beautiful house in Manila, belonging to a Mrs. Bachrach, whose husband, an automobile man in Manila before the war, had been killed by the Japs. The house had been occupied by the Japanese naval commander and had escaped the wholesale destruction that the vandals had committed when they were driven out in February.

Mrs. Bachrach heard that I wanted a place to live and promptly turned the house over to me for the duration of the war. She also rounded up her old gardeners and put them to work restoring the beauties of the shrubbery which the Jap occupant had let grow wild for the past two years. The house, which was of concrete-block construction, was a beautifully designed and furnished modern four-bedroom affair, with a huge swimming pool in the midst of an exotic setting of tropical flowering and decorative plants of a hundred types and varieties.

The house sat in the middle of a lot which was half a city block in area and was surrounded by a stone wall ten feet high and topped with barbed wire. A huge wrought-iron gate closed the driveway, just inside of which was a gatekeeper's house of six rooms. A two-car garage, a hothouse for the cultivation of orchids, and three summerhouses scattered around the estate completed the picture.

I moved in that evening and put a couple of guards in the gatehouse to make sure that I didn't get disturbed by any would-be looters or stray Japs. The latter were continually filtering into Manila from the fighting zone at Montalban, only twelve miles away. A lot of them, dressed as Filipinos, were being captured or shot almost every day. Some of them were deserters but there was also evidence that Yamashita had sent others through the lines with instructions to assassinate MacArthur and other generals they might get in contact with.

About two o'clock in the morning someone woke me up, trying to open the side door just under my bedroom. I got out of bed, picked up my pistol, and quietly moved out onto the balcony to see if I could see who was trying to get in the house. The outside door on the opposite side of the house then rattled. I ran across the hall to the balcony over that door but this time I was not as quiet and the noise must have alarmed the intruder for just as I got to the balcony I saw a man run across the lawn. He disappeared into the shrubbery before I could get a shot at him. I called to the guard at the gate, who looked all over the grounds but said he couldn't find any trace of anyone. I decided I must have been seeing things and went back to bed. Just before I went to sleep I heard a shot that sounded about a block away, but there was nothing unusual about an occasional shot around Manila at that time, so I dozed off.

The next morning the guard showed me a coat thrown over the barbed wire on the stone wall opposite my room and a lot of torn vines where someone had climbed the wall. The guard also told me that a Jap dressed in Filipino clothes had been shot by a patrol down the street about fifty yards from where the coat had been thrown over the barbed wire. The Jap was not wearing a coat at the time he had been killed, which, the patrol had reported, was just after two o'clock in the morning.

The Australians, heavily supported by bombers, strafers, and fighters of the RAAF and Thirteenth Air Force, landed at Tarakan on May 1st, against negligible opposition. Bostock radioed Wurtsmith: "Desire express my entire satisfaction with your wholehearted cooperation and excellent support which your forces rendered me in support of Oboe 1 (Tarakan) operation. I desire particularly to mention the accurate bombing achieved 13th Bomber Command."

Moreshead sent word that he was "entirely satisfied" with the support he had received and was still receiving and desired to express his thanks and appreciation also.

Tokio Radio said the United States must be running short of oil or we wouldn't be going after Tarakan. They admitted that the oil there was of a high quality and that its capture would be a great asset to us.

That afternoon Colonel Cardenas, the commander of the Mexican Expeditionary Force, landed at Manila with the 201st Mexican Fighter Squadron. After a reception at the pier I took Cardenas over to see General MacArthur, and after the official exchange of greetings, the Mexicans were officially assigned to my command. They then proceeded to Clark Field, where I turned them over to Brigadier General Freddy Smith with instructions to outfit them with P-47s and give them a course of advanced combat training before putting them into action. Both officers and enlisted men were a fine-looking lot and seemed anxious to get to work against the Japs as soon as possible.

I was up at Photo Hutchison's headquarters at Lingayen when the news came in that Germany had quit unconditionally on May 8th. Colonel Jerry Johnson, with twenty-three Nip airplanes to his credit, said, "Hell, I was hoping we'd get this war over first so we could go over to Europe and show those guys how to fight a war."

Barney Giles, who had taken Harmon's place at Guam, Major General Douglass, the commander of the Seventh Air Force, and Major General LeMay, commanding the 21st Bomber Command, the B-29s, flew over to Manila on the 11th to talk over the allocation of Army Air units as the Navy released them. We agreed that the Seventh Air Force was to join the Far East Air Forces as soon as it could be moved to Okinawa, except for the three fighter groups at Iwo Jima, which were to join the Twentieth Air Force and continue to carry out the escort of the B-29s. Giles wanted to know how we would coordinate the bombing attacks on Japan when I got my bombers operating out of Okinawa. I suggested that he hit everything east of Kobe and that I work west of that line. If for any reason either of us wanted to work in the other zone we should notify each other at least twenty-four hours ahead of time. Giles agreed to this procedure. It looked to me like another needless complication. With all our aircraft soon to be in position to hit Japan proper it would seem sensible to put all the participating air forces under one control and issue one set of orders assigning times, routes, and targets and let the kids do their stuff. Priorities on the matter of the huge tonnages of bombs, gasoline, and other supplies would be far easier to handle under one command. Under the existing system, it was almost impossible to insure that one of the two big air outfits to be set up would not have too much and the other have too little at some time or another. Giles agreed with me but there was nothing either of us could do about it. A decision of that kind had to come from Washington and we did not have a unified command there, either.

On the 14th I invited Whitehead to my house in Manila to have dinner and spend the night. After one of Sergeant Raymond's best meals, Whitey and I walked out around the swimming pool, which was empty, as the water for that section of Manila came from the Ipo Dam which the Japs held in spite of Krueger's efforts to dislodge them from the hills on each side. I told Whitehead that I wanted water in the pool so that I could invite him down for a swim before dinner sometime and suggested that he get in touch with Krueger and offer to put a couple of hundred planeloads of Napalm on the Jap positions and burn them out. I thought if the job was done on a big scale the Nips would not have time to blow up the dam and Krueger's troops could then turn the water on, Manila's water problem would be solved, and we could go swimming.

Four days later a deluge of radiograms poured into the office from the commander of the 43rd Division, General Wing, the commander of the 11th Corps, General Hall, from Krueger, and from MacArthur. They spoke of "magnificent preparation," "masterful execution of a difficult and complex mission," "outstanding support," and "fine coordination," for the job we had done. I passed them all on to Whitehead.

Following a continuous drenching of the Jap hill positions on each side of the Ipo Dam with over 200 fighters, each carrying over 300 gallons of Napalm, the troops of the 43rd Division had taken the whole position with negligible losses. Twenty-one hundred dead Japs had been buried when our troops had finished counting. The dam had been mined for demolition but the Nips evidently did not get a chance to blow it up and our troops turned on the water.

I suddenly remembered that I had left the valves open in the swimming pool. I called the house and was informed by the sergeant there that the whole place was flooded with water a foot deep all over the lawn but that he had dived into the pool and shut the valve and he thought it would soon drain off. It did drain off fairly well and evaporation took care of the excess, but for three days the place threatened to become a mud hole. I never did dare to tell Krueger what impelled me to hurry up his attack on the Ipo Dam.

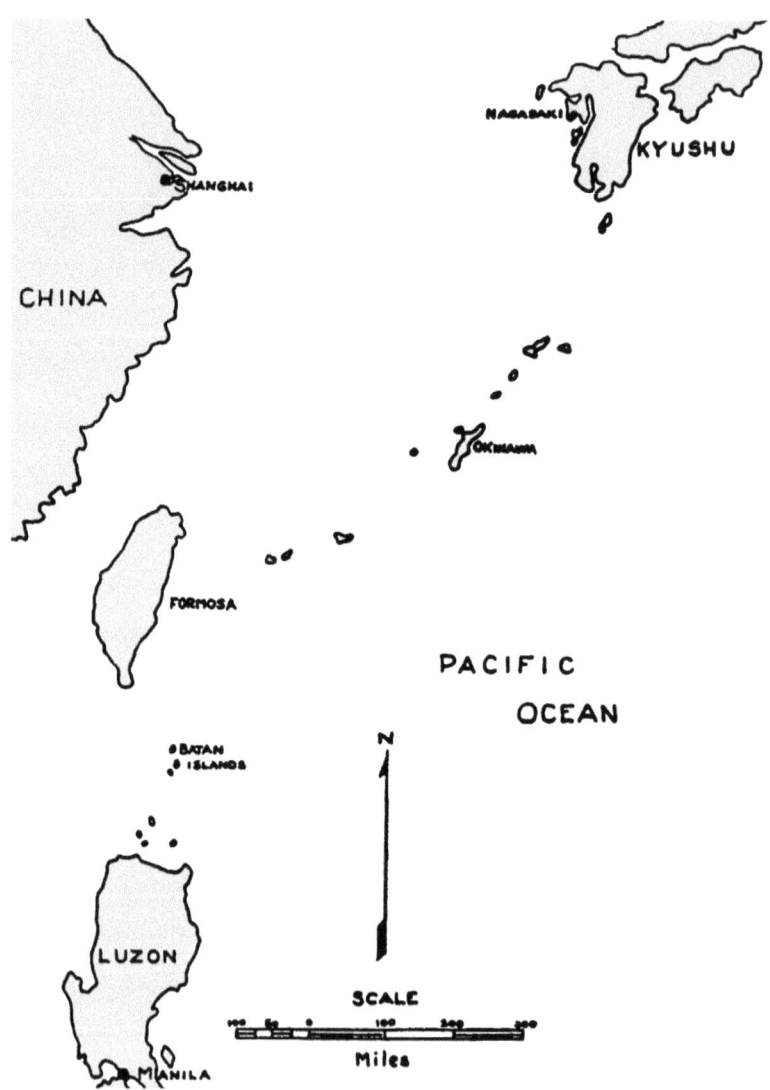

26. OKINAWA AND THE KYUSHU PLAN
May—July, 1945

On May 26th the Joint Chiefs of Staff ordered the invasion of Kyushu carried out on November 1st. MacArthur was named as land commander and chief coordinator of the plans for the transportation of the troops and the actual landing itself. This amphibious part of the show was to be commanded by Admiral Turner, with Admiral Spruance the top Naval commander of the big supporting fleet, instead of Halsey. The Twentieth Air Force was to come under my orders, in case I needed the B-29s to support the operation.

Word came in from Okinawa that airdrome construction was going slowly and, if something wasn't done about it soon, the Olympic operation would be on us before we got the Air Force in position with time to soften up Kyushu before we made the landing. The GHQ staff boys didn't think I needed the whole Fifth and Seventh Air Forces moved to Okinawa so they were not disposed to help. I talked to General MacArthur and told him that if he would let me get my airplanes up there well ahead of time I would promise him that he could land his whole initial six-division effort on the Kyushu beaches with their rifles on their backs as they had been doing for him all through the war. He said that was just what he wanted done and to go ahead on that basis. That took care of the opposition of his staff.

I then flew over to Guam for a few days and talked with the Navy crowd. Turner and Spruance both joined my plea for more shipping to carry more engineers to build more airdromes in a hurry. They were both worried about the Jap air force, particularly the kamikaze boys who had already hit so many ships during the Okinawa operation that it had the Navy worried. They wanted me to have plenty of aircraft up there to blast out the Nip airdromes and get rid of the Jap air force before Olympic started. Admiral Sherman promised to help in every way he could and said he would start moving the Seventh Air Force headquarters to Okinawa right away and the combat units as fast as airdrome construction permitted.

I wired Whitehead to send Photo Hutchison up there with his advanced headquarters right away and start planning to move his own headquarters up soon after. I also sent a message to Colonel Hewitt to go to Okinawa, pick out a place for my own advanced headquarters, and start collecting building materials for it.

Giles told me that Major General Douglass was to be ordered back to Washington on some job that Hap Arnold wanted him for. I decided to give the Seventh Air Force to Tommy White and sent word to Wurtsmith to be ready to release him about the first of July.

On June 6th, just before leaving Manila on the cruiser Boise for the Brunei Bay operation which was scheduled for the 10th, General MacArthur said he would like me to go along with him. I hesitated, as I was pretty busy, but the General said, "No, I think the rest will do you good. You'd better come along with me. We'll have a good holiday and, besides, there will be chocolate ice-cream sodas three times a day." I said I'd join him on the 8th at Palawan, as I wanted to spend some time there inspecting Brigadier General Diz Barnes' task-force wing, which had been doing some excellent bombing to prepare the way for our landings in Brunei Bay and which would be the unit I'd have to depend on if the Japs gave us any trouble in Borneo. The airdrome at Tarakan hadn't turned out so well and would only hold a small fighter force. If I needed anything heavier, Barnes would have to supply it.

I joined General MacArthur on board the Boise at Palawan on the 8th as I had promised. The ship steamed south that afternoon and the next afternoon joined the main convoy, which was coming up from Morotai, carrying the 9th Australian Division commanded by General Wooten. We made the rendezvous in the Balabac Straits between Palawan Island

and North Borneo. The weather was perfect, the mountains on either side of the straits were beautiful, I had had about nine hours' sleep the night before, and there was no sign of a Jap airplane in the skies. Our fighter cover droned back and forth over the convoy but they were just putting in flying time. It was so peaceful it didn't seem as though there was a war on, after all. The General and I had four ice-cream sodas that day.

On the morning of the 10th at six o'clock a lone Jap bomber came over, dropped one bomb, which missed a landing craft, and then flew away under a hail of antiaircraft gunfire.

We watched the Naval gunfire on the landing beach on the island of Labuan, our first objective, and after the RAAF and Thirteenth Air Force bombers got through a farewell blasting of the Jap positions, General MacArthur, General Moreshead, Admiral Royal, the Naval commander, Bostock, and myself went ashore.

The Aussie first-wave troops had landed and pushed inland from the beach about a quarter of a mile. They put out their patrols and then calmly started cooking their tea. Nothing seemed to worry this fine-looking body of troops. They were bronzed and healthy-looking, well equipped, and there was no question about their morale. They had run into only two or three Japs and the natives who greeted them at the beach said the Nips had pulled out about ten days before, when our heavy bombing started. There were originally about 3000 Japs on the island but now the natives estimated the number at less than a thousand.

The brass-hat party moved along the road paralleling the beach, about a quarter of a mile from the water, to the accompaniment of an occasional sniper's shot and a burst of machine-gun fire ahead of us and farther inland. I began to feel all over again as I had at the Leyte landing. MacArthur kept walking along, enjoying himself hugely, chatting with a patrol along the road every once in a while, and asking the men what they were shooting at. Moreshead and Bostock asked me where we were going. I shrugged my shoulders and pointed at MacArthur.

Just then a tank came lumbering along the road and we stood aside to let it pass. As the tank reached the top of a little rise perhaps fifty yards ahead of us a burst of rifle and machine-gun fire broke out and then stopped. The turret gunner looked out, said, "We got those two obscene, unmentionable so-and-so's," and the tank drove on.

We walked ahead to inspect. MacArthur commented on the good clothes and well-kept equipment the two dead Japs had and remarked that they looked like first-class troops. Probably a suicide outfit of a few hundred left to resist to the last, to cover the withdrawal of the main body. The rest of us nodded, pretending to take it nonchalantly.

Just then an Australian Army photographer came up to take pictures of the two dead Nips lying there in the ditch. His bulb flashed, and he dropped to the ground with a sniper's bullet in his shoulder. I figured it was time to do something. I walked over to General MacArthur and said that all he had to do was to hang around that place long enough and he would collect one of those bullets, too, and spoil our whole trip. It looked to me as though we had finally gotten into the Jap outpost position and, if he wanted my vote, it was for allowing the infantry to do the job they came ashore for. We certainly couldn't help them much. In addition, it was a long way back to the Boise and we just had time to walk back to the beach, get in the boat, and ride back to the cruiser, if we were going to be in time for dinner. I thought it would be discourteous in the extreme to keep dinner waiting, when, after all, we were just guests. Finally, the captain had told me we were going to have chocolate ice cream that evening. No one else said a word but it was quite evident that I had a clear majority ready to vote for my proposition.

MacArthur smiled and said, "All right, George, we'll go back. I wouldn't have you miss that ice cream for anything."

We walked back to the beach, stopping to look over the troops who were now moving forward, and talked to some of the commanders. I mentioned the fact that the sooner they fixed up the Labuan airdrome the sooner I'd have some airplanes on the island to help them clear the Nips out of North Borneo. They all assured me that they wanted to see airplanes there, too, just as quickly as I did.

We got back to the Boise in plenty of time for the ice cream. The next morning we all went over to the beach near Brooketon, on the other side of Brunei Bay, where the Aussies had landed the day before. General Wooten joined us at the beach. We waded through about a half mile of swamp to a road where a half dozen jeeps picked us up and drove us into the town of Brooketon itself. The place was completely wrecked by bombing. Wooten said they had encountered very little opposition until they got about ten miles inland, where they were in contact with about 500 Japs, who were dug in on a hill commanding the road and who had a few field pieces to help them out. He had radioed for some airplanes from Palawan to blast the artillery out of the hills so that he could use the road.

MacArthur, of course, wanted to see what was going on, so we climbed in the jeeps and headed off for more trouble. About five miles along the road we came to an overturned Jap truck. It seemed that about two hours before the truck, with twelve Nips on board, had dashed along the road with the lights turned on, the horns blowing, and the fools all yelling, "Banzai," heading for the Aussies who were marching toward them. The Aussie machine-gunners had taken care of the truck and all the Japs.

We then halted to watch a flight of B-25s come over and cheered when they took out the Nip gun position on their first pass. It was pretty bombing. MacArthur was in fine humor and said, "Let's go on." We did for a while, until an Australian colonel halted us. He was not awed a bit at MacArthur's five stars and, much to my gratification, refused to let us go forward another inch. He said that right around the bend a few yards ahead of us there was a real fight going on and he would not have it interfered with. We discussed the situation with him until some Aussie artillery came up and started getting into position to support an attack.

There wasn't anything more to see so we went back to the Boise and pulled out of the harbor for Jolo, in the Sulu archipelago, where General MacArthur was to meet General Bob Eichelberger and the Sultan of Sulu and check up on the progress of clearing the Japs out of the islands.

The next day we landed at Jolo, the capital of the island of the same name, where Eichelberger met us and took us on a tour of the island. The roads were surprisingly good and the scenery excellent, but what appealed to me the most were the Moros themselves. They looked like the unconquered people they were. Well built, intelligent-looking, and walking proudly erect, they looked you in the eye when they talked to you. You knew instinctively that here was fighting stock. The Japs never did subdue them in three years of occupation. Actually the Nips only captured the old walled city of Jolo and were besieged there the whole time. Patrols were ambushed and wiped out. The Japs then tried raiding the villages and burning houses, but the raiding parties melted away from the constant sniper fire that an unseen enemy kept pouring into them wherever they went. Even when they withdrew behind the city walls they were harassed continually by the Moro swordsmen, who would scale the walls at night, kill a Jap or two, seize their arms and ammunition, and get away. There were still about a hundred or more Japs hiding in the hills but the Moro guerrillas were whittling the number down day by day.

Since time immemorial they had been pirates, roaming all over the East Indies and raiding the coastal settlements from Singapore to Manila and from Ambon to the China coast. They got most of their wives that way and they must have picked some good-looking ones, for the Moros were the best-looking natives I saw in the Southwest Pacific. When piracy went out of fashion about the time we took over the Philippines, they turned to fishing and cattle raising but they still preserved a large measure of their independence and considered that they were only nominally a part of the Philippine Republic. They are Mohammedans and consider the Sultan of Sulu their spiritual head. During the war they recognized him as their military head as well and took orders from him as the leader in the continuous campaign against the Japs.

The Sultan came to Jolo while we were there to pay his respects to General MacArthur and presented him as well as Bob Eichelberger and myself with krises, the traditional wavy swords of the Moro. They were beautiful weapons, sharp as razors, with handles inlaid with mother-of-pearl and, according to the interpreter, about a hundred years old. They were in excellent shape but showed signs of having been used and I wondered as I thanked the Sultan for the gift how many skulls that wicked-looking blade had split before it had been retired from service.

Eichelberger told me that in his mopping-up campaign they had buried 27,000 Japs in Leyte since taking over from Krueger. This brought the total to around 210,000 Nips that had been killed in the Leyte campaign since we landed there in October 1944. In the whole Philippines since that time, we had buried over half a million of the enemy. The total still remaining was probably not over 75,000 and most of those were with Yamashita in the Cagayan Valley in northwestern Luzon.

I left General MacArthur at Davao in Mindanao and flew back to Manila on the 13th. I told the General that I was worried about Balikpapan, the next Australian objective, which the 7th Division was scheduled to take on July 1st. That division would be making its first amphibious landing and they had not been in combat for over a year. Vasey was dead. The Japs were supposed to have around 7000 troops there and, if all the antiaircraft guns that we had run into were also capable of shooting at surface targets, there needed to be a lot of preparatory work done if we were to keep the casualties down. I told MacArthur I wanted to smother the whole Balikpapan defense system and would make the whole Far East Air Forces available if necessary. The General said to go ahead. He then decided he was going to go on that landing, too, although up to that time he had not intended to.

On the way I stopped in at Tolosa, where Wurtsmith now had his headquarters, and told him to get together with Bostock on a plan for knocking out every Jap gun that could bring fire to bear on the proposed landing area and to keep me posted on his progress. I told him to call on Whitehead for any extra effort that was necessary to aid the RAAF and the Thirteenth Air Force to accomplish the mission. My estimate was that it would need the application of at least 3000 tons of bombs.

General Arnold arrived in Manila on the 16th and, with General Stratemeyer, who had flown in from China, we discussed plans for utilization of the air forces now available for deployment from Europe. Lieutenant General Jimmy Doolittle was to set up his headquarters in Okinawa, where a newly constituted Eighth Air Force would be accommodated, equipped with B-29s. The Eighth and the Twentieth, under LeMay, would constitute the Strategic Air Force, still under Arnold's control as an agent of the Joint Chiefs of Staff. Arnold said he was considering sending General Spaatz to Guam to run the show for him.

Arnold wanted to know whether it would be necessary to send many ex-European groups out to me, as it looked to him as though I already had all that I could manage. I said that if the Japs didn't quit before we invaded Japan, I would have to move the whole Far East Air Forces into Kyushu, and in that case Okinawa and the Philippines would have plenty of space from which aircraft in large numbers would be used to keep Formosa beaten down, for maintaining pressure on the Jap forces in China, and for keeping up the air blockade of Jap shipping in the China Sea and the Yellow Sea. I also wanted three crews per airplane, now that there were plenty of trained men available, so that we could work the equipment oftener and also allow the tired combat men some rest once in a while. Hap agreed and sent radios to Washington to give me what I wanted.

General MacArthur returned from Davao and General Arnold had a conference with him on the 17th. Hap had discussed with me the day before a scheme of bombing Japan for the next six months, without trying to land in Kyushu as scheduled, to see if the Japs would not quit without an actual invasion. I said that, while I wouldn't be surprised if the Nips threw in the sponge any time, I thought it would be wise to go ahead with the plans to land in Kyushu in November just as a safety measure. If the Nips did quit before that time we would automatically have the troops, shipping, and supplies on hand for the occupational force, so that we couldn't lose if we kept on as originally planned. Arnold said he wanted to get an answer from General MacArthur to send to the other members of the Joint Chiefs of Staff. I wished him luck but told him that I was quite sure MacArthur would not buy his scheme.

As I expected, General MacArthur refused to have anything to do with the idea. He could see no merit in Arnold's proposal and more than hinted that the commanders in the Pacific, who had been fighting the Jap for the past three or more years, were thoroughly capable of deciding as to the best measures to undertake to defeat Japan.

Arnold finally sent a wire off to Washington recommending that the Kyushu operation stay on the schedule as planned, that bombing of Japan be intensified to the maximum, beginning as soon as possible, and that decisions as to our next objective after we took Kyushu be left until later in the fall. MacArthur agreed.

Arnold rested the next day. He was still a little weak and the interview with MacArthur had been a bit strenuous. On the 19th we flew to Clark Field and chatted with Whitehead until midnight, when Arnold and I took off for Okinawa. During the evening word came in that General Buckner, the 10th Army commander, had been killed and that Lieutenant General Roy S. Geiger, Marine Corps, had taken over. Buckner had been observing the action in a front-line position and had been hit by a Jap artillery shell. Geiger, an old Marine Corps flyer, had had a lot of amphibious landing and fighting experience during the war and was thoroughly capable of winding up the Okinawa campaign, which was almost over anyhow.

At Okinawa we met Geiger and the staff of the 10th Army, the Army and Navy construction engineers, and Photo Hutchison, and discussed the possibility of hurrying up airdrome work so that we could get started on the heavy bombing of Kyushu that I had outlined to Arnold in Manila. The Navy engineer said he could do it if about twelve battalions of engineer troops then in the Marianas could be turned over to him. Arnold promised to take the matter up with Nimitz on the return trip to Washington. The Navy agreed to turn one of their fields over to me in another week and, after inspecting the construction already in sight, I radioed Whitehead to send two fighter groups and a strafing group to Okinawa by July 1st, and sent word to Tommy White to send his two strafing groups in during the first week in July. Arnold left for Guam that evening. I stayed overnight at 10th Army Headquarters.

The Navy announced the end of Jap resistance in Okinawa that day, with the exception of a few small pockets of isolated enemy troops, which would be liquidated shortly. The operation had taken eighty-two days. Over 100,000 Japs had been killed and nearly 9000 more captured. Our losses were announced as 2573 Marines and 4417 Army troops killed

Okinawa and the Kyusha Plan

or missing and 12,265 Marines and 17,033 Army troops wounded.

During the night I found out that the war on Okinawa was not quite over. Around midnight a party of Japs blundered into a fight with the guards about fifty yards from my tent. One of the guards came to the door and said it was all right but would I please stay where I was and not turn on the lights.

I put my pistol on a chair by the side of the bed and assured him that I had no intention of going outside to have either Nips or 10th Army guards take a shot at me. The shooting died down a little later and I went to sleep.

The next morning, as I was taking off for Manila, Photo Hutchison told me that he had had another battle going on during the night around his headquarters and that his guards had killed eight Japs. He didn't know how many were involved in the fight but was sure it was at least eight.

On June 24th I said goodbye with profound regret to Sergeant Raymond. He had worked like a dog for me for over two years, a large part of the time pretty close to the equator, and a kitchen is not a good place to be in the tropics. Raymond was tired and thin and I had known for some time that I would have to let him go home before he broke down. He had trained another cook, a Filipino, who was working out quite well, but there was only one Raymond and I knew I was going to miss him and the good cooking that had kept me well fed and contented ever since he arrived in the theater. I wrote a letter to Dan London, his old boss in San Francisco, and asked him to take care of the sergeant as soon as he could get discharged from the Army. I heard later on that Raymond was back on the job, had been promoted to Executive Chef, and was running his kitchen with military precision. Dan said he almost felt like saluting every time he visited the kitchen.

On the 25th I got a wire from Cooper that he was back from Europe and had no job. I immediately wired Arnold to send him out to me with top priority for air travel. He evidently got it, as he came into my office on the 28th. I made him Deputy Chief of Staff and told him that in addition to his other duties I expected him to make certain that the people back home found out that we were still at war in the Pacific.

Whitehead's promotion that day to the rank of lieutenant general was the first thing for him to publicize, with special attention to Kansas newspapers.

The landing at Balikpapan went off as scheduled on July 1st. A radio that evening said that the 7th Division had gone ashore with no casualties and that at six o'clock that evening the casualty list was eleven men wounded, one of them seriously. The dispatch said the Jap defenses were completely wrecked. They should have been. We slugged the place with 4000 tons of bombs before the Aussies went ashore. The usual congratulatory messages came in from the Australians and the Naval commander, and when General MacArthur arrived back in Manila on the 3rd, he said the bombing just before the troops went in was the most accurate he had ever seen.

One of MacArthur's aides, who went ashore with the General, told me a story that was even worse than what he did at Labuan.

About a half mile inland from the landing beach, MacArthur and his party of generals and admirals were standing on a hill looking out over the rolling country toward the town of Balikpapan. An Australian brigadier handed MacArthur a map and proceeded to point out the various points of interest and the locations of the troops. Suddenly a machine gun opened fire on the hill. It was at extreme range, but leaves were being snipped off the bushes and little spurts of dust being kicked up around the place. Everyone except the Aussie and MacArthur hit the dirt and started sliding back down the slope. After the brigadier had finished, MacArthur folded up the map, handed it back, and thanked him for his information. Then, pointing to another hill about a quarter of a mile away, he said, "Let's go over there and see what's going on. By the way, Brigadier, I think it would be a good idea to have a patrol take out that Jap machine gun before someone gets hurt."

That was the man that you used to hear called "Dug-out Doug." Perhaps it was meant to emphasize an entirely opposite characteristic of the man, like calling a fat man "Skinny."

The first of my fighter groups to arrive in Okinawa, the old 35th, began operations out of their new location on July 3rd, with 48 planes making a sweep along the west coast of Kyushu as far north as the Jap naval base at Sasebo. Only three Jap aircraft, all floatplanes, were seen in the air and these were added to the 35th Group score. We had no losses. Captain Richard Celia got credit for the first official victory over Japan proper.

That evening the Jap radio announced that we were so hard up for airplanes and pilots that we were using flimsy training type P-51 Mustangs piloted by young American girls in our operations against Formosa. I don't know what the purpose was of so preposterous a lie but we all got a good laugh out of it.

Giles wired me from Guam that "now that the cat is out of the bag, in the interests of good liaison" would I let him know where I had gotten my last replacement of fighter pilots.

I spent the 6th and 7th at Okinawa, where I selected a site for my headquarters and told Colonel Hewitt to get busy fixing it up for occupancy about August 1st. I saw General Stilwell, who had relieved General Geiger in command of the 10th Army, and told him of the huge bomb stockage that would be on its way soon and which I wanted unloading priority for so that we could get going on the softening up of Kyushu. He promised to take care of it. Stilwell was extremely cordial to me and seemed as keen and sharp-tempered when talking about someone or something that annoyed him as I had imagined from all the stories about the man.

The 35th Group shot down four more Nip planes over Kyushu on the 7th. I went over to the airdrome and decorated the lads who had gotten the first three victories a few days before as well as three other pilots who had accounted for the latest four. The kids were disgusted over the fact that in five days of flying all over Kyushu they had only knocked down seven Jap planes. They asked me to tell the 49th Group, who were all disappointed at not being the first fighter group to go to Okinawa, that they might just as well stay in the Philippines, as there were already more than enough fighters up there to take care of what was left of the Jap Air Force.

That night I stayed over at Photo Hutchison's headquarters. Somehow or other he had managed to get two Quonset huts from the Navy and had another building going up for a mess hall that looked like part of a standard Army portable hut. Additions were being constructed as fast as lumber became available.

Just after dark a truck drove up with a load of lumber. I pretended I was looking the other way but I could have sworn I saw a couple of bottles passed to the driver while a detail of men from Hutchison's headquarters was unloading the stuff. According to current rumors, two bottles of almost any liquor was good for a load of lumber, and for a case a Quonset hut might be unloaded by mistake. Not a big Quonset—those were worth two cases.

I was awakened about one o'clock the next morning by sounds of lumber being loaded into a truck. I went to the door and saw Hutch bossing the job. When he came back, I asked where he was going to erect the new building. Photo grinned rather sheepishly and said, "Dammit, I just found out that we hijacked a load of lumber that was supposed to go to General Whitehead's camp up the road. Now I'm out two bottles of whiskey."

The truck delivered the lumber before daybreak, unloaded quietly, and got away without anyone discovering whose truck had delivered the load. As far as I know Whitehead never did find out about it.

On July 10th it was announced from Washington that the B-29s in the Marianas would form the Twentieth Air Force, under General Twining, and that those operating from Okinawa would form the Eighth Air Force, under General Jimmy Doolittle. The Eighth and Twentieth would together be called the United States Strategic Air Force, with General Spaatz in command, Giles his Deputy, and LeMay his Chief of Staff. Strategic control was still to remain with the Joint Chiefs of Staff, with General Arnold as their agent.

On the same day Nimitz turned over control of the Seventh Air Force to the Far East Air Forces and told the Marine Fighter Wing at Okinawa to operate in conjunction with our show there by arrangement with my local commander.

On the 12th Lord Louis Mountbatten and a few members of his staff flew over from India to Manila for a visit and conference with MacArthur. We briefed him on the coming Olympic operation, and his staff in turn gave us the details of the proposed British operation to recapture Singapore. Mountbatten wanted some bombing assistance at that time, if we had any to spare. MacArthur asked me what I could do. I said the Australians had three squadrons of B-24s which would be able to operate from the Borneo fields by that time, and just before the date of the Singapore landing I could send down a couple of groups of B-24s from the Far East Air Forces for two or three days' final preparation work. Mountbatten was quite pleased and thanked us most heartily. I told my staff to work out the details with his air officer, who had accompanied him to Manila.

The next evening I invited Mountbatten to dinner at my house in Manila. Whitehead, Wurtsmith, Fighter Hutchinson, who was now my Chief of Staff, and Jimmy Crabb, who had the 5th Bomber Command under Whitehead, were the other guests.

Before dinner we all went swimming in the pool, which Lord Louis said he wished he could take back to India with him. He turned out to be a most enjoyable guest. A bit to my surprise it wasn't over ten minutes before I heard him talking to "Whitey," "Squeeze," "Hutch," and "Jimmy" as though they had been college chums together. In answer, I heard "Louis," "Dicky," and "Admiral" from my gang.

About midnight the party broke up and Lord Louis suddenly discovered he had lost his ring. He said it must have slipped off his finger when he was in the pool. He hated to lose it, as it had been given to him on his twenty-first birthday by the Prince of Wales. It was just a simple signet ring, but he had become rather silly and sentimental about the thing.

I told him not to worry about it. As soon as I got up the next morning we would find it for him if we had to drain the pool and I would send it up to Clark Field, where he was going to have lunch with Whitehead before taking off for India. He didn't think I would find it but thanked me just the same and drove in town to his billet.

When I went up to bed I picked up a towel, lying on the bathroom floor, that Mountbatten had used when he came out of the pool. The ring fell on the floor as I lifted the towel.

The next morning I sent it to Clark Field by Fighter Hutchinson. That afternoon Hutch returned with a note of thanks from Lord Louis and two bottles of excellent Scotch, which Mountbatten had said were overcrowding his bag. Hutch thought it would be a good idea to invite him to return to Manila often for a swim in my pool, on condition that he always wear that ring.

All through July we kept moving aircraft into Okinawa from both the Fifth and Seventh Air Forces. Whitehead and Tommy White set up their headquarters on the island and began the final sweep of Jap shipping from the Yellow Sea and the Straits of Tsushima, between Japan and Korea. That target dried up quickly and by the first of August a group of strafers during the daytime and a squadron of radar-equipped B-24s were sufficient to maintain the blockade. By this time whole days would go by without sighting a single vessel. Most of Japan's shipping was on the bottom.

In conjunction with the B-29s from the Marianas, who were battering the big cities of Japan apart and burning them down, we concentrated our attacks on the island of Kyushu, smashing airdromes, burning up gasoline stocks, and wrecking the railway centers, bridges, and marshalling yards. The attacks were being made with an ever-increasing weight, as airdromes were being finished on Okinawa, allowing us to move the aircraft forward from the Philippines and the Marianas. By the end of July, on an average day when the weather permitted large operations, there would be over 1500 of my airplanes operating along the line from Japan to Formosa to Shanghai to Borneo and the Netherlands East Indies. Of this number around 600 bombers, strafers, and fighters would be attacking targets in Japan itself. It was a far cry from the days back in 1942, when a raid of fifty or sixty planes was such big news that we had boasted about it for days.

On August 2nd General Spaatz paid me a visit in Manila. We spent a couple of days working out the details of coordinating the efforts of our two air forces prior to the date of the Olympic operation and the procedure from that time on. During August and September Spaatz was to concentrate his attacks on Japan from around Osaka, east, while I took care of Kyushu and the western half of the main island of Honshu and maintained the shipping blockade.

We decided to keep each other informed daily as to the next day's scheduled operations and to get together frequently for adjustments in the plan. Doolittle's Eighth Air Force, which would get its planes about the end of August, would coordinate its attacks with mine on the western part of Honshu. Later on, a week or so prior to our landing on Kyushu on November 1st, Spaatz said for me to assign the missions to him for the final crushing of any possible opposition to the actual invasion itself. We both agreed that, after Kyushu was occupied and our air forces were operating from airdromes in Japan itself, the difficulty of coordination would almost demand a single air force command but decided to discuss that point later on.

27. THE JAPANESE SURRENDER
August—September, 1945

On the 4th, the word came in that the first atomic bomb would be dropped on Hiroshima the morning of the 6th. No one really knew what this new weapon would do, so the whole area, for fifty miles around the target city, was declared off limits for all aircraft on that day, except, of course, the one doing the bombing.

I flew to Okinawa the next day and gave the news to Whitehead, who rearranged several of his missions that had been scheduled to be in the Hiroshima area on the 6th. Changing the missions involved changing some of the bomb-loads after they had already been put in the bomb-bays. Whitehead's comment was priceless: "These newfangled gadgets are certainly raising hell with my operations."

Whitehead had a job to do, and this new earth-shaking development, which some people said would cause an explosion that would be felt halfway around the world, didn't cause him anything but annoyance if it interfered with carrying out his mission.

The next day we held our breaths and waited for the big event. The first news came from the Jap radio, which reported that they were out of communication with Hiroshima, following a tremendous shock from some new type of explosive. The crew of the B-29 that had dropped the bomb reported a successful strike, with the cloud of smoke and dust rising to over 20,000 feet. Much to our gratification we felt no shock at Okinawa. The Japs got it all and later radio reports began to tell the story. The city was wrecked and the indications were that around a hundred thousand people had perished.

That afternoon the Fifth Air Force, from Okinawa to Manila, for a few hours stopped thinking about the atomic bomb or even about the war. Something else had happened that somehow touched them more intimately and more deeply. They had lost someone that everyone, from general to private, admired, constantly boasted about, and loved—someone they had thought was safe and out of harm's way. Dick Bong had been killed in the crash of a new jet fighter plane at Los Angeles.

On the 8th Russia declared war on Japan and the next morning her forces started crossing the Manchurian borders and driving south and southwest on Harbin, Mukden, and Port Arthur.

The second atomic bomb struck Nagasaki on the 9th. Although the clouds had interfered with aiming at the center of the city, the results were comparable to what had happened at Hiroshima. Everything in the path of the bomb had been wiped out. There was no question about the destructive capabilities of this new and terrifying weapon.

The radio was crackling now from Tokio to Switzerland, from Switzerland to Washington, London, and Moscow. On the 10th, Domei, the official Jap news agency, broadcast a statement that Japan was willing to surrender, providing that the Emperor's prerogatives were not impaired by the terms. There was no argument now about what territory Japan could keep. They were willing to accept about what the Allies were demanding—unconditional surrender, the terms of total and acknowledged defeat. I returned to Manila.

On the 11th the Swiss legation in Washington received a note from our State Department clarifying our position in regard to Emperor Hirohito. The note was immediately transmitted to Tokio. It stated that the Allied military commander of the occupation forces would run Japan through the Emperor. Hirohito was to be spared, but the living god, the head of the Shinto faith, was reduced to the role of mouthpiece for MacArthur. The General ordered the 11th Airborne Division moved from Luzon to Okinawa, to be ready to go in by air to the Tokio area as the spearhead of the occupation forces, and we waited for the Japanese answer and called off the bombing so that the Nips would have a chance to think it over.

Nothing was heard from Tokio the 12th and 13th. The orders went out for the resumption of full-scale attacks. This meant that 7000 tons of bombs a day on the Jap homeland was what they could look forward to until they were ready to quit. We loaded up and briefed the crews.

On the 14th in Washington, which was our 15th, the official surrender text from Tokio, which had been transmitted to the Swiss legation, was given to our State Department and carried to the White House by Secretary Byrnes. President Truman at once announced the end of the war.

As soon as the radio flash came in I got in touch with all my commanders and told them to call off the attack. If any aircraft were already in the air they were to be recalled by radio at once. Whitehead had his whole force in motion, in the air or getting ready to take off. He got in touch with all of them except about twenty strafers, who were on the way to sweep the Tsushima Straits for shipping. They found no targets and returned that afternoon to their base with their bombs. Wurtsmith's Thirteenth, White's Seventh, and Bostock's Australians were in about the same position but managed to get everyone back without getting involved in attack or combat. The shooting part of the war was over.

We celebrated a little that night. Not much, as everyone seemed a bit tired and wanted to get some sleep. The 49th Fighter Group opened the magnum of brandy. They said the next day that it was not bad.

Now that the war was over, we added up the scores of the fighter groups to see what the kids were to brag about for he rest of their lives. The 49th Group led the field with a final total of 677 confirmed air victories. The 475th with 545, the 8th with 443, the 35th with 397, and the 348th with 356 followed in order as the Fifth Air Force's contribution. The two fighter groups of the Thirteenth Air Force, the 18th and 347th, finished in that order with scores of 274 and 246.

Bong, with forty victories, and McGuire, close behind with thirty-eight, were well ahead of their closest rival, Colonel Charles MacDonald, the commander of the 475th Group, the same outfit that McGuire had belonged to. MacDonald got twenty-seven. Colonel Jerry Johnson, the crack shot from the 49th Group, finished with twenty-four. Colonel Neel Kearby of the 348th, Major Jay Robbins of the 8th, and Lieutenant Colonel Robert Westbrook of the 347th, all tied at twenty-two.

Lieutenant Colonel Thomas Lynch, with an even twenty, completed the list of those who had scored a score or more confirmed victories in the war against Japan.

On the 15th MacArthur ordered the Japs to send their envoys to Manila for a conference. They were told to fly their own airplane painted white, with green crosses on the fuselage and wings, and using the call letters B-A-T-A-A-N, to Ie Shima, an island just off the northwest tip of Okinawa where we had a large fighter base. There our transport aircraft would bring them to Manila.

As the distance from Okinawa to Atsugi airdrome, ten miles west of Yokohama, where we expected to land the 11th Airborne Division, was too great for a round trip by our C-47 and C-46 type transports without refueling, I asked Arnold to let me have about 250 C-54s, which could carry over double the load of my transports and had enough fuel capacity to make the round trip without taking on gas at Yokohama. I was quite sure that the Japs could not furnish us any aviation gasoline or that, if they did have any, it would not be good enough to use in our engines. Arnold told the Air Transport Command, which was operating the C-54s in the Pacific, to furnish me the number I wanted and ordered them to start concentrating in Okinawa and await my instructions.

On the 17th the Japs replied, saying that they didn't understand whether their envoys were to negotiate for or sign surrender terms. MacArthur answered that the language of his message was plain and he wanted his instructions carried out at once.

The next day the Nips radioed that they would land at Ie Shima on the 19th. In the message they said that their call letters would be JNP. MacArthur's reply emphasized that the call letters would be BATAAN, as he had originally directed.

The Jap envoys arrived in Manila at 7:00 p.m. the 19th at Nichols Field, where they were met by General Willoughby, MacArthur's Chief of Intelligence, and taken to a hotel in Manila. The talks started that evening and continued until the Nips left the next day. They were to fix up Atsugi airdrome, about ten miles west of Yokohama, for our air invasion force, take the propellers off their airplanes or otherwise render them unflyable, furnish us transportation and fuel, set up a schedule of withdrawal of troops from the Tokio Bay area, and start disarmament right away. We were to land at Atsugi about the 28th of August and the surrender ceremony would be held on the battleship Missouri in Tokio Bay two days later. The Nips said there was a lot of work to be done repairing the airdrome at Atsugi but they would do the best they could to meet the date.

When the envoys arrived at the hotel in Manila a few aides and orderlies were assigned to look after them. One of the Japanese generals asked a GI orderly if he could get him a couple of cartons of cigarettes and handed him two five-dollar bills. The orderly returned a few minutes later with the cigarettes and handed the Nip general his change. It was in the Jap-invasion paper currency that the Nips had printed by the carload in the Philippines. Even the grade of paper was poor. The Jap general took it without saying a word.

On the 21st I came down with a strep throat that put me in the hospital for four days while the doctors pumped penicillin into me. I told them that they'd better do the job right as I intended to land in Japan with MacArthur on Invasion Day if I had to go in on a stretcher. The colonel in charge at the hospital said I ought to take it easy for a month, but I asked him who gave him orders, anyhow. He shrugged his shoulders and gave me another shot of penicillin.

On the 25th I got a radio from Whitehead which I rather enjoyed and, reading between the lines, I believe he did, too. That morning Colonel Clay Tice and Flight Officer Hall, both of the 49th Fighter Group, flying two P-38s, had left Okinawa on a reconnaissance flight over Kyushu. While circling around the southern part of the island, Hall "noticed undue gas consumption" and, after talking it over with Tice, the two landed on a Jap field at Nittagahara. Jap infantry officers and men welcomed them with open arms, refueled Hall's airplane, gave them candy and refreshment, and bowed them all over the place. The two planes then took off and returned to Okinawa without further incident. The 49th Group had been the first to land in Japan. The story was a bit thin but it could have been on the level. I didn't ask any questions.

Jackie Cochran, the record-holding woman aviator and head of the Wasps—the woman's organization she had organized to ferry aircraft from the factories and modification centers in the United States during the war—visited me on the 26th and had dinner with me at Fort McKinley. Jackie had gotten into the theater as a magazine correspondent and was still annoyed that she had not been able to get there before the war was over. I was still pretty weak but enjoyed the visit immensely. I not only like Jackie as a fine person but she is the only woman flyer that I wouldn't mind flying with if she were at the controls.

On the 28th I flew to Okinawa. The date of our air invasion had been postponed to August 30th and the surrender ceremony on the deck of the Missouri was set for September 2nd. I sent word to Bostock the Australian, Isitt the New Zealand air commander, Wurtsmith, White, and Whitehead that I wanted them to accompany me when we landed at Atsugi and be present with me at the surrender ceremony. They all arrived on the 29th, when General MacArthur and Spaatz also came up to be ready for the take-off the next morning. MacArthur said he was going to land in Japan about noon. I told Whitehead to send in an escort of old-timers from the 49th Fighter Group and the 3rd Attack Group. They would have to remain there until gasoline was available before they could be of any use in the event of an emergency, but I wanted the kids to have the honor and satisfaction of being in at the finish.

On the 30th in ideal flying weather we flew to Atsugi. The south coast of Japan was beautiful in the warm sunlight as we approached the land looking for the snow-capped peak of Fujiyama so familiar to everyone from photography and picture postcards. But Fujiyama didn't have any snow on it. The mountain is beautiful without snow but the white cap makes it a truly magnificent sight. We felt let down. Someone said they would bet the Nips melted the snow off on purpose.

Less than 500 airborne infantry were on the ground when MacArthur and all his top generals of the Southwest Pacific landed just after noon. Jap automobiles with Jap drivers took us into Yokohama. Just before leaving the airdrome someone thought it would be a good idea for MacArthur to have a guard of honor or at least someone to do guard duty around the hotel when we arrived in the city. Twenty good-looking, tall Yank soldiers were hustled into a truck and joined the parade.

Just before we left, MacArthur had noticed that most of us were carrying pistols in shoulder holsters. It had become a habit during the past three years. He said we had better leave the pistols behind. There were fifteen fully armed Jap divisions within ten miles of us. If they decided to start anything, those toy cannon of ours wouldn't do any good. We left our weapons in the airplanes. I found out afterward that it was excellent psychology and made a tremendous impression on the Japs to see us walking around in their country unarmed and seemingly with utter disregard of danger from the nation of 70,000,000 people we had defeated. To them it meant that there was no doubt about it. They had lost.

All along both sides of the road to Yokohama, stationed about a hundred yards apart, were Jap soldiers, fully armed and with their backs to the line of automobiles as we passed. It was partly a token of submission but was also meant to insure against any possibility of sniping by any Jap who didn't agree with the imperial edict calling the war off.

Yokohama was practically a shell of what had once been a thriving city of a million people. The fire bombs of the B-29 attacks had burned out whole blocks, and piles of rubbish had overflowed into the streets. In preparation for our arrival, an attempt had been made to clear the streets, and electric current and water service had been restored in the buildings earmarked for the occupation forces, but Yokohama had a job on its hands that would take years to accomplish. The waterfront was in fairly good shape and the park in front of the New Grand Hotel, which was to be our temporary headquarters, had been damaged only slightly, but everywhere else the eye met scenes of desolation and destruction. The fury of modern warfare was an appalling thing to think about.

I wanted my Chinese boy Foo to look after me. I had brought in my airplane ten days' rations for my party, who were all quartered in the New Grand Hotel, and I intended Foo to do the cooking and serve our meals in the suite that had been assigned to me. No one was supposed to live in that hotel except generals and very important colonels. I told the manager of the place that I wanted a room next to my suite assigned to General Foo from the island of Hainan. Foo moved in. As a matter of fact, Foo did come from Hainan.

On September 2nd, with all my Air Force generals who were to take part in the ceremony, I boarded a destroyer at Yokohama and went out to the battleship Missouri, which was anchored in the middle of the southern end of Tokio Bay.

All the top Army, Navy, and Air commanders of the Pacific were there. After the commanders had greeted and congratulated each other for a half hour or so, the eleven Japs who were to sign the surrender documents came aboard and formed a little square in front of the table on which the documents awaited them. We lined up to watch the ceremony and MacArthur walked to the microphone on the opposite side of the table facing the Japs. Back of him were the representatives and alternates of the Allied nations, who were to sign the documents for their respective governments.

MacArthur made a short speech and then told the Japs to sign. Prince Shigemitsu, Minister of Foreign Affairs, dressed in a frock coat and striped trousers, stepped forward and sat down, pushing his wooden leg under the table. He took off his silk hat, laid it on the table, took off his white gloves, put his hat back on his head, and finally put both the hat and gloves down. He was quite visibly nervous about something. It was the first time he had done anything like that and he was probably afflicted with stage fright. He looked from one paper to another and seemed to be puzzled as to which one he should look at first.

"Sutherland," MacArthur's voice punctuated the dead silence like a pistol shot. General Sutherland, his Chief of Staff, stepped forward.

"Show him where to sign." It still sounded explosive. Sutherland stepped to the table and pointed to the line where Shigemitsu was to sign his name.

The Minister of Foreign Affairs picked up the pen and, scarcely glancing at the text of the surrender document, signed for the Emperor of Japan and the Japanese Government. General Yoshijiro Umezo, Chief of the Army General Staff, signed for the Japanese Imperial General Headquarters. Umezo glanced briefly at the last sentence: ". . . The authority of the Emperor and the Japanese Government to rule the state shall be subject to the Supreme Commander for the Allied Powers, who will take such steps as he deems proper to effectuate these terms of surrender." He then stepped back and took his place alongside Shigemitsu with the rest of the Japanese delegation.

It was now the turn of the Allied Nations. As Supreme Commander, MacArthur stepped forward, calling on General Wainwright, formerly of Bataan and Corregidor, and General Percival, the British commander who had surrendered at Singapore, to stand behind him. We had flown the two generals in from the Jap prison camp at Mukden just three days before.

MacArthur sat down and pulled five pens from his pocket. He picked up the first one, wrote a few letters of his signature, and handed the pen to Wainwright. The next pen went to Percival. Then one for the archives of the Military Academy at West Point and one for the Naval Academy at Annapolis. The fifth was his own.

He then reached into another pocket of his shirt, pulled out a little red-barreled fountain pen, finished signing, and put the pen back in his pocket. I recognized it. It belonged to Jean MacArthur, his wife, who was back in Manila listening like the rest of the world to the broadcast of this last act of World War II.

Admiral Nimitz, with Halsey and Sherman standing behind him, signed for the United States. China, Great Britain, the USSR, France, Australia, Canada, Holland, and New Zealand followed. The ceremony was over. The Japs left the Missouri and I joined MacArthur on the way back to Yokohama.

That night I listened to a rebroadcast of MacArthur's speech addressed to the people of the United States. I had read it and heard it before. I wanted to hear it again.

"Today the guns are silent. A great tragedy has ended. A great victory has been won. The skies no longer rain death—the seas bear only commerce—men everywhere walk upright in the sunlight. The entire world lies quietly at Peace. . . .

"A new era is upon us. Even the lesson of Victory itself brings with it profound concern, both for our future security, and the survival of civilization. . . .

"Men since the beginning of time have sought Peace. Various methods through the ages have been attempted, to devise an international process to prevent or settle disputes between nations. . . . Military alliances, balances of power, Leagues of Nations, all in turn, failed, leaving the only path to be by way of the crucible of War. The utter destructiveness of war now blots out this alternative. We have had our last chance. If we do not devise some greater and more equitable system, Armageddon will be at our door. . . .

"My fellow countrymen, today I report to you that your sons and daughters have served you well and faithfully. . . . They are homeward-bound—take care of them."

www.ingramcontent.com/pod-product-compliance
Lightning Source LLC
Chambersburg PA
CBHW041124110526
44592CB00020B/2685